Fitting Models to Biological Data Using Linear and Nonlinear Regression

A practical guide to curve fitting

Harvey Motulsky &

Arthur Christopoulos

OXFORD
UNIVERSITY PRESS

2004

OXFORD
UNIVERSITY PRESS

Oxford New York
Auckland Bangkok Buenos Aires Cape Town Chennai
Dar es Salaam Delhi Hong Kong Istanbul Karachi Kolkata
Kuala Lumpur Madrid Melbourne Mexico City Mumbai Nairobi
São Paulo Shanghai Taipei Tokyo Toronto

Published by Oxford University Press, Inc.
198 Madison Avenue, New York, New York 10016

www.oup.com

Library of Congress Cataloging-in-Publication Data
Motulsky, Harvey.
Fitting models to biological data using linear and nonlinear regression :
a practical guide to curve fitting / by Harvey Motulsky, Arthur Christopoulos.
p. cm.
Includes bibliographical references.
ISBN-13 978-0-19-517179-2; 978-0-19-517180-8 (pbk.)
ISBN 0-19-517179-9; 0-19-517180-2 (pbk.)
1. Biology—Mathematical models. 2. Regression analysis. 3. Nonlinear theories.
4. Curve fitting. I. Christopoulos, Arthur. II. Title.
QH323.5.M68 2003
570'.1'5118—dc22 2003057957

Contents at a Glance

Contents

Preface

Regression analysis, especially nonlinear regression, is an essential tool to analyze biological (and other) data. Many researchers use nonlinear regression more than any other statistical tool. Despite this popularity, there are few places to learn about nonlinear regression. Most introductory statistics books focus only on linear regression, and entirely ignore nonlinear regression. The advanced statistics books that do discuss nonlinear regression tend to be written for statisticians, exceed the mathematical sophistication of many scientists, and lack a practical discussion of biological problems.

We wrote this book to help biologists learn about models and regression. It is a practical book to help biologists analyze data and make sense of the results. Beyond showing some simple algebra associated with the derivation of some common biological models, we do not attempt to explain the mathematics of nonlinear regression.

The book begins with an example of curve fitting, followed immediately by a discussion of how to prepare your data for nonlinear regression, the choices you need to make to run a nonlinear regression program, and how to interpret the results and troubleshoot problems. Once you have completed this first section, you'll be ready to analyze your own data and can refer to the rest of this book as needed.

This book was written as a companion to the computer program, GraphPad Prism (version 4), available for both Windows and Macintosh. Prism combines scientific graphics, basic biostatistics, and nonlinear regression. You can learn more at www.graphpad.com. However, almost all of the book will also be useful to those who use other programs for nonlinear regression, especially those that can handle global curve fitting. All the information that is specific to Prism is contained in the last section and in boxed paragraphs labeled "GraphPad notes".

We thank Ron Brown, Rick Neubig, John Pezzullo, Paige Searle, and James Wells for helpful comments.

Visit this book's companion web site at www.curvefit.com. You can download or view this entire book as a pdf file. We'll also post any errors discovered after printing, links to other web sites, and discussion of related topics. Send your comments and suggestions to Hmotulsky@graphpad.com .

Harvey Motulsky
President, GraphPad Software
GraphPad Software Inc.
Hmotulsky@graphpad.com

Arthur Christopoulos
Dept. of Pharmacology
University of Melbourne
arthurc1@unimelb.edu.au

A. Fitting data with nonlinear regression

1. An example of nonlinear regression

As a way to get you started thinking about curve fitting, this first chapter presents a complete example of nonlinear regression. This example is designed to introduce you to the problems of fitting curves to data, so it leaves out many details that will be described in greater depth elsewhere in this book.

> GraphPad note: You'll find several step-by-step tutorials on how to fit curves with Prism in the companion tutorial book, also posted at www.graphpad.com.

Example data

Various doses of a drug were injected into three animals, and the change in blood pressure for each dose in each animal was recorded. We want to analyze these data.

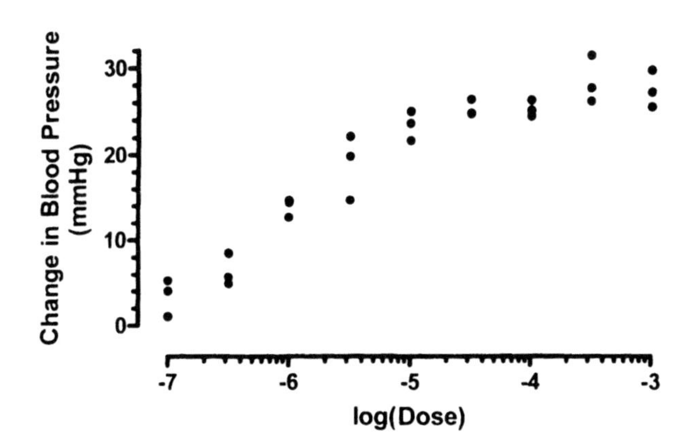

log(dose)	Y1	Y2	Y3
-7.0	1	4	5
-6.5	5	4	8
-6.0	14	14	12
-5.5	19	14	22
-5.0	23	24	21
-4.5	26	24	24
-4.0	26	25	24
-3.5	27	31	26
-3.0	27	29	25

Step 1: Clarify your goal. Is nonlinear regression the appropriate analysis?

Nonlinear regression is used to fit data to a model that defines Y as a function of X. Y must be a variable like weight, enzyme activity, blood pressure or temperature. Some books refer to these kinds of variables, which are measured on a continuous scale, as "interval" variables. For this example, nonlinear regression will be used to quantify the potency of the drug by determining the dose of drug that causes a response halfway between the minimum and maximum responses. We'll do this by fitting a model to the data.

Three notes on choosing nonlinear regression:

- With some data, you may not be interested in determining the best-fit values of parameters that define a model. You may not even care about models at all. All you may care about is generating a standard curve that you can use to interpolate unknown values. If this is your goal, you can still use nonlinear regression. But you won't have to be so careful about picking a model or interpreting the results. All you care about is that the curve is smooth and comes close to your data.

- If your outcome is a binomial outcome (for example male vs. female, pass vs. fail, viable vs. not viable) linear and nonlinear regression are not appropriate. Instead, you need to use a special method such as *logistic regression* or *probit* analysis. But it is appropriate to use nonlinear regression to analyze outcomes such as receptor binding or enzyme activity, even though each receptor is either occupied or not, and each molecule of enzyme is either bound to a substrate or not. At a deep level, binding and enzyme activity can be considered to be binary variables. But you measure binding of lots of receptors, and measure enzyme activity as the sum of lots of the activities of lots of individual enzyme molecules, so the outcome is really more like a measured variable.

- If your outcome is a survival time, you won't find linear or nonlinear regression helpful. Instead, you should use a special regression method designed for survival analysis known as *proportional hazards regression* or *Cox regression*. This method can compare survival for two (or more) groups, after adjusting for other differences such as the proportions of males and females or age. It can also be used to analyze survival data where subjects in the treatment groups are matched. Other special methods fit curves to survival data assuming a theoretical model (for example the Weibull or exponential distributions) for how survival changes over time.

GraphPad note: No GraphPad program performs logistic regression, probit analysis, or proportional hazards regression (as of 2003).

Step 2: Prepare your data and enter it into the program

For this example, we don't have to do anything special to prepare our data for nonlinear regression. See Chapter 2 for comments on transforming and normalizing your data prior to fitting a curve.

Entering the data into a program is straightforward. With Prism, the X and Y columns are labeled. With other programs, you may have to specify which column is which.

Step 3: Choose your model

Choose or enter a model that defines Y as a function of X and one or more parameters. Section C explains how to pick a model. This is an important decision, which cannot usually be relegated to a computer (see page 66).

For this example, we applied various doses of a drug and measured the response, so we want to fit a "dose-response model". There are lots of ways to fit dose-response data (see I. Fitting dose-response curves). In this example, we'll just fit a standard model that is alternatively referred to as the *Hill equation*, the *four-parameter logistic equation,n* or the *variable slope sigmoidal* equation (these three names all refer to exactly the same model). This model can be written as an equation that defines the response (also called the dependent variable Y) as a function of dose (also called the independent variable, X) and four parameters:

$$Y=Bottom+\frac{Top-Bottom}{1+\left(\frac{10^{LogEC_{50}}}{10^{X}}\right)^{HillSlope}}$$

The model parameters are *Bottom*, which denotes the value of Y for the minimal curve asymptote (theoretically, the level of response, if any, in the absence of drug), *Top*, which denotes the value of Y for the maximal curve asymptote (theoretically, the level of response produced by an infinitely high concentration of drug), $LogEC_{50}$, which denotes the logarithm of drug dose (or concentration) that produces the response halfway between the *Bottom* and *Top* response levels (commonly used as a measure of a drug's potency), and the *Hill Slope*, which denotes the steepness of the dose-response curve (often used as a measure of the sensitivity of the system to increments in drug concentrations or doses). The independent variable, X, is the logarithm of the drug dose. Here is one way that the equation can be entered into a nonlinear regression program:

```
Y = Bottom + (Top-Bottom)/(1+10^(LogEC50-X)*HillSlope)
```

> Note: Before nonlinear regression was readily available, scientists commonly transformed data to create a linear graph. They would then use linear regression on the transformed results, and then transform the best-fit values of slope and intercept to find the parameters they really cared about. This approach is outdated. It is harder to use than fitting curves directly, and the results are less accurate. See page 19.

Step 4: Decide which model parameters to fit and which to constrain

It is not enough to pick a model. You also must define which of the parameters, if any, should be fixed to constant values. This is an important step, which often gets skipped. You don't have to ask the program to find best-fit values for all the parameters in the model. Instead, you can fix one (or more) parameter to a constant value based on controls or theory. For example, if you are measuring the concentration of drug in blood plasma over time, you know that eventually the concentration will equal zero. Therefore you probably won't want to ask the program to find the best-fit value of the bottom plateau of the curve, but instead should set that parameter to a constant value of zero.

Let's consider the four parameters of the sigmoidal curve model with respect to the specific example above.

Parameter	Discussion
Bottom	The first dose of drug used in this example already has an effect. There is no lower plateau. We are plotting the change in blood pressure, so we know that very low doses won't change blood pressure at all. Therefore, we won't ask the program to find a best-fit value for the Bottom parameter. Instead, we'll fix that to a constant value of zero, as this makes biological sense.
Top	We have no external control or reference value to assign as Top. We don't know what it should be. But we have plenty of data to define the top plateau (maximum response). We'll therefore ask the program to find the best-fit value of Top.
LogEC50	Of course, we'll ask the program to fit the $logEC_{50}$. Often, the main reason for fitting a curve through dose-response data is to obtain this measure of drug potency.
HillSlope	Many kinds of dose-response curves have a standard Hill slope of 1.0, and you might be able to justify fixing the slope to 1.0 in this example. But we don't have strong theoretical reasons to insist on a slope of 1.0, and we have plenty of data points to define the slope, so we'll ask the program to find a best-fit value for the Hill Slope.

> Tip: Your decisions about which parameters to fit and which to fix to constant values can have a large impact on the results.

You may also want to define constraints on the values of the parameters. For example, you might constrain a rate constant to have a value greater than zero. This step is optional, and is not needed for our example.

Step 5: Choose a weighting scheme

If you assume that the average scatter of data around the curve is the same all the way along the curve, you should instruct the nonlinear program to minimize the sum of the squared distances of the points from the curve. If the average scatter varies, you'll want to instruct the program to minimize some weighted sum-of-squares. Most commonly, you'll choose weighting when the average amount of scatter increases as the Y values increase, so the relative distance of the data from the curve is more consistent than the absolute distances. This topic is discussed in Chapter 14. Your choice here will rarely have a huge impact on the results.

For this example, we choose to minimize the sum-of-squares with no weighting.

Step 6: Choose initial values

Nonlinear regression is an iterative procedure. Before the procedure can begin, you need to define initial values for each parameter. If you choose a standard equation, your program may provide the initial values automatically. If you choose a user-defined equation that someone else wrote for you, that equation may be stored in your program along with initial values (or rules to generate initial values from the X and Y ranges of your data).

If you are using a new model and aren't sure the initial values are correct, you should instruct the nonlinear regression program to graph the curve defined by the initial values. With Prism, this is a choice on the first tab of the nonlinear regression dialog. If the resulting curve comes close to the data, you are ready to proceed. If this curve does not come near your data, go back and alter the initial values.

Step 7: Perform the curve fit and interpret the best-fit parameter values

Here are the results of the nonlinear regression as a table and a graph.

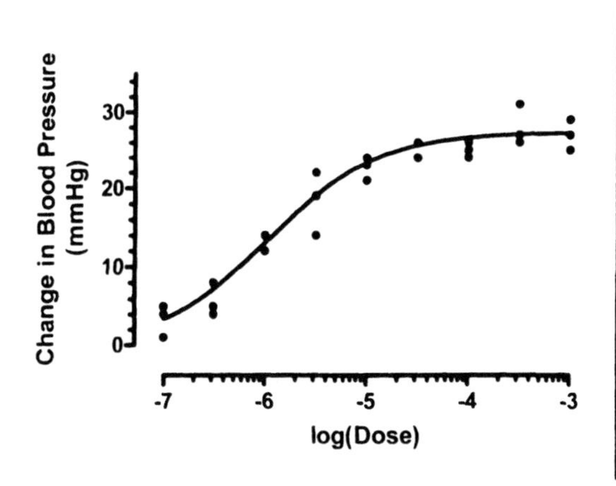

Best-fit values	
BOTTOM	0.0
TOP	27.36
LOGEC50	-5.946
HILLSLOPE	0.8078
EC50	1.1323e-006
Std. Error	
TOP	0.7377
LOGEC50	0.06859
HILLSLOPE	0.09351
95% Confidence Intervals	
TOP	25.83 to 28.88
LOGEC50	-6.088 to -5.804
HILLSLOPE	0.6148 to 1.001
EC50	8.1733e-007 to 1.5688e-006
Goodness of Fit	
Degrees of Freedom	24
R^2	0.9547
Absolute Sum of Squares	96.71
Sy.x	2.007

When evaluating the results, first ask yourself these five questions (see Chapter 4):

1. Does the curve come close to the data? You can see that it does by looking at the graph. Accordingly the R^2 value is high (see page 34).

2. Are the best-fit values scientifically plausible? In this example, all the best-fit values are sensible. The best-fit value of the $logEC_{50}$ is -5.9, right in the middle of your data. The best-fit value for the top plateau is 26.9, which looks about right by inspecting the graph. The best-fit value for the Hill slope is 0.84, close to the value of 1.0 you often expect to see.

3. How precise are the best-fit parameter values? You don't just want to know what the best-fit value is for each parameter. You also want to know how certain that value is. It isn't enough to look at the best-fit value. You should also look at the 95% confidence interval (or the SE values, from which the 95% CIs are calculated) to see how well you have determined the best-fit values. In this example, the 95% confidence intervals for all three fitted parameters are reasonably narrow (considering the number and scatter of the data points).

4. Would another model be more appropriate? Nonlinear regression finds parameters that make a model fit the data as closely as possible (given some assumptions). It does not automatically ask whether another model might work better. You can compare the fit of models as explained beginning in Chapter 21.

5. Have you violated any assumptions of nonlinear regression? The assumptions are discussed on page 30. Briefly, nonlinear regression assumes that you know X precisely, and that the variability in Y is random, Gaussian, and consistent all the way along the curve (unless you did special weighting). Furthermore, you assume that each data point contributes independent information.

See chapters 4 and 5 to learn more about interpreting the results of nonlinear regression.

2. Preparing data for nonlinear regression

Avoid Scatchard, Lineweaver-Burk, and similar transforms whose goal is to create a straight line

Before nonlinear regression was readily available, shortcuts were developed to analyze nonlinear data. The idea was to transform the data to create a linear graph, and then analyze the transformed data with linear regression. Examples include Lineweaver-Burk plots of enzyme kinetic data, Scatchard plots of binding data, and logarithmic plots of kinetic data.

> Tip: Scatchard, Lineweaver-Burk, and related plots are outdated. Don't use them to analyze data.

The problem with these methods is that they cause some assumptions of linear regression to be violated. For example, transformation distorts the experimental error. Linear regression assumes that the scatter of points around the line follows a Gaussian distribution and that the standard deviation is the same at every value of X. These assumptions are rarely true after transforming data. Furthermore, some transformations alter the relationship between X and Y. For example, when you create a Scatchard plot the measured value of Bound winds up on both the X axis(which plots Bound) and the Y axis (which plots Bound/Free). This grossly violates the assumption of linear regression that all uncertainty is in Y, while X is known precisely. It doesn't make sense to minimize the sum-of-squares of the vertical distances of points from the line if the same experimental error appears in both X and Y directions.

Since the assumptions of linear regression are violated, the values derived from the slope and intercept of the regression line are not the most accurate determinations of the parameters in the model. The figure below shows the problem of transforming data. The left panel shows data that follows a rectangular hyperbola (binding isotherm). The right panel is a Scatchard plot of the same data (see "Scatchard plots" on page 205). The solid curve on the left was determined by nonlinear regression. The solid line on the right shows how that same curve would look after a Scatchard transformation. The dotted line shows the linear regression fit of the transformed data. Scatchard plots can be used to determine the receptor number (B_{max}, determined as the X-intercept of the linear regression line) and dissociation constant (K_d, determined as the negative reciprocal of the slope). Since the Scatchard transformation amplified and distorted the scatter, the linear regression fit does not yield the most accurate values for B_{max} and K_d.

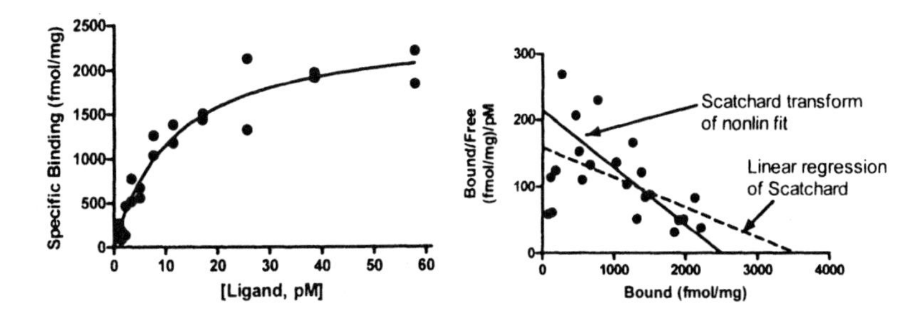

Don't use linear regression just to avoid using nonlinear regression. Fitting curves with nonlinear regression is not difficult. Considering all the time and effort you put into collecting data, you want to use the best possible technique for analyzing your data. Nonlinear regression produces the most accurate results.

Although it is usually inappropriate to *analyze* transformed data, it is often helpful to *display* data after a linear transform. Many people find it easier to visually interpret transformed data. This makes sense because the human eye and brain evolved to detect edges (lines), not to detect rectangular hyperbolas or exponential decay curves.

Transforming X values

Nonlinear regression minimizes the sum of the square of the *vertical* distances of the data points from the curve. Transforming X values only slides the data back and forth horizontally, and won't change the vertical distance between the data point and the curve. Transforming X values, therefore, won't change the best-fit values of the parameters, or their standard errors or confidence intervals.

In some cases, you'll need to adjust the model to match the transform in X values. For example, here is the equation for a dose-response curve when X is the logarithm of concentration.

```
Y = Bottom + (Top-Bottom)/(1+10^((LogEC50-X)*HillSlope))
```

If we transform the X values to be concentration, rather than the logarithm of concentration, we also need to adapt the equation. Here is one way to do it:

```
Y = Bottom + (Top-Bottom)/(1+10^((LogEC50-log(X))*HillSlope))
```

Note that both equations fit the logarithm of the EC50, and not the EC50 itself. See "Why you should fit the $logEC_{50}$ rather than EC_{50}" on page 263. Also see page 100 for a discussion of rate constants vs. time constants when fitting kinetic data. Transforming parameters can make a big difference in the reported confidence intervals. Transforming X values will not.

Don't smooth your data

Smoothing takes out some of the erratic scatter to show the overall trend of the data. This can be useful for data presentation, and is customary in some fields. Some instruments can smooth data as they are acquired.

Avoid smoothing prior to curve fitting. The problem is that the smoothed data violate some of the assumptions of nonlinear regression. Following smoothing, the residuals are no longer independent. You expect smoothed data points to be clustered above and below the curve. Furthermore, the distances of the smoothed points from the curve will not be Gaussian, and the computed sum-of-squares will underestimate the true amount of scatter. Accordingly, nonlinear regression of smoothed data will determine standard errors for the parameters that are too small, so the confidence intervals will be too narrow. You'll be misled about the precision of the parameters. Any attempt to compare models will be invalid because these methods are based on comparing sum-of-squares, which will be wrong.

The graphs below show original data (left) and the data after smoothing (right).

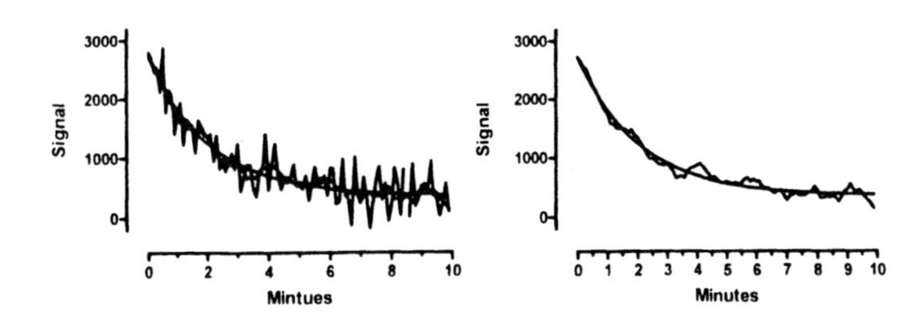

Parameter	Fit actual data Best-fit	SE	Fit smoothed data Best-fit	SE
Y_0	2362	120.5	2363	38.81
k	0.4902	0.05577	0.4869	0.01791
Plateau	366.4	56.70	362.4	18.43
Sum-of-squares	7,811,565		814,520	

The best-fit values of the parameters are almost the same when you fit smoothed data as when you fit the actual data. But the standard errors are very different. Smoothing the data reduced the sum-of-squares by about a factor of nine, and so reduced the reported standard errors by about a factor of three. But smoothing the data doesn't make the estimates of the parameter values more precise. The standard errors and confidence intervals are simply wrong since the curve fitting program was mislead about the scatter in the data.

Transforming Y values

Dividing, or multiplying, all Y values by a constant is OK.

Multiplying or dividing all Y values by a constant does not change the best-fit curve. You'll get the same (or equivalent) best-fit values of the parameters with the same confidence intervals. It is a good idea to transform to change units to prevent your Y values from being very high or very low (see next section). And it is OK to multiply or divide by a constant to make the units more convenient.

Note that some parameters are expressed in units of the Y axis, so if you change the Y values, you'll also change the units of the parameter. For example, in a saturation binding study, the units of the B_{max} (maximum binding) are the same as the units of the Y axis. If you transform the Y values from counts/minute to sites/cell, you'll also change the units of the B_{max} parameter. The best-fit value of that parameter will change its numerical value, but remain equivalent.

Subtracting a constant is OK

Subtracting a constant from all Y values will not change the distance of the points from the best-fit curve, so will not affect which curve gets selected by nonlinear regression.

Think carefully about nonlinear transforms

As mentioned above, transforming Y values with a linear transformation (such as dividing all values by a constant, or subtracting a constant from all values), won't change the nature of the best-fit curve. In contrast, nonlinear transformations (such as converting Y values to their logarithms, square roots, or reciprocals) will change the relative position of data points from the curve and cause a different curve to minimize the sum-of-squares. Therefore, a nonlinear transformation of Y values will lead to different best-fit parameter values. Depending on the data, this can be good or bad.

Nonlinear regression is based on the assumption that the scatter of data around the curve follows a Gaussian distribution. If the scatter of your data is in fact Gaussian, performing a nonlinear transform will invalidate your assumption. If you have data with Gaussian scatter, avoid nonlinear Y transforms. If, however, your scatter is not Gaussian, a nonlinear transformation might make the scatter more Gaussian. In this case, it is a good idea to apply nonlinear transforms to your Y values.

Change units to avoid tiny or huge values

In pure math, it makes no difference what units you use to express your data. When you analyze data with a computer, however, it can matter. Computers can get confused by very small or very large numbers, and round-off errors can result in misleading results.

When possible, try to keep your Y values between about 10^{-9} and 10^{9}, changing units if necessary. The scale of the X values usually matters less, but we'd suggest keeping X values within that range as well.

> Note: This guideline is just that. Most computer programs will work fine with numbers much larger and much smaller. It depends on which program and which analysis.

Normalizing

One common way to normalize data is to subtract off a baseline and then divide by a *constant*. The goal is to make all the Y values range between 0.0 and 1.0 or 0% and 100%. Normalizing your data using this method, by itself, will not affect the results of nonlinear regression. You'll get the same best-fit curve, and equivalent best-fit parameters and confidence intervals.

If you normalize from 0% to 100%, some points may end up with normalized values less than 0% or greater than 100. What should you do with such points? Your first reaction might be that these values are clearly erroneous, and should be deleted. This is not a good idea. The values you used to define 0% and 100% are not completely accurate. And even if they were, you expect random scatter. So some points will end up higher than 100% and some points will end up lower than 0%. Leave those points in your analysis; don't eliminate them.

Don't confuse two related decisions:

- Should you normalize the data? Normalizing, by itself, will *not* change the best-fit parameters (unless your original data had such huge or tiny Y values that curve fitting encountered computer round-off or overflow errors). If you normalize your data, you may find it easier to see what happened in the experiment and to compare results with other experiments.

- Should you constrain the parameters that define bottom and/or top plateaus to constant values? Fixing parameters to constant values *will* change the best-fit results of the remaining parameters, as shown in the example below.

Averaging replicates

If you collected replicate Y values (say triplicates) at each value of X, you may be tempted to average those replicates and only enter the mean values into the nonlinear regression program. Hold that urge! In most situations, you should enter the raw replicate data. See page 87.

Consider removing outliers

When analyzing data, you'll sometimes find that one value is far from the others. Such a value is called an *outlier*, a term that is usually not defined rigorously. When you encounter an outlier, you may be tempted to delete it from the analyses. First, ask yourself these questions:

- Was the value entered into the computer correctly? If there was an error in data entry, fix it.

- Were there any experimental problems with that value? For example, if you noted that one tube looked funny, you have justification to exclude the value resulting from that tube without needing to perform any calculations.

- Could the outlier be caused by biological diversity? If each value comes from a different person or animal, the outlier may be a correct value. It is an outlier not because of an experimental mistake, but rather because that individual may be different from the others. This may be the most exciting finding in your data!

After answering "no" to those three questions, you have to decide what to do with the outlier. There are two possibilities.

- One possibility is that the outlier was due to chance. In this case, you should keep the value in your analyses. The value came from the same distribution as the other values, so it should be included.

- The other possibility is that the outlier was due to a mistake - bad pipetting, voltage spike, holes in filters, etc. Since including an erroneous value in your analyses will give invalid results, you should remove it. In other words, the value comes from a different population than the others and is misleading.

The problem, of course, is that you can never be sure which of these possibilities is correct. Statistical calculations can quantify the probabilities.

Statisticians have devised several methods for detecting outliers. All the methods first quantify how far the outlier is from the other values. This can be the difference between the outlier and the mean of all points, the difference between the outlier and the mean of the remaining values, or the difference between the outlier and the next closest value. Next, standardize this value by dividing by some measure of scatter, such as the SD of all values, the SD of the remaining values, or the range of the data. Finally, compute a P value answering this question: If all the values were really sampled from a Gaussian population, what is the chance of randomly obtaining an outlier so far from the other values? If this probability is small, then you will conclude that the outlier is likely to be an erroneous value, and you have justification to exclude it from your analyses.

One method of outlier detection (Grubbs' method) is described in the companion book, *Prism 4 Statistics Guide*. No outlier test will be very useful unless you have lots (say a dozen or more) replicates.

Be wary of removing outliers simply because they seem "too far" from the rest. All the data points in the figure below were generated from a Gaussian distribution with a mean of 100 and a SD of 15. Data sets B and D have obvious "outliers". Yet these points came from the same Gaussian distribution as the rest. If you removed these values as outliers, the mean of the remaining points would be further from the true value (100) rather than closer to it. Removing those points as "outliers" makes any further analysis less accurate.

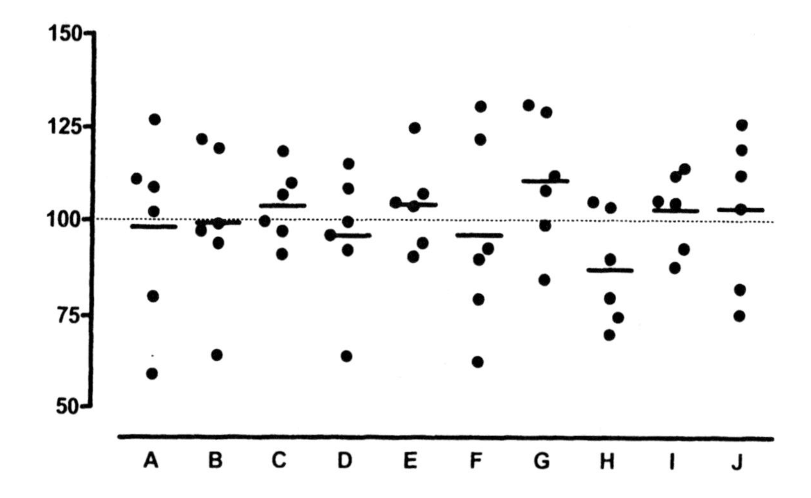

Tip: Be wary when removing a point that is obviously an outlier.

3. Nonlinear regression choices

Choose a model for how Y varies with X

Nonlinear regression fits a model to your data. In most cases, your goal is to get back the best-fit values of the parameters in that model. If so, it is crucial that you pick a sensible model. If the model makes no sense, even if it fits the data well, you won't be able to interpret the best-fit values.

In other situations, your goal is just to get a smooth curve to use for graphing or for interpolating unknown values. In these cases, you need a model that generates a curve that goes near your points, and you won't care whether the model makes scientific sense.

Much of this book discusses how to pick a model, and explains models commonly used in biology.

If you want to fit a global model, you must specify which parameters are shared among data sets and which are fit individually to each data set. This works differently for different programs. See page 70.

Fix parameters to a constant value?

As part of picking a model, you need to decide which parameters in the model you will set to a constant value based on control data. For example, if you are fitting an exponential decay, you need to decide if the program will find a best-fit value for the bottom plateau or whether you will set that to a constant value (perhaps zero). This is an important decision, that is best explained through example.

Here are data to be fit to an exponential decay.

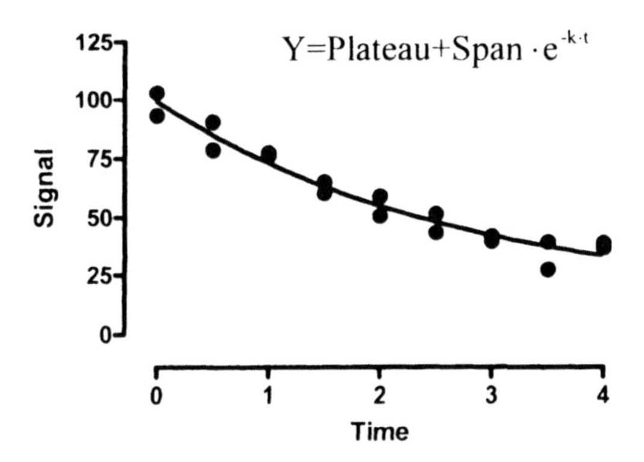

We fit the data to an exponential decay, asking the computer to fit three parameters, the starting point, the rate constant, and the bottom plateau. The best-fit curve shown above looks fine. But as you can see below, many exponential decay curves fit your data almost equally well.

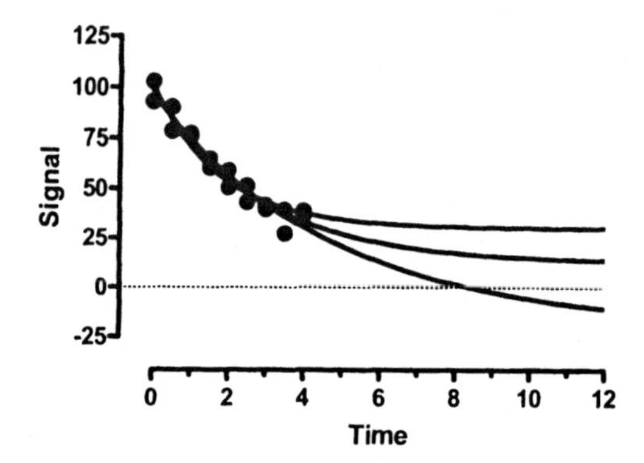

The data simply don't define all three parameters in the model. You didn't collect data out to long enough time points, so your data simply don't define the bottom plateau of the curve. If you ask the program to fit all the parameters, the confidence intervals will be very wide – you won't determine the rate constant with reasonable precision. If, instead, you fix the bottom parameter to 0.0 (assuming you have normalized the data so it has to end up at zero), then you'll be able to determine the rate constant with far more precision.

> Tip: If your data don't define all the parameters in a model, try to constrain one or more parameters to constant values.

It can be a mistake to fix a parameter to a constant value, when that constant value isn't quite correct. For example, consider the graph below with dose-response data fit two ways. First, we asked the program to fit all four parameters: bottom plateau, top plateau, $\log EC_{50}$ (the middle of the curve) and the slope. The solid curve shows that fit. Next, we used the mean of the duplicates of the lowest concentration to define the bottom plateau (1668), and the mean of the duplicates of the highest concentration to define the top plateau (4801). We asked the program to fix those parameters to constant values, and only fit logEC50 and slope. The dashed curve shows the results. By chance, the response at the lowest concentration was a bit higher than it was for the next two concentrations. By forcing the curve to start at this (higher) value, we pushed the $\log EC_{50}$ to the right. When we fit all four parameters, the best-fit value of the $\log EC_{50}$ was -6.00. When we fixed the top and bottom plateaus to constant values, the best-fit value of the $\log EC_{50}$ was -5.58.

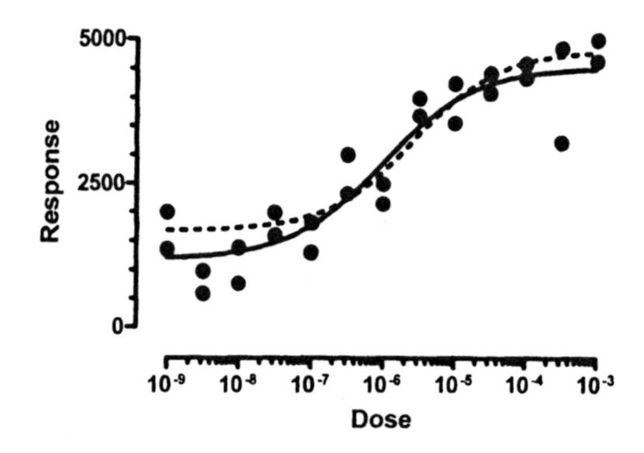

In this example, fixing parameters to constant values (based on control measurements) was a mistake. There are plenty of data points to define all parts of the curve. There is no need to define the plateaus based on duplicate determinations of response at the lowest and highest dose.

> Tip: If you are going to constrain a parameter to a constant value, make sure you know that value quite accurately.

Initial values

Nonlinear regression is an iterative procedure. The program must start with estimated initial values for each parameter. It then adjusts these initial values to improve the fit.

If you pick an equation built into your nonlinear regression program, it will probably provide initial values for you. If you write your own model, you'll need to provide initial values or, better, rules for computing the initial values from the range of the data.

You'll find it easy to estimate initial values if you have looked at a graph of the data, understand the model, and understand the meaning of all the parameters in the equation. Remember that you just need estimated values. They don't have to be very accurate.

If you are having problems estimating initial values, set aside your data and simulate a family of curves (in Prism, use the analysis "Create a family of theoretical curves"). Once you have a better feel for how the parameters influence the curve, you might find it easier to go back to nonlinear regression and estimate initial values.

Another approach to finding initial values is to analyze your data using a linearizing method such as a Scatchard or Lineweaver-Burk plot. While these methods are obsolete for analyzing data (see page 19), they are reasonable methods for generating initial values for nonlinear regression. If you are fitting exponential models with multiple phases, you can obtain initial values via curve stripping (see a textbook of pharmacokinetics for details).

Page 43 shows an example where poor initial values prevent nonlinear regression from finding best-fit values of the parameters.

Weighting

The goal of regression is to find best-fit values for the parameters of the model (e.g., slope and intercept for linear regression, other parameters, such as rate constants, for nonlinear regression). More precisely, the goal is the find values for the parameters that are most likely to be correct. It turns out that you can't decide which parameter values are most likely to be correct without first making an assumption about how the data are scattered around the line or curve.

Most commonly, linear and nonlinear regression make two assumptions:

- The scatter follows a Gaussian (also called a "normal") distribution.
- The standard deviation of the scatter (the average amount of scatter) is the same for all values of X.

Given these two assumptions, how does nonlinear regression decide which parameter values are most likely to be correct? The answer, discussed in more detail in Chapter 13, is that the parameters that are most likely to be correct are those that generate a curve that minimizes the sum of the squares of the vertical distance between data points and curve. In other words, least-squares regression minimizes:

$$\sum \left(Y_{Data} - Y_{Curve}\right)^2$$

If the scatter really is Gaussian and uniform, least-squares regression finds the parameter values that are most likely to be correct.

It is possible to change the first assumptions, and make some other assumption about the scatter. But the Gaussian assumption works well, and is used for almost all curve fitting. The only real exception is robust nonlinear regression, designed to reduce the influence of outliers, which uses different assumptions.

> GraphPad note: Prism 4 does not perform robust nonlinear regression, but the next version almost certainly will.

The second assumption is that the average amount of scatter is the same all the way along the curve, at all values of X. This assumption is often not true. Instead, the average amount of scatter often increases as Y increases. With this kind of data, a least-squares method tends to give undue weight to the points with large Y values, and ignores points with low Y values. To prevent this, it is common to apply a weighting scheme.

The most common alternative to minimizing the sum of the squares of the vertical distances of the points from the curve, is to minimize the sum of the squares of the relative distances of the points from the curve. In other words, minimize this quantity:

$$\sum \left(\frac{Y_{Data} - Y_{Curve}}{Y_{Data}}\right)^2$$

This is called relative weighting or weighting by $1/Y^2$.

If you know that the scatter is not uniform, you should choose an appropriate weighting scheme. Often it will be hard to know what weighting scheme is appropriate. Fortunately, it doesn't matter too much. Simulations have shown that picking the wrong weighting scheme, while making the best-fit values less accurate, doesn't make as large an impact as you might guess.

If you collected replicate values at each value of X, you might be tempted to weight by the standard deviation of the replicates. When replicates are close together with a small standard deviation, give that point lots of weight. When replicates are far apart, and have a large standard deviation, give that point little weight. While this sounds sensible, it actually doesn't work very well. To see why, read the section starting on page 87.

See Chapter 14 for more information on unequal weighting of data points.

Other choices

Your program will give you lots of choices for additional calculations, and for how you want to format the results. For GraphPad Prism, see Chapters 45 to 47. If you are new to nonlinear regression, leave all these settings to their default values, and spend your time learning about models and about interpreting the results. Then go back and learn about the various options.

4. The first five questions to ask about nonlinear regression results

Nonlinear regression programs can produce lots of output, and it can be hard to know what to look at first. This chapter explains the first five questions you should ask when interpreting nonlinear regression results. The next chapter explains the rest of the results.

Does the curve go near your data?

The whole point of nonlinear regression is to fit a model to your data – to find parameter values that make the curve come near your data points. The first step in evaluating the results, therefore, is to look at a graph of your data with the superimposed curve.

> Tip: Don't neglect this first step. Look at your data with the superimposed curve.

If your goal is simply to create a standard curve from which to interpolate unknown values, it may be enough to look at the curve. You may not care about the best-fit values of the parameters and their standard errors and confidence intervals. More commonly, however, you will be fitting curves to your data in order to understand your system. Under this circumstance, you need to look carefully at the nonlinear regression results. Nonlinear regression generates quite a few results, and the rest of this chapter is guide to interpreting them.

Are the best-fit parameter values plausible?

When evaluating the parameter values reported by nonlinear regression, the first thing to do is check that the results are scientifically plausible. When a computer program fits a model to your data, it can't know what the parameters mean. Therefore, it can report best-fit values of the parameters that make no scientific sense. For example, make sure that parameters don't have impossible values (rate constants simply cannot be negative). Check that EC_{50} values are within the range of your data. Check that maximum plateaus aren't too much higher than your highest data point.

If the best-fit values are not scientifically sensible, then the fit is no good. Consider constraining the parameters to a sensible range, and trying again.

For help with troubleshooting results that just don't make sense, see Chapter 6.

How precise are the best-fit parameter values?

You don't just want to know what the best-fit value is for each parameter. You also want to know how certain that value is.

Nonlinear regression programs report the standard error of each parameter, as well as the 95% confidence interval. If your program doesn't report a 95% confidence interval, you can calculate one using the equation on page 103. Or you can quickly estimate the confidence interval as the best-fit value plus or minus two standard errors.

If all the assumptions of nonlinear regression are true, there is a 95% chance that the interval contains the true value of the parameter. If the confidence interval is reasonably narrow, you've accomplished what you wanted to do – found the best fit value of the parameter with reasonable certainty. If the confidence interval is really wide, then you've got a problem. The parameter could have a wide range of values. You haven't nailed it down. See Chapter 6 for troubleshooting tips.

Note that the confidence interval and standard error are calculated from one experiment based on the scatter of the points around the curve. This gives you a sense of how well you have determined a parameter. But before making any substantial conclusions, you'll want to repeat the experiment. Then you can see how the parameter varies from experiment to experiment. This includes more sources of experimental variation than you see in a single experiment.

For details, see part E. Confidence intervals of the parameters.

> Tip: The standard errors and confidence intervals are important results of regression. Don't overlook them.

Would another model be more appropriate?

Nonlinear regression finds parameters that make a model fit the data as closely as possible (given some assumptions). It does not automatically ask whether another model might work better. You can compare the fit of models as explained beginning in Chapter 21.

Even though a model fits your data well, it may not be the best, or most correct, model. You should always be alert to the possibility that a different model might work better. In some cases, you can't distinguish between models without collecting data over a wider range of X. In other cases, you would need to collect data under different experimental conditions. This is how science moves forward. You consider alternative explanations (models) for your data, and then design experiments to distinguish between them.

Have you violated any of the assumptions of nonlinear regression?

Nonlinear regression is based on a set of assumptions. When reviewing your results, therefore, you should review the assumptions and make sure you have not violated them.

- X is known precisely. All the "error" (scatter) is in Y.
- The variability of Y values at any particular X value follows a known distribution. Almost always, this is assumed to be a Gaussian (normal) bell-shaped distribution.
- Standard nonlinear regression assumes that the amount of scatter (the SD of the residuals) is the same all the way along the curve. This assumption of uniform variance is called *homoscedasticity.* Weighted nonlinear regression assumes that the scatter is predictably related to the Y value. See Chapter 14.
- The observations are independent. When you collect a Y value at a particular value of X, it might be higher or lower than the average of all Y values at that X value. Regression assumes that this is entirely random. If one point

happens to have a value a bit too high, the next value is equally likely to be too high or too low.

If you choose a global model, you also are making two additional assumptions (see page 69):

- All data sets are expressed in same units. If different data sets were expressed in different units, you could give different data sets different weights just by expressing the values in different units.

- The scatter is the same for each data set. At each X value, for each data set, the scatter should be the same (unweighted fit) or should vary predictably with Y (and be accounted for by weighting).

5. The results of nonlinear regression

The previous chapter discussed the five most important questions to ask when reviewing results of nonlinear regression. This chapter explains the rest of the results.

Confidence and prediction bands

The plot of the best-fit curve can include the 95% confidence band of the best-fit curve, or the 95% prediction band. The two are very different.

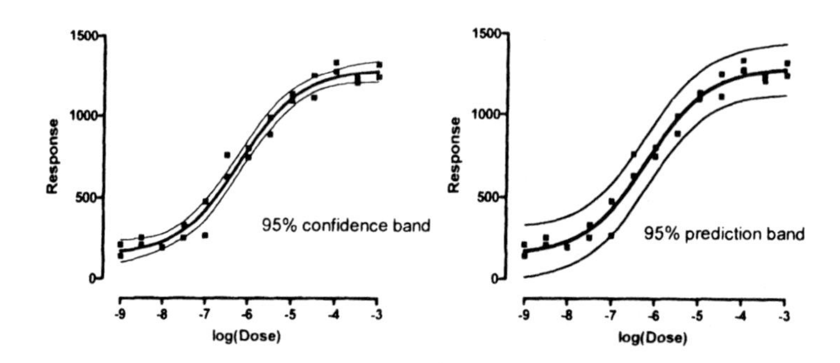

The confidence band tells you how well you know the curve. You can be 95% sure that the true best-fit curve (which you could only know if you had an infinite number of data points) lies within the confidence band.

The prediction band shows the scatter of the data. If you collected many more data points, you would expect 95% to fall within the prediction band.

Since the prediction band has to account for uncertainty in the curve itself as well as scatter around the curve, the prediction band is much wider than the confidence band. As you increase the number of data points, the confidence band gets closer and closer to the best-fit curve, while the prediction band doesn't change predictably. In the example below, note that the confidence bands (shown as solid) contain a minority of the data points. That's ok. The confidence bands have a 95% chance of containing the true best-fit curve, and with so much data these bands contain far fewer than half the data points. In contrast, the dashed prediction bands include 95% of the data points.

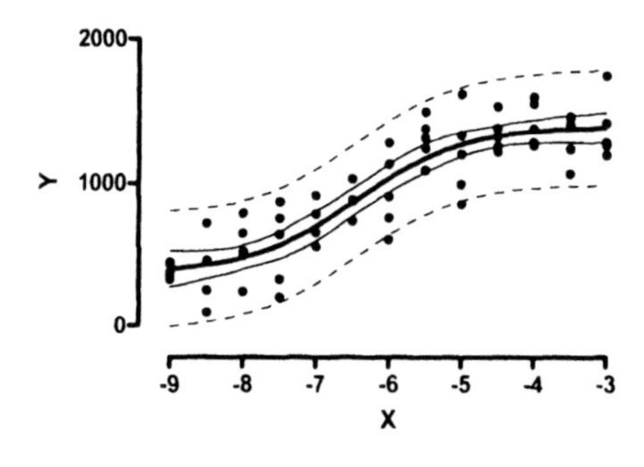

Correlation matrix

If the standard errors are large (so the confidence intervals are wide), you'll want to investigate further. One possibility is that the model is somewhat redundant, so two parameters are linked. Some programs report a correlation matrix to help diagnose this problem.

For every pair of parameters, the correlation matrix reports a value showing you how tightly those two parameters are linked. This value is like a correlation coefficient, ranging in value from -1 to +1.

The program finds the best fit values of each parameter. That means that if you change the value of any parameter (without changing the rest), the sum of squares will go up (the fit will be worse). But what if you change the value of one parameter and fix it, and then ask the program to find a new best-fit value for another parameter? One extreme is when the parameters are completely unlinked. Changing one parameter makes the fit worse, and you can't compensate at all by changing the other. In this extreme case, the value reported by the correlation matrix would be zero. The other extreme is when the two parameters are completely intertwined. Changing one parameter makes the fit worse, but this can be completely compensated for by changing the other. The value reported in the correlation matrix would be +1 (if you compensate for an increase in one parameter by increasing the other) or -1 (if you compensate for an increase in one parameter by decreasing the other).

Parameters in most models are somewhat related, so values in the range -0.8 to + 0.8 are seen often. Higher correlations (greater than 0.9, and especially greater than 0.99, or lower than -0.9 or -0.99) mean that data simply don't define the model unambiguously. One way to solve this problem is to simplify the model, perhaps by fixing a parameter to a constant value. Another solution is to collect more data (either use a wider range of X values, or include data from another kind of experiment in a global fit).

> GraphPad note: Prism does not report the correlation matrix. Let us know if you'd like to see the covariance matrix in future versions.

Sum-of-squares

Sum-of-squares from least squares nonlinear regression

The *sum-of-squares* (SS) is the sum of the squares of the vertical distances of the points from the curve. It is expressed in the units used for the Y values, squared. Standard (least squares) nonlinear regression works by varying the values of the model parameters to minimize SS. If you chose to weight the values and minimize the relative distance squared (or some other weighting function), goodness-of-fit is quantified with the weighted sum-of-squares.

You'll only find the SS value useful if you do additional calculations to compare models.

$S_{y.x}$ (Root mean square)

The equation below calculates the value, $s_{y.x}$, from the sum-of-squares and degrees of freedom:

$$s_{y \cdot x} = \sqrt{\frac{SS}{N-P}}$$

$S_{y.x}$ is the standard deviation of the vertical distances of the points from the line, N is the number of data points, and P is the number of parameters. Since the distances of the points from the line are called residuals, $s_{y.x}$ is the *standard deviation of the residuals*. Its value is expressed in the same units as Y. Some programs call this value s_e.

R^2 (coefficient of determination)

The value R^2 quantifies goodness of fit. R^2 is a fraction between 0.0 and 1.0, and has no units. Higher values indicate that the curve comes closer to the data. You can interpret R^2 from nonlinear regression very much like you interpret r^2 from linear regression. By tradition, statisticians use uppercase (R^2) for the results of nonlinear and multiple regressions and lowercase (r^2) for the results of linear regression, but this is a distinction without a difference.

When R^2 equals 0.0, the best-fit curve fits the data no better than a horizontal line going through the mean of all Y values. In this case, knowing X does not help you predict Y. When R^2=1.0, all points lie exactly on the curve with no scatter. If you know X, you can calculate Y exactly. You can think of R^2 as the fraction of the total variance of Y that is explained by the model (equation).

> Tip: Don't make the mistake of using R^2 as your main criterion for whether a fit is reasonable. A high R^2 tells you that the curve came very close to the points, but doesn't tell you that the fit is sensible in other ways. The best-fit values of the parameters may have values that make no sense (for example, negative rate constants) or the confidence intervals may be very wide.

R^2 is computed from the sum of the squares of the distances of the points from the best-fit curve determined by nonlinear regression. This sum-of-squares value is called SS_{reg}, which is in the units of the Y-axis squared. To turn R^2 into a fraction, the results are normalized to the sum of the square of the distances of the points from a horizontal line through the mean of all Y values. This value is called SS_{tot}. If the curve fits the data well, SS_{reg} will be much smaller than SS_{tot}.

The figure below illustrates the calculation of R^2. Both panels show the same data and best-fit curve. The left panel also shows a horizontal line at the mean of all Y values, and vertical lines show how far each point is from the mean of all Y values. The sum of the square of these distances (SS_{tot}) equals 62735. The right panel shows the vertical distance of each point from the best-fit curve. The sum of squares of these distances (SS_{reg}) equals 4165.

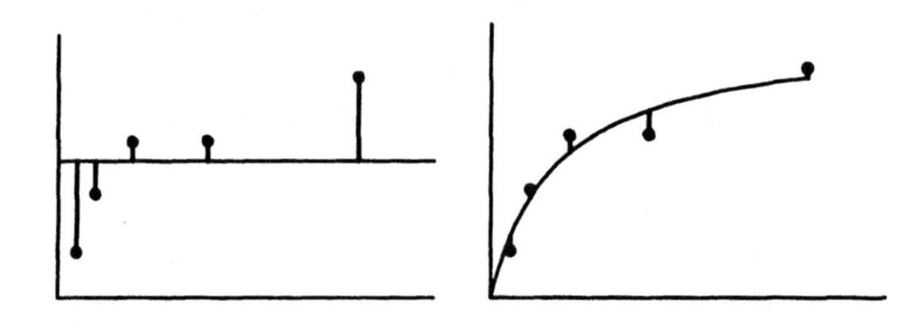

R^2 is calculated using this equation:

$$R^2 = 1.0 - \frac{SS_{reg}}{SS_{tot}} = 1.0 - \frac{4165}{62735} = 1.0 - 0.0664 = 0.9336$$

Note that R^2 is not really the square of anything. If SS_{reg} is larger than SS_{tot}, R^2 will be negative. While it is surprising to see something called "squared" have a negative value, it is not impossible (since R^2 is not actually the square of R). R^2 will be negative when the best-fit curve fits the data worse than a horizontal line at the mean Y value. This could happen if you pick an inappropriate model (maybe you picked an exponential association model rather than an exponential dissociation model), or enforce an inappropriate constraint (for example, if you fix the Hill slope of a dose-response curve to 1.0 when the curve goes downhill).

> Warning: If you want to compare the fit of two equations, don't just compare R^2 values. Comparing curves is more complicated than that.

Does the curve systematically deviate from the data?

If you've picked an appropriate model, and your data follow the assumptions of nonlinear regression, the data will be randomly distributed around the best-fit curve. You can assess this in three ways:

- The distribution of points around the curve should be Gaussian.
- The average distance of points from curve should be the same for all parts of the curve (unless you used weighting).
- Points should not be clustered. Whether each point is above or below the curve should be random.

Residuals and *runs* help you evaluate whether the curve deviates systematically from your data.

Residuals from nonlinear regression

A *residual* is the distance of a point from the curve. A residual is positive when the point is above the curve, and is negative when the point is below the curve. If you listed the residuals of a particular curve fit in the form of a table, you would use the same X values as the original data, but the Y values would be the vertical distances of each corresponding point from the curve.

A sample residual plot is shown below. If you look carefully at the curve on the left, you'll see that the data points are not randomly distributed above and below the curve. There are clusters of points at early and late times that are below the curve, and a cluster of points at middle time points that are above the curve. This is much easier to see on the graph of the residuals in the inset. The data are not randomly scattered above and below the X-axis.

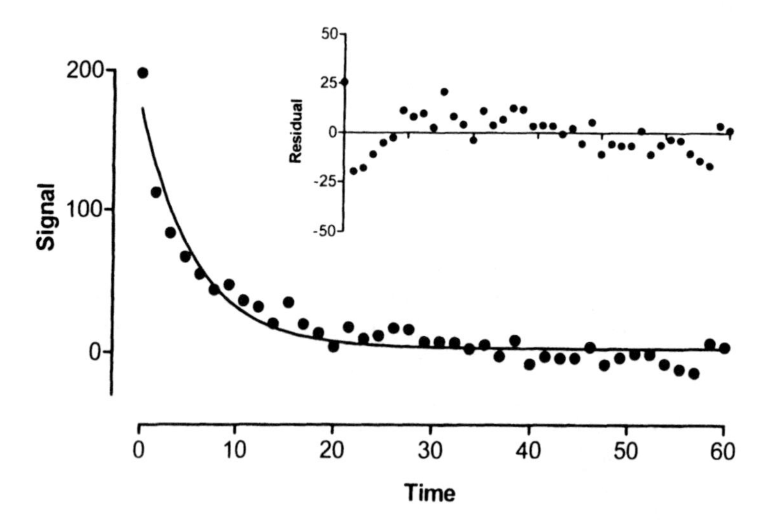

Runs test from nonlinear regression

The *runs test* determines whether the curve deviates systematically from your data. A run is a series of consecutive points that are either all above or all below the regression curve. Another way of saying this is that a run is a consecutive series of points whose residuals are either all positive or all negative.

If the data points are randomly distributed above and below the regression curve, it is possible to calculate the expected number of runs. If there are N_a points above the curve and N_b points below the curve, the number of runs you expect to see equals $[(2N_aN_b)/(N_a+N_b)]+1$. In most cases, you'll find about equal number of points above and below the curve, so N_a and N_b each equal N/2, where N is the total number of points. With this approximation, the rule is that you expect to see about 1+ N/2 runs.

If you observe fewer runs than expected, it may be a coincidence or it may mean that you picked an inappropriate regression model and the curve systematically deviates from your data. The P value from the runs test answers this question: If the data really follow the model you selected, what is the chance that you would obtain as few (or fewer) runs as observed in this experiment?

The P values are always one-tailed, asking about the probability of observing as few runs (or fewer) than observed. If you observe more runs than expected, the P value will be higher than 0.50.

If the runs test reports a low P value, conclude that the data don't really follow the equation you have selected.

In the example above, you expect 21 runs. There are 13 runs, and the P value for the runs test is 0.0077. If the data were randomly scattered above and below the curve, there is less than a 1% chance of observing so few runs. The data systematically deviate from the curve. Most likely, the data were fit to the wrong equation.

Testing whether the residuals are Gaussian

Least-squares nonlinear regression (as well as linear regression) assumes that the distribution of residuals follows a Gaussian distribution (robust nonlinear regression does not make this assumption). You can test this assumption by using *a normality test* on the residuals. This test is usually available in a many statistical and some curve-fitting programs.

> GraphPad note: Prism allows you to test whether your data or residuals follow a Gaussian distribution by performing a normality test on the residuals. From the residual table, click Analyze and choose Column Statistics. Then choose a normality test as part of the Column Statistics analysis.

Could the fit be a local minimum?

The nonlinear regression procedure adjusts the parameters in small steps in order to improve the goodness-of-fit. If the procedure converges on an answer, you can be sure that altering any of the variables a little bit will make the fit worse. But it is theoretically possible that large changes in the variables might lead to a much better goodness-of-fit. Thus the curve that nonlinear regression decides is the "best" may really not be the best.

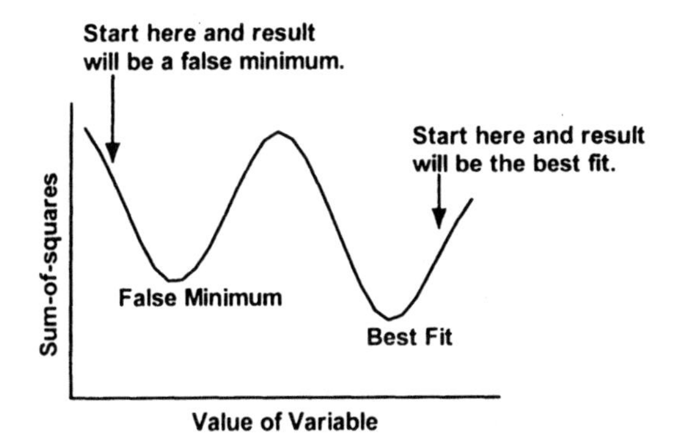

Think of latitude and longitude as representing two parameters you are trying to fit. Now think of altitude as the sum-of-squares. Nonlinear regression works iteratively to reduce the sum-of-squares. This is like walking downhill to find the bottom of the valley. See page 91. When nonlinear regression has converged, changing any parameter increases the sum-of-squares. When you are at the bottom of the valley, every direction leads uphill. But there may be a much deeper valley over the ridge that you are unaware of. In nonlinear regression, large changes in parameters might decrease the sum-of-squares.

This problem (called finding a local minimum) is intrinsic to nonlinear regression, no matter what program you use. You will rarely encounter a local minimum if you have collected data with little scatter and over an appropriate range of X values, have chosen an appropriate equation, and have provided a sensible initial value for each parameter.

To test for the presence of a false minimum, run nonlinear regression several times. Each time, provide a different set of initial values. If these fits all generate the same parameter values, with the same sum-of-squares, you can be confident you have not encountered a false minimum.

6. Troubleshooting “bad” fits

This chapter shows you five examples of bad fits, and explains what went wrong in each one and how the analysis can be salvaged.

Poorly defined parameters

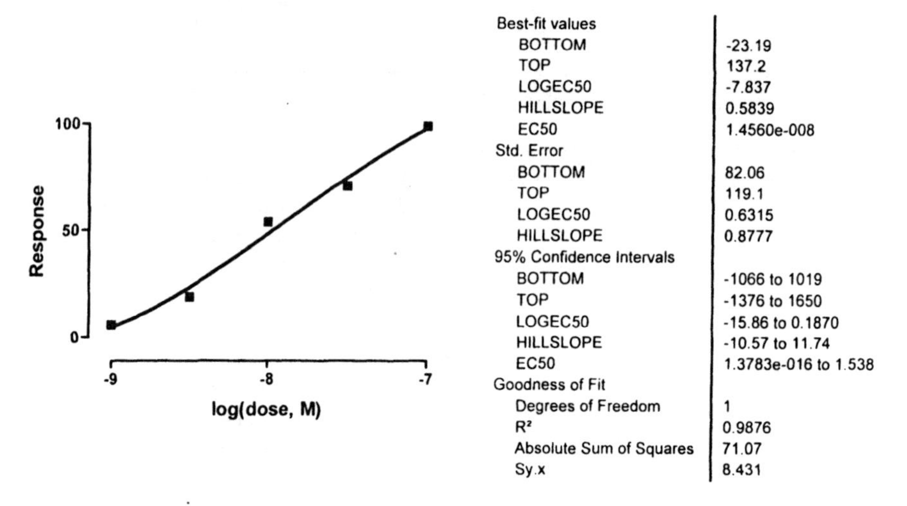

Best-fit values	
BOTTOM	-23.19
TOP	137.2
LOGEC50	-7.837
HILLSLOPE	0.5839
EC50	1.4560e-008
Std. Error	
BOTTOM	82.06
TOP	119.1
LOGEC50	0.6315
HILLSLOPE	0.8777
95% Confidence Intervals	
BOTTOM	-1066 to 1019
TOP	-1376 to 1650
LOGEC50	-15.86 to 0.1870
HILLSLOPE	-10.57 to 11.74
EC50	1.3783e-016 to 1.538
Goodness of Fit	
Degrees of Freedom	1
R^2	0.9876
Absolute Sum of Squares	71.07
Sy.x	8.431

In this example, a response was measured at five concentrations, and the results were fit to a sigmoidal dose-response curve with variable slope, also called a four-parameter logistic equation or the Hill equation.

When you first look at the results, all may seem fine. The R^2 is high, and both the $\log EC_{50}$ and the Hill slope have reasonable best-fit values.

A closer look shows you the trouble. The confidence interval of the EC_{50} spans 15 orders of magnitude (ranging from homeopathy to insolubility). The confidence interval for the Hill Slope is also extremely wide, extending from -11 (a steep curve heading down as you go from left to right) to +12 (a steep curve heading up hill). With such wide confidence intervals, the curve fit is not at all helpful.

> Tip: Always look at the confidence intervals for best-fit values. If the confidence intervals are extremely wide, the fit is not likely to be very helpful.

The problem is clear once you think about it. Nonlinear regression was asked to fit four parameters - the bottom plateau, the top plateau, $\log EC_{50}$ and Hill slope. But the data don’t extend far enough to define the bottom and top plateaus. Since the top and bottom are not defined precisely, neither is the middle, so the $\log EC_{50}$ is very imprecise.

It is easy to salvage this experiment and get useful results. Since the data were normalized to run from 0 to 100, there is no need to ask the nonlinear regression program to fit the best-fit values of the Bottom and Top. Instead, constrain Bottom to equal 0 and Top to equal 100. Now you only have to ask the program to fit two parameters ($\log EC_{50}$ and slope). Here are the revised results.

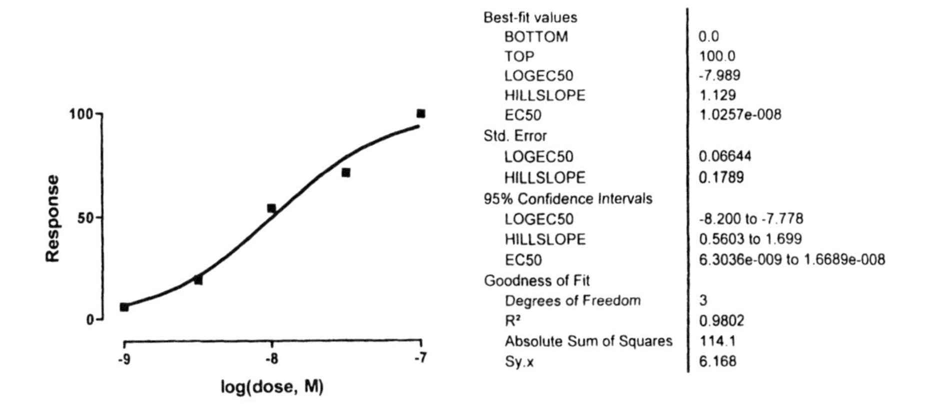

Best-fit values	
BOTTOM	0.0
TOP	100.0
LOGEC50	-7.989
HILLSLOPE	1.129
EC50	1.0257e-008
Std. Error	
LOGEC50	0.06644
HILLSLOPE	0.1789
95% Confidence Intervals	
LOGEC50	-8.200 to -7.778
HILLSLOPE	0.5603 to 1.699
EC50	6.3036e-009 to 1.6689e-008
Goodness of Fit	
Degrees of Freedom	3
R^2	0.9802
Absolute Sum of Squares	114.1
Sy.x	6.168

The confidence interval of the $\log EC_{50}$ extends plus or minus 0.2 log units, so the EC50 is known within a factor of 1.6 (the antilog of 0.2) . Since the curve is defined by so few points, the Hill Slope is still not defined very well, but the confidence interval is much narrower than it was before.

> Tip: If your data are normalized in any way, think carefully about whether any of the parameters in the equation you selected can be fixed to a constant value.

Model too complicated

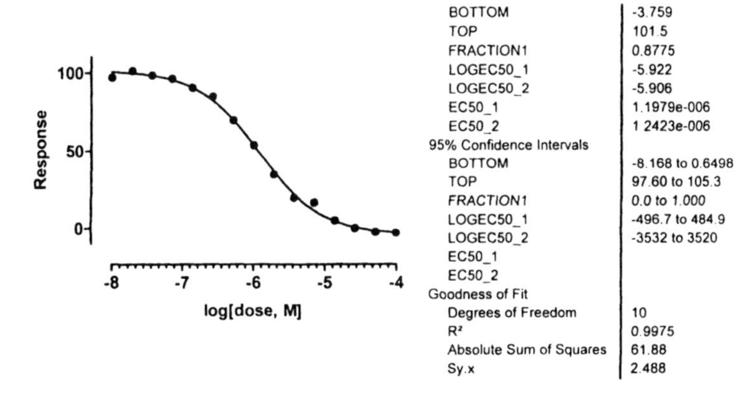

BOTTOM	-3.759
TOP	101.5
FRACTION1	0.8775
LOGEC50_1	-5.922
LOGEC50_2	-5.906
EC50_1	1.1979e-006
EC50_2	1.2423e-006
95% Confidence Intervals	
BOTTOM	-8.168 to 0.6498
TOP	97.60 to 105.3
FRACTION1	0.0 to 1.000
LOGEC50_1	-496.7 to 484.9
LOGEC50_2	-3532 to 3520
EC50_1	
EC50_2	
Goodness of Fit	
Degrees of Freedom	10
R^2	0.9975
Absolute Sum of Squares	61.88
Sy.x	2.488

This example shows results of a competitive binding assay. The nonlinear regression fit a two-component competitive binding curve in order to find the two $\log EC_{50}$ values and the relative fraction of each component.

The curve follows the data nicely, so R^2 is high. But look closer and you'll see some major problems. The confidence interval for the parameter *fraction* (the fraction of sites that are of high-affinity) ranges from 0 to 1 – all possible values. The two log EC_{50} values are almost identical with extremely wide confidence intervals. With such wide confidence intervals, this fit is not helpful.

Part of the problem is the same as with example 1. These data were normalized to run from 0 to 100, so the values of Bottom and Top ought to be constrained to those values. Here are the revised results.

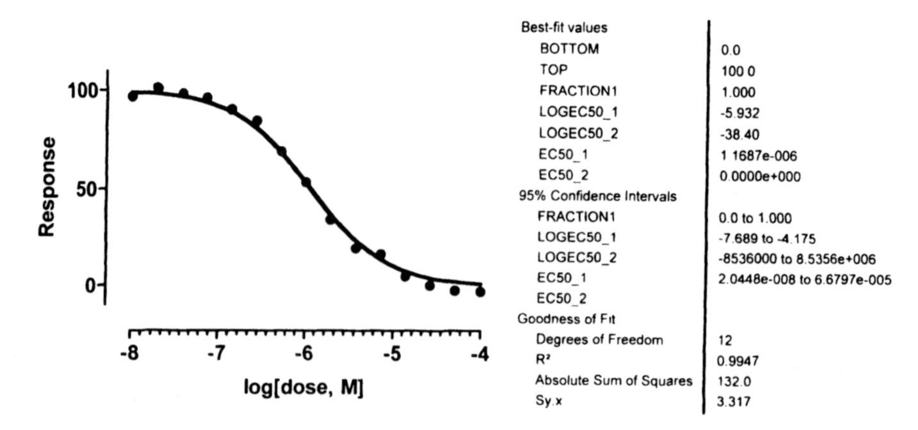

Best-fit values	
BOTTOM	0.0
TOP	100.0
FRACTION1	1.000
LOGEC50_1	-5.932
LOGEC50_2	-38.40
EC50_1	1.1687e-006
EC50_2	0.0000e+000
95% Confidence Intervals	
FRACTION1	0.0 to 1.000
LOGEC50_1	-7.689 to -4.175
LOGEC50_2	-8536000 to 8.5356e+006
EC50_1	2.0448e-008 to 6.6797e-005
EC50_2	
Goodness of Fit	
Degrees of Freedom	12
R^2	0.9947
Absolute Sum of Squares	132.0
Sy.x	3.317

The curve still looks fine, and now the results look better, at first glance. But the confidence interval for the $\log EC_{50}$ of the second site is extremely wide. This is because the best-fit value of the parameter *fraction* is 100%, meaning that all the sites are of the first type. Of course the confidence interval for the second site is wide, if the program isn't convinced that that site even exists.

Is there any evidence for the existence of this second site? Here are the results fitting it to a one-site competitive binding model, constraining the Bottom and Top to 0 and 100.

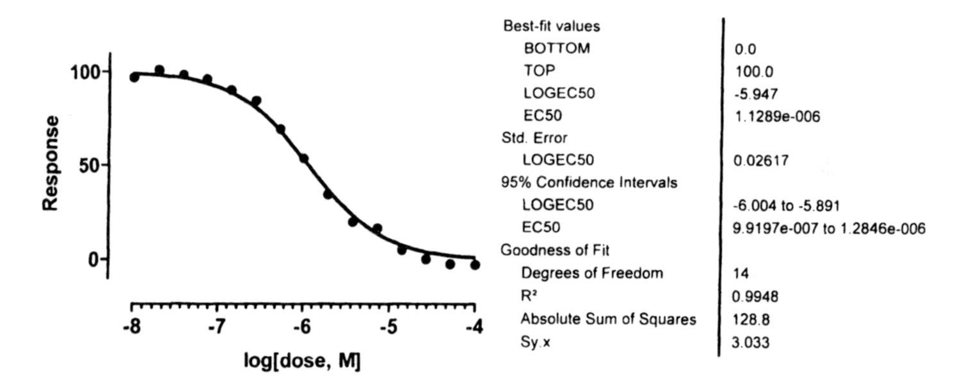

Best-fit values	
BOTTOM	0.0
TOP	100.0
LOGEC50	-5.947
EC50	1.1289e-006
Std. Error	
LOGEC50	0.02617
95% Confidence Intervals	
LOGEC50	-6.004 to -5.891
EC50	9.9197e-007 to 1.2846e-006
Goodness of Fit	
Degrees of Freedom	14
R^2	0.9948
Absolute Sum of Squares	128.8
Sy.x	3.033

The sum-of-squares is a bit lower with the one-site fit than two two-site fit. And the one-site fit has a $\log EC_{50}$ value with a narrow (useful) confidence interval. There simply is nothing in the data to suggest the presence of a second site. When you fit two sites, the program complies with your request, but the results are nonsense.

It is possible to compare two models statistically (see Chapter 21), but there is no point in a statistical comparison when the fit of one of the models (the two-site model in this example) simply makes no sense.

Tip: Don't fit a complicated (i.e. two sites or phases) model if a simpler model fits the data just fine.

The model is ambiguous unless you share a parameter

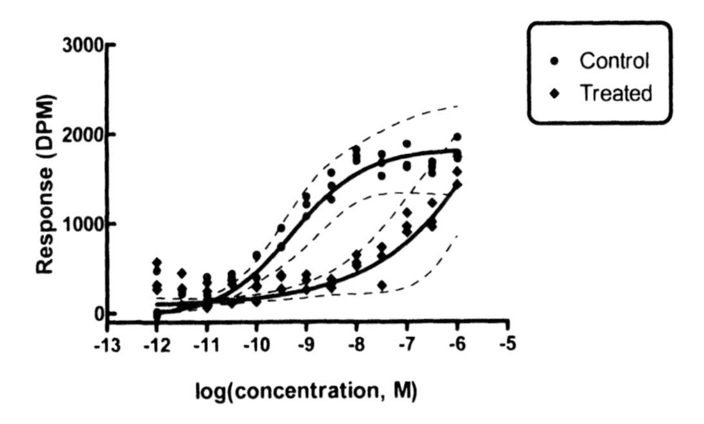

This example shows dose-response data for two experimental conditions, where the results were measured as the generation of a radiolabeled intracellular second messenger. Here are the results of fitting each curve to a variable slope dose-response curve.

	Control	Treated
Best-fit values		
Bottom	183.2	270.4
Top	1739	3637
Logec50	-9.323	-5.548
Hillslope	0.8244	0.4677
Ec50	4.7520e-010	2.8310e-006
95% Confidence Intervals		
Bottom	70.26 to 296.1	179.5 to 361.2
Top	1648 to 1829	-4541 to 11815
Logec50	-9.506 to -9.141	-9.414 to -1.682
Hillslope	0.5626 to 1.086	0.09042 to 0.8450
Ec50	3.1220e-010 to 7.2350e-010	3.8550e-010 to 0.02079

The fit of the control data is OK, although we'd prefer a narrower confidence interval for the bottom plateau. And since the bottom is a bit uncertain, so is the middle, resulting in a confidence interval for the EC_{50} that is wider than we'd like.

The fit of the treated data is not satisfactory at all. The confidence intervals are so wide that you really can't conclude anything. This is not surprising. The treated data don't begin to plateau at high concentrations, so the best-fit value of the top plateau is very uncertain. The middle of the curve is defined by the top and bottom, so the EC_{50} is very uncertain as well.

One solution would be to normalize all the data from 0 to 100 and fix the top and bottom plateaus at those values. But in these experiments, we don't have good control data defining the bottom and top. A better alternative is to use global fitting. If we are willing to assume that both curves have the same bottom, top and slope, we can use nonlinear regression to find one best-fit value for those parameters for both data sets, and to find individual EC_{50} values for each data set. This assumes that while the treatment may affect the EC_{50} (this is what we want to find out), the treatment does not affect the basal or maximum responses or the steepness (slope) of the curve.

Here are the results.

	Control	Treated
Best-fit values		
Bottom (Shared)	267.5	267.5
Top (Shared)	1748	1748
LogEC50	-9.228	-6.844
Hillslope (Shared)	0.8162	0.8162
EC50	5.9170e-010	1.4310e-007
95% Confidence Intervals		
Bottom (Shared)	211.7 to 323.3	211.7 to 323.3
Top (Shared)	1656 to 1840	1656 to 1840
LogEC50	-9.409 to -9.047	-7.026 to -6.662
Hillslope (Shared)	0.6254 to 1.007	0.6254 to 1.007
EC50	3.9000e-010 to 8.9780e-010	9.4110e-008 to 2.1770e-007

The parameter *Bottom* is now determined by both data sets, so has a narrower confidence interval. The parameter *Top* is also shared, so the nonlinear regression algorithm is able to come up with a reasonable best-fit value and confidence interval. Since both the top and bottom are now determined with a reasonable degree of certainty, the middle of the curve is also determined reasonably well. The program was able to fit EC_{50} values for both curves, with reasonable values and acceptably narrow confidence intervals.

This analysis only worked because we assumed that the treatment changed the $\log EC_{50}$ without changing the bottom, top or slope of the curve. Whether this assumption is reasonable depends on the scientific context of the experiment.

The key decision here was to share the value of the parameter *Top*. Sharing the *Bottom* and *HillSlope* helped narrow the confidence intervals, but was not essential.

> Tip: When fitting a family of curves, think carefully about whether it makes sense to share the value of some parameters among data sets.

Bad initial values

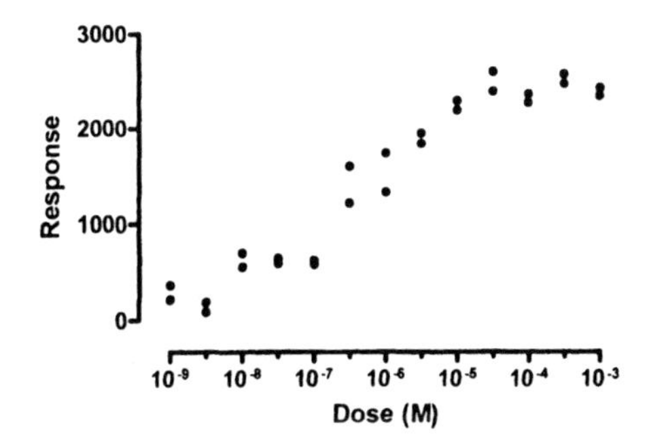

This dose response curve was fit to the equation below, a sigmoidal dose-response curve that fit the parameter pEC_{50} rather than $logEC_{50}$. The pEC_{50} is an alternative way of expressing the concentration giving the middle of the curve. It equals -1 times the $logEC_{50}$).

```
Y = Bottom + (Top-Bottom)/(1+10^((-pEC50 - X)*HillSlope))
```

Nonlinear regression failed to fit the model to the data. Instead it reports "Does not converge".

The data look fine. The equation was entered properly. What could be wrong? To help diagnose the problem, plot the curve generated by the initial values.

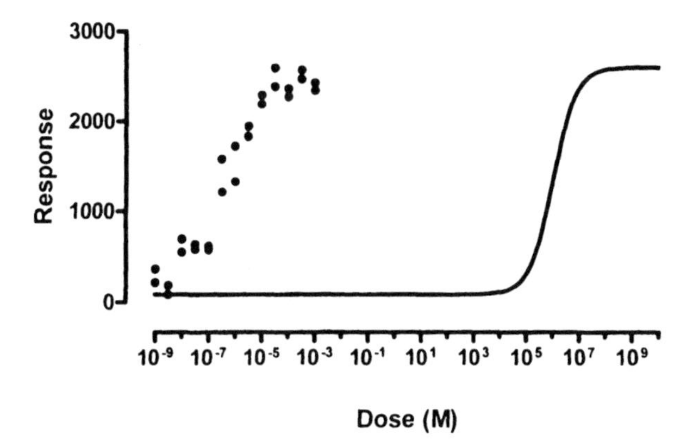

Oops. Why is the curve so far from the points? Review the rules entered for initial values.

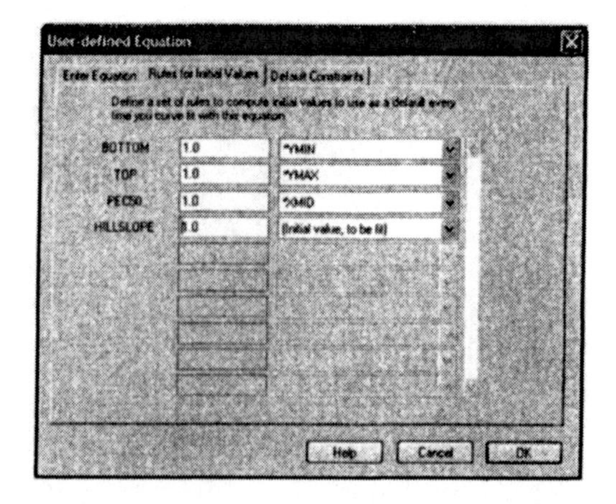

Now the problem is apparent. The initial value of the pEC_{50} was set equal to the average of the lowest (-9) and highest (-3) X values, or -6. But the equation is written in terms of the pEC_{50}, not the $logEC_{50}$. The initial value for the pEC_{50} should have been about 6, not -6. This mistake ensured that the curve generated by the initial values is quite far from the data points, so far that the program is stuck. You can see why. If the program changes the value of pEC_{50} to make it a bit larger (the curve farther to the left) or a bit smaller (the curve farther to the right) the sum-of-squares doesn't change. All the data points are under the top plateau of the curve defined by the initial values. Since the sum-of-squares doesn't change with small changes in the pEC_{50} value, the program has no clue about which way it should go. It doesn't know if the initial value of pEC_{50} is too high or too low. Since it is stuck, the program reports "Does not converge". It would need to make much larger changes in the value of pEC_{50} to affect the sum-of-squares, but the nonlinear regression procedure doesn't do this.

The solution is clear – use better initial values. We change the rule for the initial value of pEC_{50} to -1 times the XMID, which is 6. Now the curve generated by the initial values is close to the points, and curve fitting proceeds without problem.

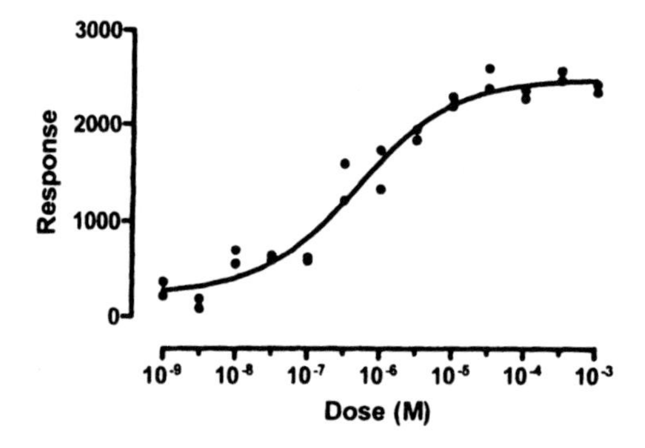

Tip: If nonlinear regression won't converge on a best-fit solution, graph the curve generated by the initial values, and change those initial values if the curve is far from the data.

Redundant parameters

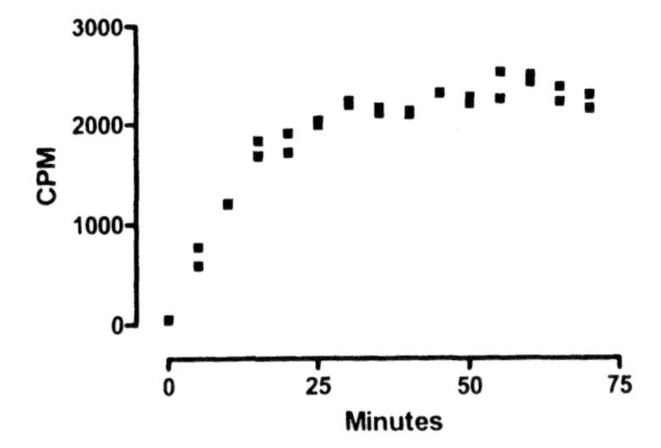

This example shows the kinetics of association binding. We attempted to fit it to this equation, where X is time. For details about this kind of experiment, see page 234.

$$Y = Y_{max}\left(1 - e^{-(Ligand \times k_{on} + k_{off}) \times X}\right)$$

The parameter Ligand is the concentration of radioactively labeled ligand used in the experiment. That is set by the experimenter, so we constrain Ligand to a constant value, and ask nonlinear regression to find best-fit values for Y_{max}, k_{on} and k_{off}.

The nonlinear regression program gives an error message such as "Bad model". But the model is correct – that equation describes the kinetics of ligand binding. To check that the problem was unreasonable initial values, let's graph the curve defined by the initial values.

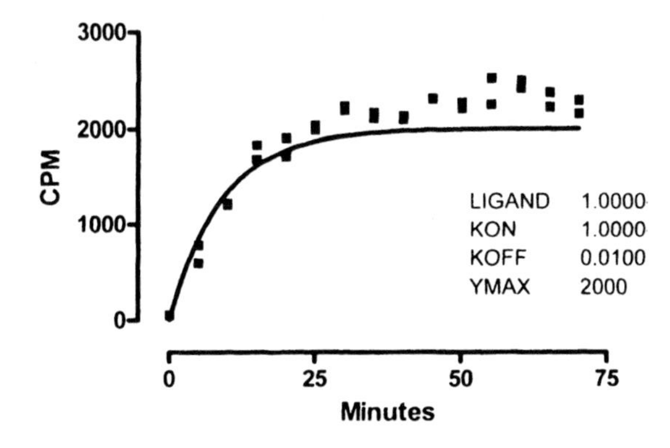

While that is hardly a best-fit curve, it does come quite close to the points, close enough for curve fitting to work. So why does nonlinear regression fail? Look more carefully at the equation. Note the term *Ligand×Kon+Koff*. *Ligand* is a constant, but we are trying to fit both *Kon* and *Koff* from one experiment. The data simply don't define both values. You can see that if the value of *Kon* is reduced a bit, the curve will be exactly the same if the value of *Koff* is increased. There is no way to fit one set of data to determine values of both *Kon* and *Koff* – the answer is ambiguous. We are trying to fit a model that simply is not entirely defined by our data.

The simplest way to solve this problem is to determine the value of *Koff* from a separate dissociation experiment. Let's say its value is 0.0945. Now we can constrain *Koff* to be a constant with that value, and the curve fitting can proceed. Here are the results:

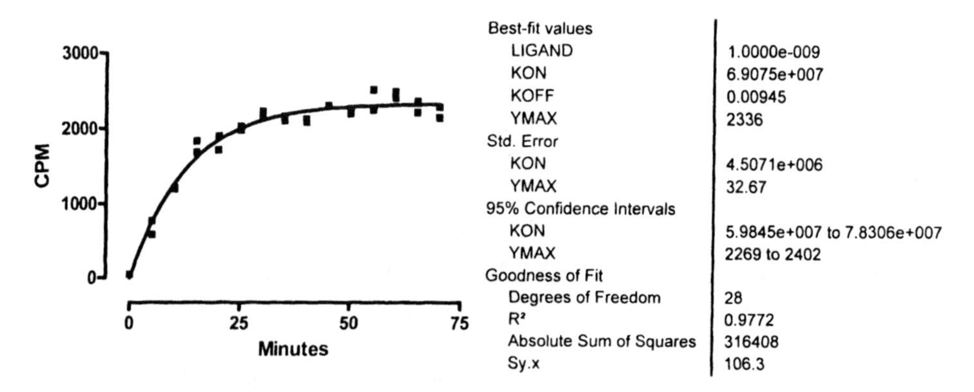

Best-fit values	
LIGAND	1.0000e-009
KON	6.9075e+007
KOFF	0.00945
YMAX	2336
Std. Error	
KON	4.5071e+006
YMAX	32.67
95% Confidence Intervals	
KON	5.9845e+007 to 7.8306e+007
YMAX	2269 to 2402
Goodness of Fit	
Degrees of Freedom	28
R^2	0.9772
Absolute Sum of Squares	316408
Sy.x	106.3

It is possible to determine the values of both *Kon* and *Koff*, just not from one data set. To do this, fit a family of association curves (each with a different concentration of radioligand) and share the values of *Kon* and *Koff* among all the data sets. See page 236. A related approach is to simultaneously analyze association and dissociation experiments. See page 238.

Tips for troubleshooting nonlinear regression

If your program generates an error message rather than curve fitting results, you probably won't be able to make sense of the exact wording of the message. It tells you *where* in the curve fitting process the error occurred, but it will rarely help you figure out *why* the error occurred.

Nonlinear regression depends on choosing a sensible model. If you are not familiar with the model (equation) you chose, take the time to learn about it. If you fit the wrong model, the results won't be helpful.

It is not enough to just pick a model. You also have to decide which parameters in the model should be fit by nonlinear regression, and which should be fixed to constant values. Failure to do this is a common cause of nonlinear regression problems.

Successful nonlinear regression requires a reasonable initial value for each parameter. In other words, the curve generated by those initial values must come close enough to the data points that the regression procedure can figure out which parameters are too high and which are too low. A vital first step in troubleshooting nonlinear regression is to plot the curve that results from those initial values. If that curve is far from the data, you'll need to change the initial values of the parameters. Even if your program picks the initial values for you, you still should review them if the fit is problematic.

If the model has several compartments or phases, see if a simpler model (fewer compartments or phases) fits your data.

B. Fitting data with linear regression

7. Choosing linear regression

Linear regression can be viewed as just a special case of nonlinear regression. Any nonlinear regression program can be used to fit to a linear model, and the results will be the same as if you had chosen linear regression. Because linear regression is usually performed separately from nonlinear regression, we explain linear regression separately here.

The linear regression model

Linear regression fits this model to your data:

$$Y = \text{intercept} + \text{slope} \times X$$

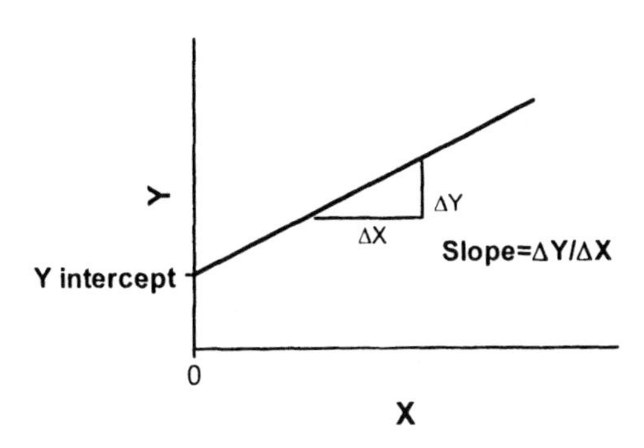

The slope quantifies the steepness of the line. It equals the change in Y for each unit change in X. It is expressed in the units of the Y axis divided by the units of the X axis. If the slope is positive, Y increases as X increases. If the slope is negative, Y decreases as X increases.

The Y intercept is the Y value of the line when X equals zero. It defines the elevation of the line.

> Note: Most programs (including GraphPad Prism) offer separate analysis choices for linear and nonlinear regression. That is why we wrote separate chapters for linear regression. But you can enter a linear model into a nonlinear regression program, and it will work fine. Linear regression is just a simpler, special, case.

Don't choose linear regression when you really want to compute a correlation coefficient

Linear regression and correlation are related, and it is easy to mix them up.

Correlation calculations are not actually a kind of regression. This is because correlation calculations do not find a best-fit line. Instead, correlation quantifies how consistently two variables vary together. When a change in one variable corresponds closely to a change in the other, statisticians say that there is a lot of *covariation* or *correlation*. The direction and magnitude of correlation is quantified by the correlation coefficient, r. If r is positive, it means that as one variable goes up so does the other. If r is negative, it means that as one variable goes up, the other variable goes down. The value of r is always between -1 and +1.

Note that the correlation coefficient quantifies the relationship between two variables. It doesn't matter which you call X and which you call Y – the correlation coefficient will be the same either way. In contrast, linear regression finds the line that best predicts Y from X, so your choices of which variable to label X and which to label Y will impact the results.

Analysis choices in linear regression

Force the line through the origin?

If you choose regression, you may force the line to go through a particular point such as the origin. In this case, linear regression will determine only the best-fit slope, as the intercept will be fixed. Use this option when scientific theory tells you that the line must go through a particular point (usually the origin, X=0, Y=0) and you only want to know the slope. This situation arises rarely.

Use common sense when making your decision. For example, consider a protein assay. You measure optical density (Y) for several known concentrations of protein in order to create a standard curve. You then want to interpolate unknown protein concentrations from that standard curve. When performing the assay, you adjusted the spectrophotometer so that it reads zero with zero protein. Therefore you might be tempted to force the regression line through the origin. But this constraint may result in a line that doesn't fit the data so well. Since you really care that the line fits the standards very well near the unknowns, you will probably get a better fit by not constraining the line.

Most often, you should let linear regression find the best-fit line without any constraints.

Fit linear regression to individual replicates or to means?

If you collected replicate Y values at every value of X, there are two ways to calculate linear regression. You can treat each replicate as a separate point, or you can average the replicate Y values, to determine the mean Y value at each X, and do the linear regression calculations using the means.

You should consider each replicate a separate point when the sources of experimental error are the same for each data point. If one value happens to be a bit high, there is no reason to expect the other replicates to be high as well. The errors are independent.

Average the replicates and treat the mean as a single value when the replicates are not independent. For example, the replicates would not be independent if they represent triplicate measurements from the same animal, with a different animal used at each value of X (dose). If one animal happens to respond more than the others, that will affect all the replicates. The replicates are not independent.

We discuss this topic in more depth on page 87.

X and Y are not interchangeable in linear regression

Standard linear regression calculations are based on the assumption that you know all the independent (X) values perfectly, and that all of the uncertainty or scatter is in the assessment of the dependent (Y) values. This is why standard linear regression (as well as nonlinear regression) minimizes the sum of the squares of the *vertical* distances of the points from the line.

While these assumptions are rarely 100% valid, linear regression is nonetheless very useful in biological (and other) research because uncertainty in X is often much smaller than the uncertainty in Y. In many cases, in fact, the experimenter controls the X values, and therefore doesn't have to worry about error in X.

Standard linear regression is not so helpful when there is substantial error in determining both X and Y. The graph below compares two assay methods. Each point on the graph represents the results of two assays of a single blood sample. The result of one assay method is plotted on the X axis, and the result of the other method is plotted on the Y axis.

The graph shows two linear regression lines. One line was created by linear regression defining the results of method A to be the independent variable, known without error. This line minimizes the sum of the square of the *vertical* distances of the points from the line; this is the standard method for finding a line of best fit. The other line was created by *inverse linear regression*. Here we define the results of method B to be the independent variable known without error, and the goal (since B is on the vertical axis) is to minimize the sum of squares of the *horizontal* distances of the points from the line.

The two lines are not the same. If you were to switch the definitions of the independent and dependent variables, you'd get a different best-fit linear regression line. Neither line is optimal for comparing the two methods. The problem is that in this situation (comparing results of two analysis methods), the concept of independent and dependent variables does not really apply. An independent variable is generally one that you control and is assumed to be error free. Neither the results of method A nor the results of method B are error free.

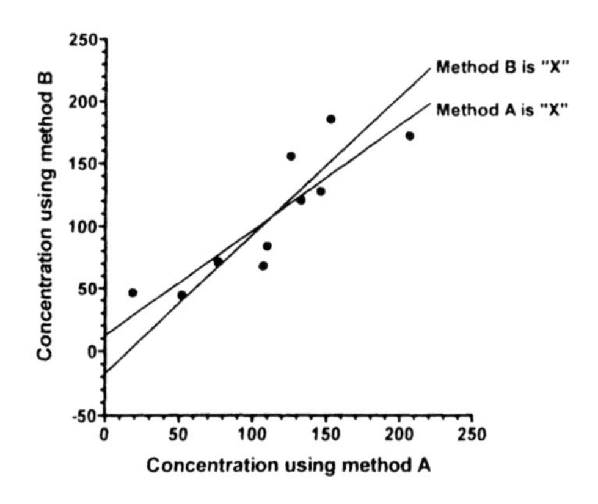

Regression with equal error in X and Y

If you assume that both X and Y variables are subject to the same amount of error, it seems logical to choose a regression method that minimizes the sum of the squares of the *perpendicular* distances between data points and line. The regression method that does this is called *Deming regression*, but its development predates Deming and was first described by Adcock in 1878. This method is also called *Model II linear regression.*

This graph shows the best-fit line created by Deming regression (assuming equal error in methods A and B) as a solid line, with dashed lines showing the two lines created by linear regression and inverse linear regression. As you'd expect, the Deming regression line is between the others.

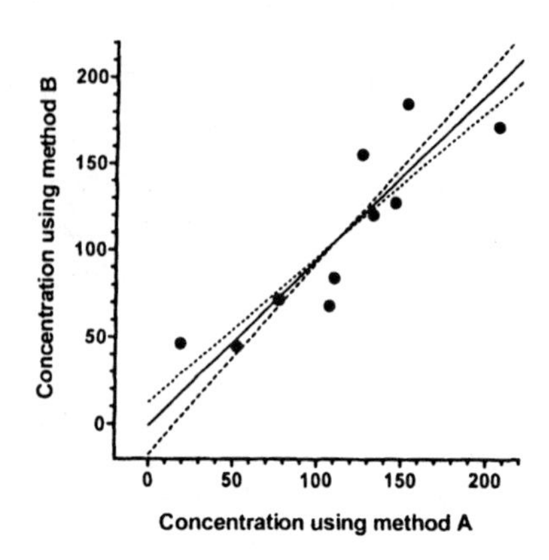

Regression with unequal error in X and Y

Deming regression can also be used when X and Y are both subject to error, but the errors are not the same. In this case, you must enter the error of each variable, expressed as a standard deviation.

How do you know what values to enter? To assess the uncertainty (error) of a method, collect duplicate measurements from a number of samples using that method. Calculate the standard deviation of the error using the equation below, where each d_i is the difference between two measurements of the same sample (or subject), and N is the number of measurements you made (N equals twice the number of samples, since each sample is measured twice).

$$SD_{error} = \sqrt{\frac{\sum d_i^2}{N}}$$

Repeat this for each method or variable (X and Y). Some programs ask you to enter the two SDerror values. Other programs ask you to enter λ (lamda), which is the square of the ratio of the two SD values.

$$\lambda = \left(\frac{SD_{X\ error}}{SD_{Y\ error}}\right)^2$$

> GraphPad note: Prism requires you to enter individual SD values, but uses these values only to calculate λ, which is then used in the Deming regression calculations. If you know λ, but not the individual SD values, enter the square root of λ as the SD of the X values, and enter 1.0 as the SD of the Y error.

8. Interpreting the results of linear regression

What is the best-fit line?

Best fit values of slope and intercept with confidence intervals

Linear regression finds the best-fit values of the slope and intercept, and reports these values along with a standard errors and confidence intervals.

The standard error values of the slope and intercept can be hard to interpret intuitively, but their main usefulness is to compute the 95% confidence intervals. If you accept the assumptions of linear regression, there is a 95% chance that the true value of the slope lies within the 95% confidence interval of the slope. Similarly, there is a 95% chance that the true value of the intercept lies within the 95% confidence interval of the intercept.

95% confidence bands of a regression line

The 95% confidence interval of the slope is a range of values, as is the 95% confidence interval of the intercept. Linear regression can also combine these uncertainties to graph a 95% confidence band of the regression line. An example is shown below. The best-fit line is solid, and the 95% confidence band is shown by two curves surrounding the best-fit line.

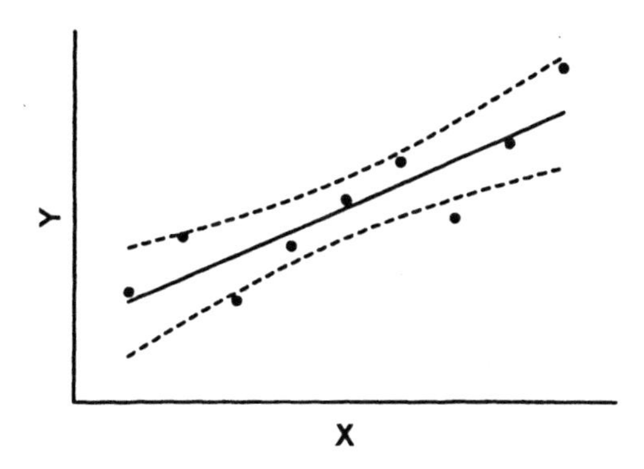

The dashed lines that demarcate the confidence interval are curved. This does not mean that the confidence interval includes the possibility of curves as well as straight lines. Rather, the curved lines are the boundaries of all possible straight lines. The figure below shows four possible linear regression lines (solid) that lie within the confidence interval (dashed).

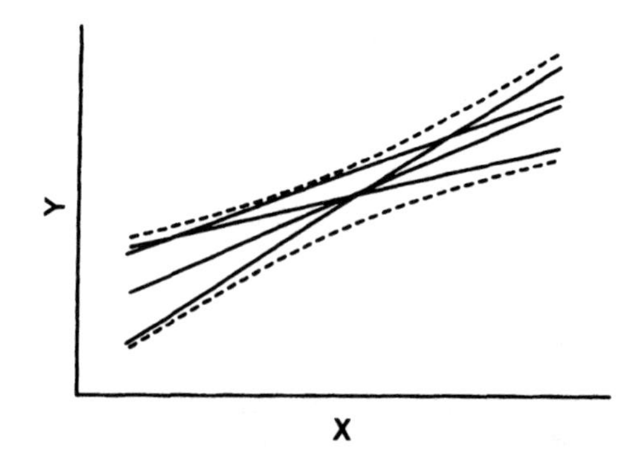

Given the assumptions of linear regression, you can be 95% confident that the two curved confidence boundaries enclose the true best-fit linear regression line, leaving a 5% chance that the true line is outside those boundaries. This is not the same as saying it will contain 95% of the data points. Many data points will be outside the 95% confidence interval boundary.

95% prediction bands of a regression line

The 95% *prediction band* is different than the 95% confidence band.

The prediction bands are farther from the best-fit line than the confidence bands, a lot farther if you have many data points. The 95% prediction interval is the area in which you expect 95% of all data points to fall. In contrast, the 95% confidence interval is the area that has a 95% chance of containing the true regression line. This graph shows both prediction and confidence intervals (the curves defining the prediction intervals are farther from the regression line).

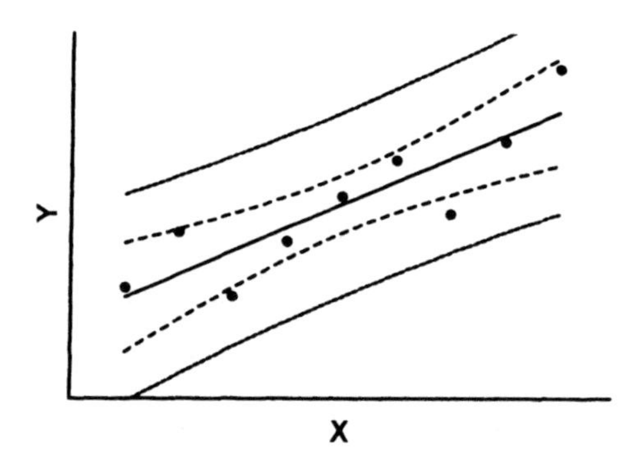

95% confidence interval of the X intercept

The graph below shows a best-fit linear regression line along with its 95% confidence band.

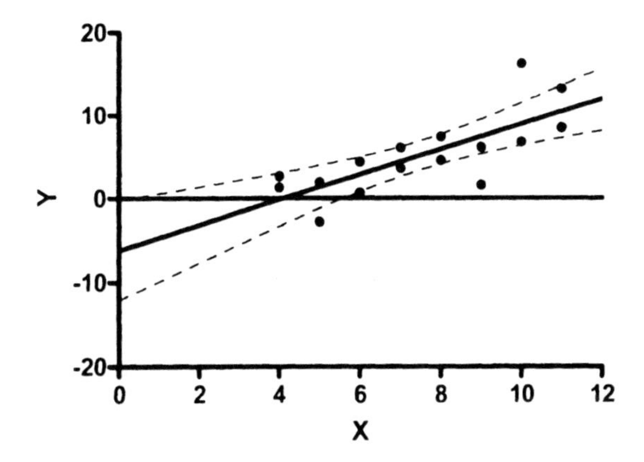

Move along the horizontal line at Y=0, and notice where it intersects the two confidence bands. It crosses the upper band just beyond X=0 and crosses the lower band near X=6. Indeed, linear regression reports that the X intercept is 4.1 with the 95% confidence interval extending from -0.1 to 5.9. Note that the confidence interval for the X intercept is not symmetrical around the best-fit value.

How good is the fit?

r^2

The value r^2 is a fraction between 0.0 and 1.0, and has no units. An r^2 value of 0.0 means that knowing X does not help you predict Y. There is no linear relationship between X and Y, and the best-fit line is a horizontal line going through the mean of all Y values. When r^2 equals 1.0, all points lie exactly on a straight line with no scatter. Knowing X lets you predict Y perfectly.

The figure below demonstrates how to compute r^2.

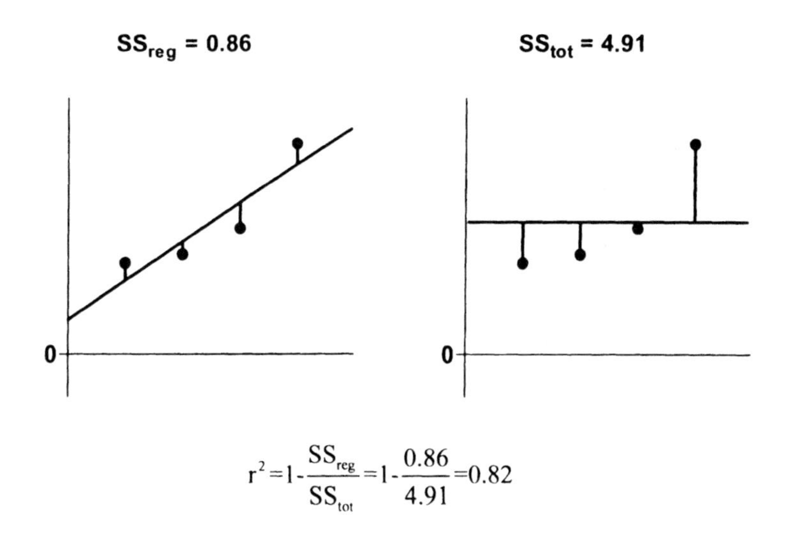

The left panel shows the best-fit linear regression line This lines minimizes the sum-of-squares of the vertical distances of the points from the line. Those vertical distances are also shown on the left panel of the figure. In this example, the sum of squares of those distances (SS_{reg}) equals 0.86. Its units are the units of the Y-axis squared. To use this value as a measure of goodness-of-fit, you must compare it to something.

The right half of the figure shows the null hypothesis -- a horizontal line through the mean of all the Y values. Goodness-of-fit of this model (SS_{tot}) is also calculated as the sum of squares of the vertical distances of the points from the line, 4.91 in this example. The ratio of the two sum-of-squares values compares the regression model with the null hypothesis model. The equation to compute r^2 is shown in the figure. In this example r^2 is 0.82. The regression model fits the data much better than the null hypothesis, so SS_{reg} is much smaller than SS_{tot}, and r^2 is near 1.0. If the regression model were not much better than the null hypothesis, r^2 would be near zero.

You can think of r^2 as the fraction of the total variance of Y that is “explained” by variation in X. The value of r^2 (unlike the regression line itself) would be the same if X and Y were swapped. So r^2 is also the fraction of the variance in X “explained” by variation in Y. In other words, r^2 is the fraction of the variation that is shared between X and Y.

In this example, 82% of the total variance in Y is “explained” by the linear regression model. The variance (SS) of the data from the linear regression model equals only 16% of the total variance of the Y values (SS_{tot})

Why r^2 is ambiguous in constrained linear regression

As you saw in the previous section, r^2 is computed by comparing the sum-of-squares from the regression line with the sum-of-squares from a model defined by the null hypothesis. There are two ways to compute r^2 when the regression line is constrained

- Compare with the sum-of-squares from a horizontal line through the mean of all Y values. This line doesn't follow the constraint -- it does not go through the origin.
- Compare with the sum-of-squares from a horizontal line through the origin. This line is usually far from most of the data, so this approach usually gives lower r^2 values.

> GraphPad note: Because r^2 is ambiguous in constrained linear regression, Prism doesn't report it. If you really want to know a value for r^2, use nonlinear regression to fit your data to the equation Y=slope*X. Prism will report r^2 defined the first way.

The standard deviation of the residuals, $s_{y.x}$

The variable $s_{y.x}$ quantifies the average size of the residuals, expressed in the same units as Y. Some books and programs refer to this value as s_e, and others as the “root mean square”. It is calculated from SS_{reg} and N (number of points) using this equation:

$$s_{y \cdot x} = \sqrt{\frac{SS_{reg}}{N-2}}$$

You'll only find this number useful if you perform additional calculations following linear regression.

Is the slope significantly different from zero?

If there is no relationship between X and Y, the linear regression line will be nearly horizontal. You can test the null hypothesis that the true slope is zero. Most programs report the F ratio and its degrees of freedom as well as a P value which answers this question: If there were no linear relationship between X and Y overall, what is the probability that randomly selected points would result in a regression line as far from horizontal (or further) than you observed? If the P value is small, you'll conclude that there is a statistically significant relationship between X and Y.

In most situations, you know that X and Y are related and you are performing linear regression to get the best-fit values of the slope and intercept. In these situations, there is no point asking whether the slope differs from 0 (you know it does), so the P value will not be informative.

Is the relationship really linear?

Residuals from a linear regression line

Residuals are the vertical distances of each point from the regression line. The X values in the residual table are identical to the X values you entered. The Y values are the residuals. A residual with a positive value means that the point is above the line; a residual with a negative value means the point is below the line.

If the assumptions of linear regression have been met, the residuals will be randomly scattered above and below the line at Y=0. The scatter should not vary with X. You also should not see large clusters of adjacent points that are all above or all below the Y=0 line. For an example, see "Residuals" on page 35.

Runs test following linear regression

The runs test determines whether your data differ significantly from a straight line.

A run is a series of consecutive points that are either all above or all below the regression line. In other words, a run is a consecutive series of points whose residuals are either all positive or all negative. A run may consist of just a single point.

If the data points are randomly distributed above and below the regression line, it is possible to calculate the expected number of runs. If there are N_a points above the curve and N_b points below the curve, the number of runs you expect to see equals $[(2N_aN_b)/(N_a+N_b)]+1$. If you observe fewer runs than expected, it may be simply a consequence of random sampling, or it may mean that your data deviate systematically from a straight line. The P value from the runs test answers this question: If the data really follow a straight line, what is the chance that you would obtain as few (or fewer) runs as observed in this experiment? If the P value is low, it means that either a rare coincidence has occurred or that your data really follow some sort of curve rather than a line.

The P values are always one-tail, asking about the probability of observing as few runs (or fewer) than observed. If you observe more runs than expected, the P value will be higher than 0.50.

If the runs test reports a low P value, conclude that the data don't really follow a straight line, and consider using nonlinear regression to fit a curve.

Comparing slopes and intercepts

If you have two or more data sets, you can test whether the slopes and intercepts of two or more data sets are significantly different. The method is discussed in detail in Chapter 18 of J Zar, *Biostatistical Analysis*, 4th edition, Prentice-Hall, 1999.

> GraphPad note: If you check the option to compare lines, Prism performs these calculations automatically.

Compares the slopes first. The P value (two-tailed) tests the null hypothesis that the slopes are all identical (the lines are parallel). The P value answers this question: If the slopes really were identical, what is the chance that randomly selected data points would yield slopes as different (or more different) than you observed. If the P value is less than 0.05, conclude that the lines are significantly different. In that case, there is no point in comparing the intercepts. The intersection point of two lines is:

$$X=\frac{\text{Intercept}_1-\text{Intercept}_2}{\text{Slope}_2-\text{Slope}_1}$$

$$Y=\text{Intercept}_1+\text{Slope}_1\cdot X$$

$$=\text{Intercept}_2+\text{Slope}_2\cdot X$$

If the P value for comparing slopes is greater than 0.05, conclude that the slopes are not significantly different and calculates a single slope for all the lines. Now the question is whether the lines are parallel or identical. Calculate a second P value testing the null hypothesis that the lines are identical. If this P value is low, conclude that the lines are not identical (they are distinct but parallel). If this second P value is high, there is no compelling evidence that the lines are different.

This method is equivalent to an Analysis of Covariance (ANCOVA), although ANCOVA can be extended to more complicated situations.

See Chapter 27 for more general comments on comparing fits.

How to think about the results of linear regression

Your approach to linear regression will depend on your goals.

If your goal is to analyze a standard curve, you won't be very interested in most of the results. Just make sure that r^2 is high and that the line goes near the points. Then go straight to the standard curve results.

In many situations, you will be most interested in the best-fit values for slope and intercept. Don't just look at the best-fit values, also look at the 95% confidence interval of the slope and intercept. If the intervals are too wide, repeat the experiment with more data.

If you forced the line through a particular point, look carefully at the graph of the data and best-fit line to make sure you picked an appropriate point.

Consider whether a linear model is appropriate for your data. Do the data seem linear? Is the P value for the runs test high? Are the residuals random? If you answered no to any of those questions, consider whether it makes sense to use nonlinear regression instead.

Checklist: Is linear regression the right analysis for these data?

To check that linear regression is an appropriate analysis for these data, ask yourself these questions. Prism cannot help answer them.

Question	Discussion
Can the relationship between X and Y be graphed as a straight line?	In many experiments the relationship between X and Y is curved, making linear regression inappropriate. Either transform the data, or use a program (such as GraphPad Prism) that can perform nonlinear curve fitting.
Is the scatter of data around the line Gaussian (at least approximately)?	Linear regression analysis assumes that the scatter is Gaussian.
Is the variability the same everywhere?	Linear regression assumes that scatter of points around the best-fit line has the same standard deviation all along the curve. The assumption is violated if the points with high or low X values tend to be farther from the best-fit line. The assumption that the standard deviation is the same everywhere is termed *homoscedasticity*.
Do you know the X values precisely?	The linear regression model assumes that X values are exactly correct, and that experimental error or biological variability only affects the Y values. This is rarely the case, but it is sufficient to assume that any imprecision in measuring X is very small compared to the variability in Y.
Are the data points independent?	Whether one point is above or below the line is a matter of chance, and does not influence whether another point is above or below the line. See "Fit linear regression to individual replicates or means?" on page 335.
Are the X and Y values intertwined?	If the value of X is used to calculate Y (or the value of Y is used to calculate X), then linear regression calculations are invalid. One example is a Scatchard plot, where the Y value (bound/free) is calculated from the X value (bound). Another example would be a graph of midterm exam scores (X) vs. total course grades (Y). Since the midterm exam score is a component of the total course grade, linear regression is not valid for these data.

C. Models

9. Introducing models

What is a model?

A model is a mathematical description of a physical, chemical or biological state or process. Using a model can help you think about such processes and their mechanisms, so you can design better experiments and comprehend the results. A model forces you to think through (and state explicitly) the assumptions behind your analyses.

Linear and nonlinear regression fit a mathematical model to your data to determine the best-fit values of the parameters of the model (e.g., potency values, rate constants, etc.).

Your goal in using a model is not necessarily to describe your system perfectly. A perfect model may have too many parameters to be useful. Rather, your goal is to find as simple a model as possible that comes close to describing your system. You want a model to be simple enough so you can fit the model to data, but complicated enough to fit your data well and give you parameters that help you understand the system and design new experiments.

You can also use a model to simulate data, and then analyze the simulated data. This can help you design better experiments.

Two interesting quotations about models:

> A mathematical model is neither an hypothesis nor a theory. Unlike scientific hypotheses, a model is not verifiable directly by an experiment. For all models are both true and false.... The validation of a model is not that it is "true" but that it generates good testable hypotheses relevant to important problems.
> -- R. Levins, Am. Scientist 54:421-31, 1966

> All models are wrong, but some are useful.
> -- George E. P. Box

Terminology

The terminology used to describe models can be confusing, as some of the words are also used in other contexts.

Regression

The term "regression" dates back to Galton's studies in the 1890s of the relationship between the heights of fathers and sons. The two were related, of course. But tall fathers tended to have sons who are shorter than they were, and short fathers tended to have sons who were taller. In other words, the height of sons tends to be somewhere between the height of his father and the mean all heights. Thus height was said to "regress to the mean". Now the term "regression" is used generally to refer to any kind of analysis that looks at the relationship between two or more variables.

Model

The regression model is an equation that defines the outcome, or dependent variable Y, as a function of an independent variable X, and one or more model parameters.

Empirical vs. mechanistic models

Empirical models simply describe the general shape of the data set that you are trying to fit a curve to. The parameters of the model don't necessarily correspond to a biological, chemical or physical process.

In contrast, *mechanistic models*, as the name implies, are specifically formulated to provide insight into a biological, chemical, or physical process that is thought to govern the phenomenon under study. Parameters derived from mechanistic models are quantitative estimates of real system properties (e.g., dissociation constants, rate constants, catalytic velocities etc.).

In general, mechanistic models are more useful in most instances because they represent quantitative formulations of a hypothesis (see A. Christopoulos, *Biomedical Applications of Computer Modeling*, 2000, CRC Press). However, if you choose the wrong mechanistic model to fit to your data, the consequences are more dire than for empirical models because you may come to inappropriate conclusions regarding the mechanism(s) you are studying.

> Note: A particular model can be either empirical *or* mechanistic, depending on the context.

Variables

A model defines an outcome (e.g., response) Y in terms of X and one or more parameters. The outcome, Y, is called the *dependent* variable, because its value depends on the value of X and the values of the parameters. X is called the *independent* variable, often time or concentration set by the experimenter.

If your model has two independent variables – say your outcome is a function of both time and concentration – then you need to use a "multiple regression" method.

Parameters

Parameters help to define the properties and behavior of the model. The regression method finds best-fit values of the parameters that make the model do as good a job as possible at predicting Y from X. In linear regression, the parameters are *slope* and *intercept*. If you fit a dose-response curve with nonlinear regression, one of the parameters is the EC_{50}. If you fit a kinetic exponential curve with nonlinear regression, one of the parameters is the rate constant.

Error

The regression method that finds the best-fit value of the parameters must be based on an assumption about how the data are scattered around the curve. Statisticians refer to scatter of data around a predicted value as "error".

Examples of simple models

We'll discuss commonly used biological models later in this book, so this section is just a preview.

Example 1. Optical density as a function of concentration.

Colorimetric chemical assays are based on a simple principle. Add appropriate reactants to your samples to initiate a chemical reaction whose product is colored. When you terminate the reaction, the concentration of colored product, and hence the optical density, is proportional to the concentration of the substance you want to assay.

$$\text{Optical Density} = Y = K \cdot [\text{substance}] = K \cdot X$$

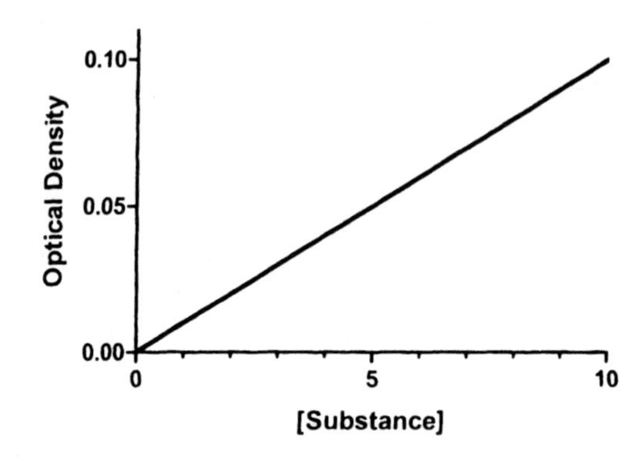

If you measured optical density at various concentrations of substrate, you could fit this model to your data to find the best-fit value of K.

Mathematically, the equation works for any value of X. However, the results only make sense with certain values. Negative X values are meaningless, as concentrations cannot be negative. The model may fail at high concentrations of substance where the reaction is no longer limited by the concentration of substance, or if the solution becomes so dark (the optical density is so high) that little light reaches the detector. At that point, the noise of the instrument may exceed the signal. It is not unusual that a model works only for a certain range of values. You just have to be aware of the limitations, and not use the model outside of its useful range.

Example 2. Dissociation of a radioligand from a receptor.

After allowing a radioactively labeled drug to bind to its receptors, wash away the free drug and then measure binding at various times thereafter. The rate of dissociation is proportional to how much is left behind. As will be explained later (see page 233), this means the dissociation follows an exponential decay. The model is:

$$Y = Y_0 \cdot e^{-kX}$$

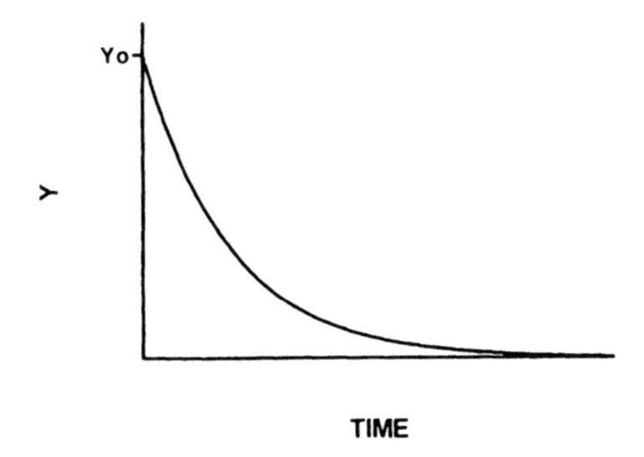

If you collect the amount of binding remaining (Y) at various times (X), you can fit to the exponential model to find the best fit values of Y_0 and K. We'll discuss this kind of experiment in more detail in chapter 39.

Example 3. Enzyme velocity as a function of substrate concentration.

If you measure enzyme velocity at many different concentrations of substrate, the graph generally looks like this:

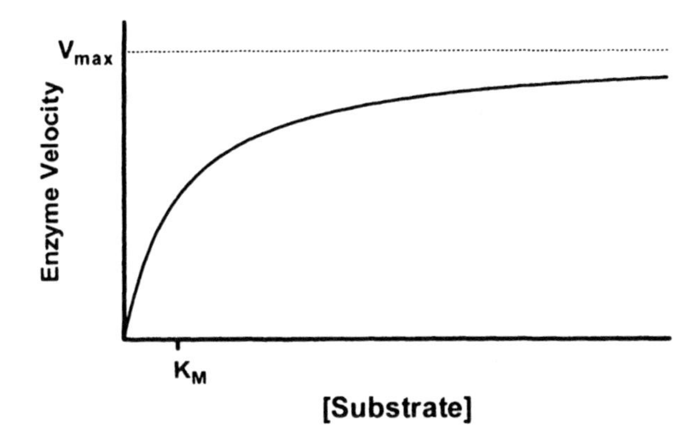

Enzyme velocity as a function of substrate concentration often follows the Michaelis-Menten equation:

$$\text{Velocity} = Y = \frac{V_{max} \cdot X}{X + K_m}$$

V_{max} is the limiting velocity as substrate concentrations get very large. V_{max} (and V) are expressed in units of product formed per time. K_m is expressed in units of substrate concentration, usually in Molar units. K_m is the concentration of substrate that leads to half-maximal velocity.

If you measure enzyme velocity (Y) at various concentrations of substrate (X), you can fit to this model to find the best-fit values of V_{max} and K_m. We'll discuss enzyme kinetics in more detail in Chapter 40.

10. Tips on choosing a model

Overview

In many cases, picking a model will be easy, as you will likely be choosing a model that is conventionally used in your field. Later in the book, we discuss in detail some of the models commonly used in experimental biology. If you aren't sure about the model, take the time to learn. If you pick a model that isn't right for your experimental situation, the regression results won't be helpful.

Don't choose a linear model just because linear regression seems simpler than nonlinear regression

The goal of linear and nonlinear regression is the same: to fit a model to your data by finding the best-fit values of one or more parameters. With linear regression, the parameters are the slope and intercept of a line. With nonlinear regression, the parameters depend on the model (e.g. the K_d and B_{max} of equilibrium saturation binding, or the rate constant and plateau of dissociation kinetics).

Linear regression is described in every statistics book, and is performed by every statistics program. Nonlinear regression is mentioned in only a few statistics books, and is not performed by all statistics programs. From a mathematician's point of view, the two procedures are vastly different, and linear regression is much simpler. From a scientist's point of view, however, the two procedures are fairly similar. Nonlinear regression is more general, as it can fit any model, including a linear one, to your data.

Your choice of linear or nonlinear regression should be based on the model that makes the most sense for your data. Don't use linear regression just to avoid using nonlinear regression.

> Tip: Most models in biology are nonlinear, so many biologists use nonlinear regression more often than linear regression. You should not use linear regression just to avoid learning about nonlinear regression.

Don't go out of your way to choose a polynomial model

The polynomial model is shown below:

$$Y=A+BX+CX^2+DX^3+EX^4...$$

You can include any number of terms. If you stop at the second (B) term, it is called a first-order polynomial equation, which is identical to the equation for a straight line. If you stop after the third (C) term, it is called a second-order, or quadratic, equation. If you stop after the fourth term, it is called a third-order, or cubic, equation.

From a scientist's point-of-view, there is nothing really special about the polynomial model. You should pick it when it is the right equation to model your data. However, few biological or chemical models are described by polynomial equations. If you fit a polynomial model to your data, you might get a nice-looking curve, but you probably won't be able to interpret the best-fit values of A, B, C, etc. in terms of biology or chemistry.

> Note: If your goal is just to get a curve to use for graphing or interpolating unknown values, polynomial regression can be a reasonable choice. But beware of extrapolating beyond the range of your data, as polynomial models can change direction suddenly beyond the range of your data.

Polynomial regression is only special from the point-of-view a mathematician or computer programmer. Even though the graph of Y vs. X is curved (assuming you have chosen at least a second order equation, and no parameter equals zero), mathematicians consider the polynomial equation to be a *linear* equation. This is because a graph of any parameter (A, B, C...) vs. Y would be linear (holding X and the other parameters constant). From a programming point of view, this means it is fairly easy to write a program to fit the model to data.

> Tip: You should pick a model that makes sense for your data and experimental design. Don't pick a polynomial model just because it is mathematically simpler to fit.

Consider global models

Global curve fitting fits several data sets at once, sharing some parameters between data sets. In some situations, this is the only way to determine reasonable values for the parameters of the models. In other situations, it helps you get much more reliable parameter values. Read about global fitting in Chapter 11.

Graph a model to understand its parameters

To make sense of nonlinear regression results, you really need to understand the model you have chosen. This is especially important if you enter your own model or choose a model you haven't used before.

A great way to learn about a model is to simulate a family of curves based on the model, and change the values of the model parameters. Then you can see how the curve changes when a parameter changes.

To generate curves, you'll need to specify the range of X values to use, as well as specifying values for all the parameters in the model. Some programs call this "plotting a function".

For example, the graph below shows a exponential decay model plotted with a single *Plateau* (0) and *Span* (2000), but with four different rate constants (*K*) (left to right: 0.7, 0.3, 0.1, and 0.05). The value of X is time, in minutes.

$$Y = Span \cdot e^{-k \cdot X} + Plateau$$

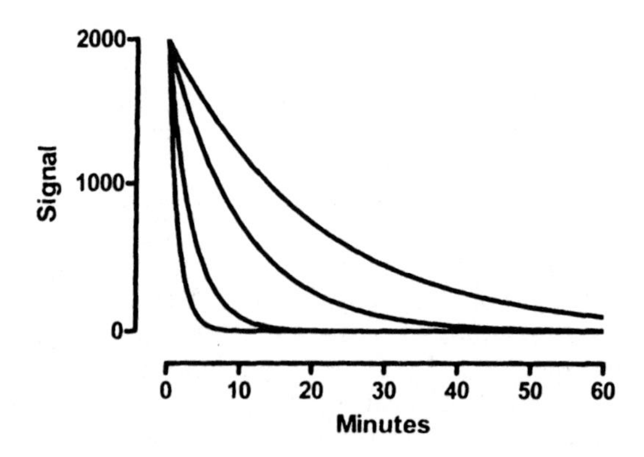

Don't hesitate to adapt a standard model to fit your needs

Most nonlinear regression programs include a list of standard models. In some cases, these won't be set up quite the way you want. In that case, it is easy to rewrite the equation to match your needs.

Here is a standard exponential decay equation and its graph:

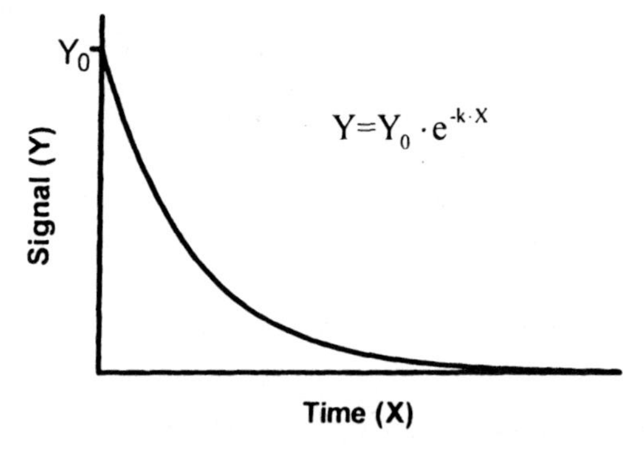

With the exponential decay model shown above, Y approaches zero at high values of X. This makes sense when you are measuring radioactive decay or metabolism of a drug. But what if you are measuring something like dissociation of a receptor from a ligand where there is a nonspecific binding? In this case, the curve plateaus at some value other than zero.

It is easy to adapt the equation to plateau at a value other than zero. Here is one way to adapt the equation.

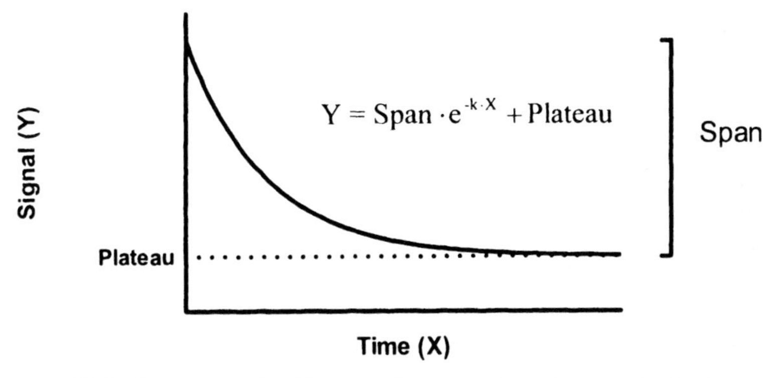

That equation will fit the span – the distance from starting place to the bottom plateau. But you might prefer to fit the starting Y value (Top) directly. No problem. Rearrange the equation like this:

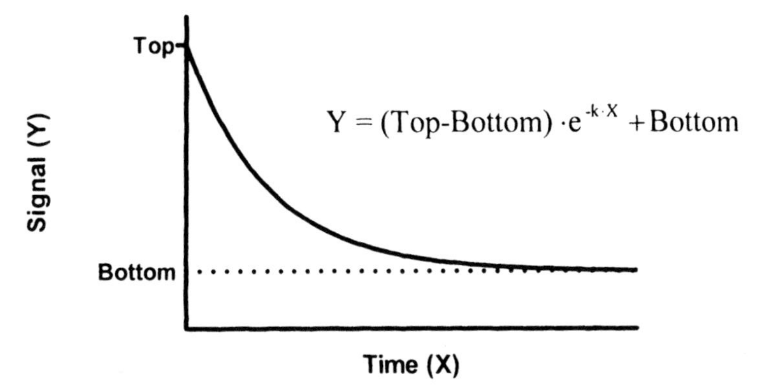

So far, we've written the exponential equation in a form such that the parameter (we call it K) that determines how rapidly the curve descends is a *rate constant*, expressed in units of inverse time. If the X values are seconds, then the rate constant is expressed in units of sec^{-1}. It is also possible to write the equations in a different form, so the parameter that determines how rapidly the curve descends is a *time constant* (τ), in units of time. If the X values are in seconds, the time constant is also expressed in units of seconds. Since the time constant is simply the inverse of the rate constant, the dissociation equation can be written in one of these forms:

$$Y=Y_0 \cdot e^{-X/\tau}$$

$$Y=(\text{Top-Bottom}) \cdot e^{-X/\tau} + \text{Bottom}$$

The details of this example will only be useful to you if you are fitting data that follow an exponential decay model. But the ideas are more general. It is easy to adapt models to fit your circumstances.

> Tip: It is easy to rearrange standard equations to meet the needs of your experimental situation and preferences. Don't feel constrained to use the equations in the form provided by someone else.

Be cautious about letting a computer pick a model for you

If you find the job of picking a model to be daunting, you may be asking: Why not just let the computer figure out which model to fit? In fact, the program TableCurve (available from www.systat.com) automatically fits data to thousands of equations and then presents you with the equation(s) that fit the data best. Using such a program is appealing because it frees you from the need to choose an equation. The problem is that the program has no understanding of the scientific context of your experiment. The equations that fit the data best usually do not correspond to scientifically meaningful models, and therefore should be considered empirical, rather than mechanistic, models. You will, therefore, probably not be able to interpret the best-fit values of the parameters in terms of biological mechanisms, and the results are unlikely to be useful for data analysis.

This approach can be useful when you just want a smooth curve for interpolation or graphing, and don't care about models or parameters. Don't use this approach when the goal of curve fitting is to fit the data to a model based on chemical, physical, or biological principles. Don't use a computer program as a way to avoid understanding your experimental system, or to avoid making scientific decisions.

> Tip: In most cases, choosing a model should be a scientific, not a statistical or mathematical, decision. Don't delegate the decision to a computer program or consulting statistician.

Choose which parameters, if any, should be constrained to a constant value

It is not enough to pick a model. You also need to decide whether any of these parameters can be constrained to constant values. Nonlinear regression will find best-fit values for the rest.

You must make this decision based on how (if) you have normalized the data and based on controls (for example, controls may define the bottom plateau of a dose-response or kinetic curve).

> Tip: One of the most common errors in curve fitting is to ask the program to fit all the model parameters in situations where some parameters should clearly be fixed to a constant value. No program can make this decision for you.

11. Global models

What are global models?

The example models in the previous chapters each define a single curve. Routinely, therefore, you would fit those models to a single set of data. This is the way curve fitting is usually done – fitting a model to a single set of data.

An alternative approach is to create a global model that defines a family of curves, rather than just a single curve. Then use a nonlinear regression program that can perform global fitting to fit a family of data sets at once. The important point is that when you fit the global model, you tell the program to share some parameters between data sets. For each shared parameter, the program finds one (global) best-fit value that applies to all the data sets. For each nonshared parameter, the program finds a separate (local) best-fit value for each data set.

Fitting data to global models is extremely useful in many contexts. Later in this book, we'll present many situations where global curve fitting is the best approach to analyzing data. For now, we'll just present two simple examples to illustrate the basic idea of global curve fitting.

Example 1. Fitting incomplete data sets.

The graph below shows two dose-response curves with two curve fits. These were fit individually, each curve fit to one of the data sets. It also shows, as horizontal lines, the 95% confidence interval of the EC_{50}.

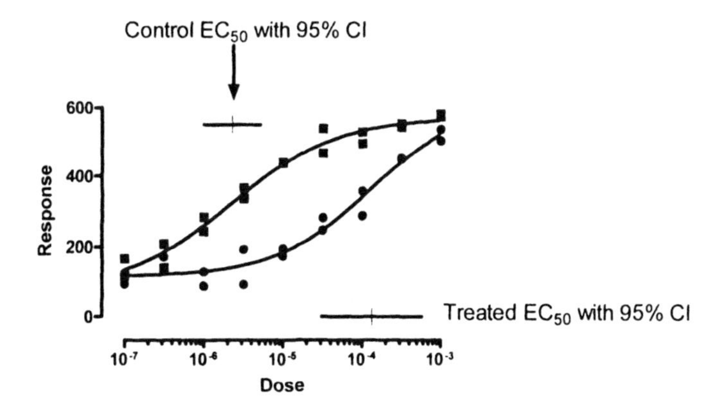

While the curves look nice, the best-fit parameters are quite uncertain. The problem is that the control data don't really define the bottom plateau of the curve, and the treated data don't really define the top plateau of the curve.

The goal of fitting a curve to dose-response data like this is to find the EC_{50}, the concentration (dose) that gives a response half-way between the minimum and maximum responses. If your data don't define the minimum and maximum responses very well, it is hard to define "halfway", so hard to define the EC_{50}. Therefore, when we fit the sample data above (fitting each curve individually), the program finds best-fit values for each EC_{50} but also presents confidence intervals for each EC_{50} that extend over more than an order of magnitude. The whole point of the experiment was to determine the two EC_{50} values, and (with this analysis) the results were disappointing. There is an unacceptable amount of uncertainty in the value of the best-fit values of the EC_{50}.

One way to determine the EC_{50} values with less uncertainty is to redo the experiment, collecting data over a wider range of doses. But we don't have to redo the experiment. We can get much better results from the original set of experiments by using global curve fitting. When you use global curve fitting, you have to tell the program which parameters to share between data sets and which to fit individually. For this example, we'll instruct the program to find one best-fit value of the top plateau that applies to both data sets, one best-fit value of the bottom plateau that applies to both data sets, and one best-fit value of the slope factor (how steep is the curve) that applies to both data sets. Of course, we won't ask the program to share the EC_{50} value. We want the program to determine the EC_{50} separately for control and treated data.

Here are the results.

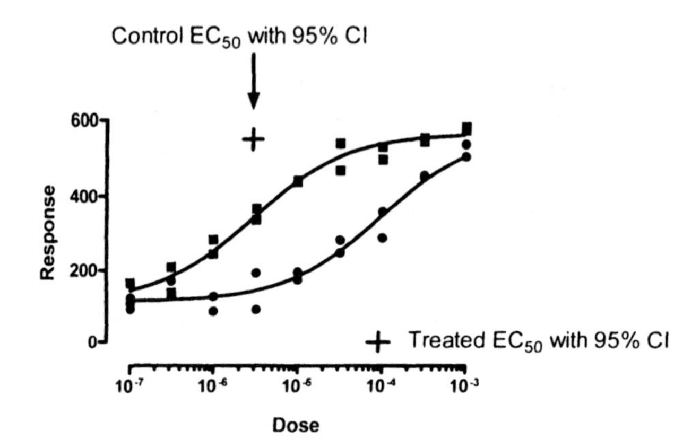

The graph of the curves looks only slightly different. But now the program finds the best-fit parameters with great confidence. Each EC_{50} value is determined, with 95% confidence, within a factor of two (compared to a factor of ten or more when the curves were fit individually). We've accomplished the goal of the experiment, to determine the two EC_{50} values with reasonable certainty.

The control data define the top of the curve pretty well, but not the bottom. The treated data define the bottom of the curve pretty well, but not the top. By fitting both data sets at once, using a global model, we are able to determine both EC_{50} values with reasonable certainty.

Example 2. The parameters you care about cannot be determined from one data set.

The graph below shows results from a homologous competitive binding experiment. This kind of experiment, and this particular example, will be explored in much more detail in Chapter 38.

The idea is pretty simple. You want to know how many of a particular kind of receptor your tissue sample has, and how tightly a particular drug (ligand) binds to those receptors. You add a single concentration of a radioactively labeled drug to all the tubes, and also add various amounts of the same drug that is not radioactively labeled. You assume that the two forms of the ligand bind identically to the receptors. As you add more of the unlabeled ligand, it binds to the receptors so less of the radioactive ligand binds. So you see downhill binding curves.

If you do such an experiment with a single concentration of radioactively labeled drug, the results are not usually acceptable. You get a nice looking curve, but the confidence intervals for both parameters you want to know (how many receptors and how tightly they bind) are very wide. You get almost the same results when you have lots of receptors that bind your drug weakly or a few receptors that bind your drug tightly. Given the unavoidable experimental error, any one experiment simply gives ambiguous results.

If you use two different concentrations of labeled drug, as we did in our sample experiment, the situation is quite different. If you fit each of the two experiments individually, the results are unacceptable. The confidence intervals are extremely wide. But fit the two curves globally, telling the program to fit one value for receptor number and one value for receptor affinity that applies to both data sets, and now the results are quite acceptable.

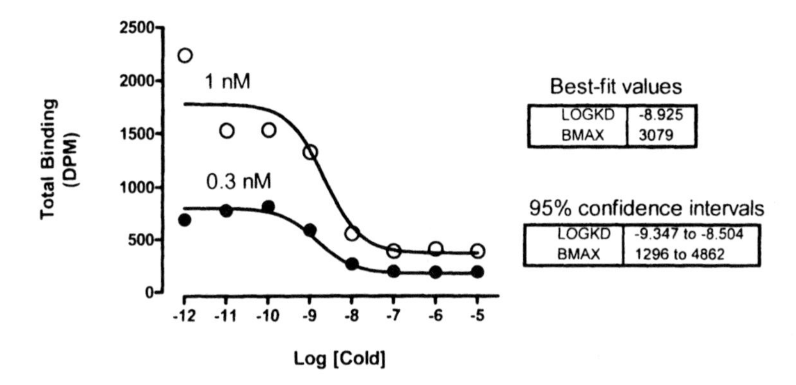

Assumptions of global models

When you fit a global model to a family of data sets, you are asking the program to find values of the parameters that minimize the grand sum-of-squares for all data sets. This makes sense only when:

- All data sets are expressed in same units. If different data sets were expressed in different units, you could give different data sets different weights just by expressing the values in different units. For example, imagine what happens if you change one data set from expressing weight in grams to expressing weight in milligrams. All the values are now increased by a factor of 1000, and the sum-of-squares for that data set is increased by a factor of one million (1000^2). Compared to other data sets, expressed in different units, this data set now has a much greater impact on the fit.

> Tip: If you really want to do global fit to data sets using different units, first normalize the data so they are comparable.

- The scatter is the same for each data set. At each X value, for each data set, the scatter should be the same (unweighted fit) or should vary predictably with Y (and be accounted for by weighting). If one data set has very little scatter and another has lots of scatter, global fitting would not make sense.

How to specify a global model

The graph below shows an example of fitting a global model to data. We shared the top, bottom, and Hill slope and fit separate values for the LogEC50.

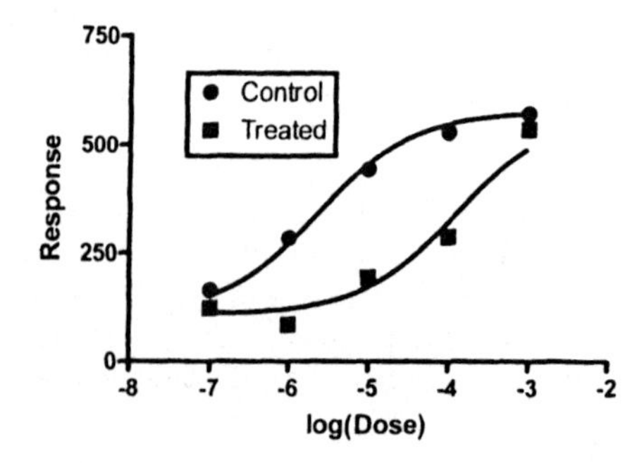

With Prism, you enter the data sets in side-by-side columns.

	X Values	A	B
	log(Dose)	Control	Treated
	X	Y	Y
1	-7.0	165	124
2	-6.0	284	87
3	-5.0	442	195
4	-4.0	530	288
5	-3.0	573	536

For this example, the built-in dose-response curve (variable slope) model is appropriate.

```
Y=Bottom + (Top-Bottom)/(1+10^((LogEC50-X)*HillSlope))
```

With Prism, use the Constraints tab of the nonlinear regression dialog to specify which parameters to share. For this example, we share Bottom, Top and HillSlope, but not LogEC50. If you share any parameter or parameters, Prism uses global curve fitting.

Fix, constrain or share a parameter

Parameter	Constraint
BOTTOM	Shared value for all data sets
TOP	Shared value for all data sets
LOGEC50	No constraint
HILLSLOPE	Shared value for all data sets

Nonlinear regression fit five best-fit values: Bottom, Top, HillSlope, LogEC50 for control and LogEC50 for treated.

Parameter	Best-fit value	SE	95% CI
Bottom	110.0	27.86	38.33 to 181.6
Top	578.6	34.26	490.6 to 666.7
HillSlope	0.7264	0.1857	0.2491 to 1.204
LogEC50 (control)	-5.618	0.07584	-5.813 to -5.423
LogEC50 (treated)	-3.883	0.1767	-4.338 to -3.429

The sum-of-squares is 5957 with 5 degrees of freedom (ten data points minus five parameters).

If you use a nonlinear regression program other than Prism, you'll need to read its documentation carefully to learn how to set up a global model. Most likely, you'll need to enter the data with a grouping variable:

log(Dose)	Response	Group
-7	165	1
-6	284	1
-5	442	1
-4	530	1
-3	573	1
-7	124	2
-6	87	2
-5	195	2
-4	288	2
-3	536	2

Enter the model, using IF-THEN statements to define a different logEC50 for each group. For our example, the first line of the equation could look something like this:

```
If (Group = 1)     {LogEC50=LogEC50C} Else     {LogEC50=LogEC50T}
```

Or this:

```
LogEC50 = IF(Group=1, LogEC50C, LogEC50T)
```

In those equivalent equations, we define the intermediate variable LogEC50 to equal either the parameter LogEC50C (control) or logEC50T (treated) depending on the value of Group. The rest of the equation (not shown) would then define Y as a function of LogEC50.

Each program has its own way of dealing with IF-THEN relationships, but you probably won't find it hard to adapt the equations shown above.

12. Compartmental models and defining a model with a differential equation

What is a compartmental model? What is a differential equation?

Let's create a pharmacokinetic model that explains how the concentration of a drug in blood plasma declines over time. You inject a bolus of drug, and measure plasma drug levels at various times thereafter. The drug is metabolized by a liver enzyme that is far from fully saturated. In this situation, the rate of drug metabolism is proportional to the concentration of drug in the plasma. This is called *first-order kinetics*. When the plasma concentration is cut in half, so is the rate of drug metabolism.

This model can be shown as this compartmental diagram:

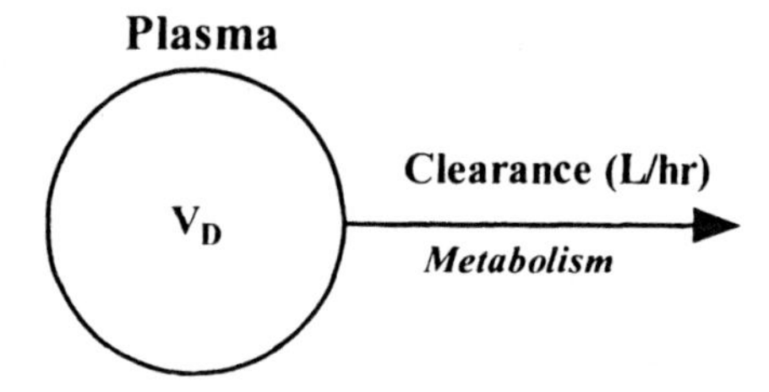

In this model, there is only one compartment labeled "plasma". In fact, this compartment may be larger than the plasma volume. It will include all other spaces that mix rapidly with the plasma. The volume of distribution, V_D, is the volume of this entire compartment. In other words, it is the volume of blood plasma (liters) it would take to hold all the drug in the body if it were equally distributed. Clearance (in L/hr) is the volume of blood plasma totally cleared of drug per minute.

Drawing a compartmental diagram helps clarify what is going on. From the diagram (some think of it as a map), you can easily create differential equations. A differential equation defines how one variable changes as another variable changes.

In this case, we measure the concentration in the plasma, abbreviated C_{plasma} . Drug is metabolized by a rate equal to the Clearance (liters/hour) times the concentration in the plasma (mg/liter). The product of clearance times concentration gives the rate of drug metabolism in mg/hr. Divide by the volume of the plasma compartment (V_D, in liters) to compute the change in the concentration in the plasma.

Written as a differential equation, our model is:

$$\frac{dC_{plasma}}{dt} = -\frac{\text{Clearance}}{V_D} \cdot C_{plasma}$$

Since we measure the concentration in the plasma, we call this Y. So the equation is equivalent to this equation and graph.

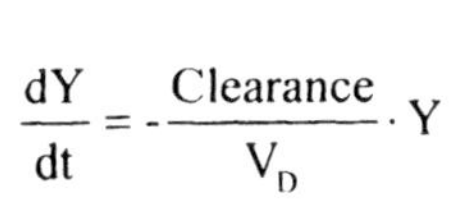

$$\frac{dY}{dt} = -\frac{Clearance}{V_D} \cdot Y$$

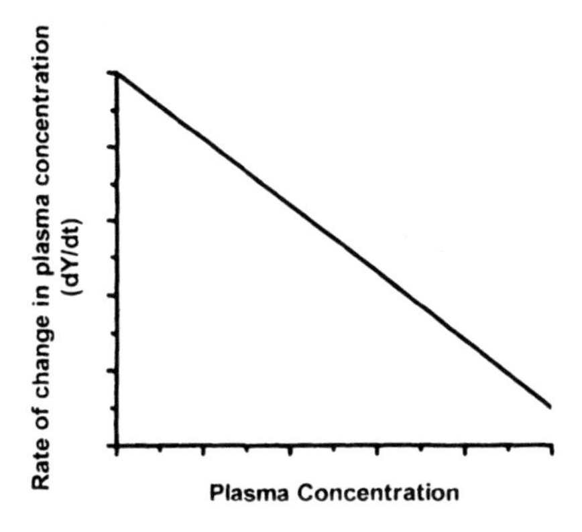

Your goal is to fit a curve to the data to learn the volume of distribution and clearance of the drug.

Integrating a differential equation

Using calculus, you (or someone you delegate this job to) can integrate the equation to form a standard model that defines Y as a function of t:

$$Y_t = Y_0 \cdot e^{-\frac{Clearance}{V_D} \cdot t} = Y_0 \cdot exp(-Clearance \cdot t/V_D)$$

At time zero, the concentration of drug (Y_o) equals the dose you injected (D in mg) divided by the volume of distribution (V_o in mL). So the equation can be rewritten like this:

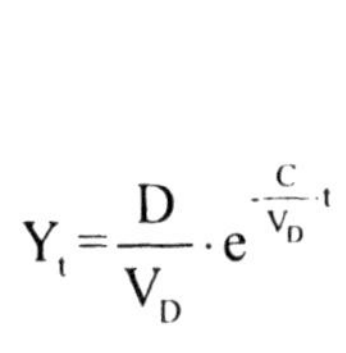

$$Y_t = \frac{D}{V_D} \cdot e^{-\frac{C}{V_D} \cdot t}$$

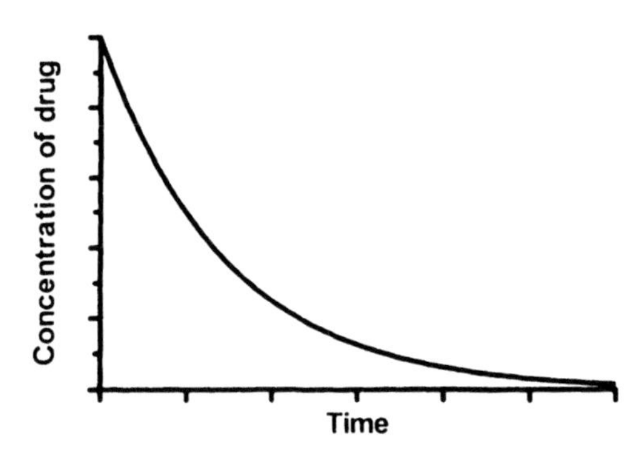

At any time t, the Y value (drug concentration) can be computed from the clearance, the volume of distribution, and the time t. Set the dose to a constant, and use nonlinear regression to fit data to find the best-fit value of clearance and volume of distribution.

> Note: With most nonlinear regression programs (including Prism), you'll need to rewrite the model so the independent variable (time) is X rather than t.

The idea of numerical integration

In the previous example, it was quite simple to integrate the differential equation to obtain an equation that defines Y (concentration) as a function of X (time). This is not always possible.

Here are some data showing the metabolism of phenytoin over time following an intravenous injection of 300 mg.

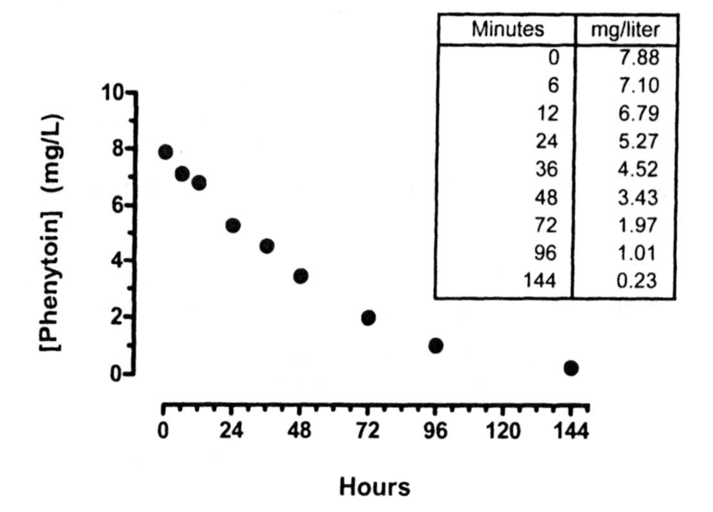

Minutes	mg/liter
0	7.88
6	7.10
12	6.79
24	5.27
36	4.52
48	3.43
72	1.97
96	1.01
144	0.23

Phenytoin saturates a large fraction of the enzyme that metabolizes it. This means that the rate of metabolism is not proportional to drug concentration. The graph below compares the first-order model discussed in the previous section with the nonlinear (Michaelis-Menten) model appropriate for phenytoin metabolism. At very low drug concentrations, the two models are indistinguishable. But at higher drug concentrations, the two diverge considerably.

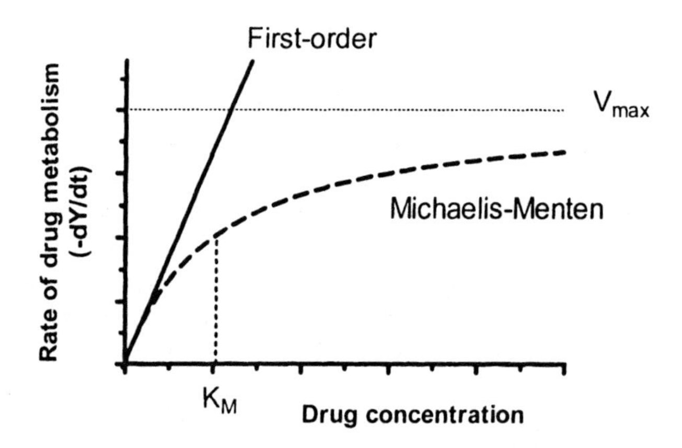

The rate of drug metabolism as a function of drug concentration follows the Michaelis-Menten relationship according to the equation below. V_{max} is the maximum velocity of the enzyme (in mg/hr), when extrapolated to very high concentrations of drug that saturate the enzyme. This is not related to V_D, the volume of distribution in mL. The use of the variable "V" is a coincidence and V_{max} and V_D are not even measured in the same units. K_m is the concentration of phenytoin (in mg/liter) where drug metabolism is half-maximal.

$$\frac{dY}{dt} = -\frac{V_{max} \cdot Y}{V_D(Km+Y)}$$

This differential equation appears to be only slightly more complicated than the one in the previous section. Instead of one parameter for drug clearance, we have two parameters (V_{max} and K_m) that quantify the enzyme that degrades the drug. But this differential equation is fundamentally different than the previous one. This differential equation cannot be integrated to form an equation that defines Y as a function of time, V_{max}, V_D, and K_m. We think that it is simply impossible to integrate, no matter how adept you are at wrangling differential equations. It certainly is difficult.

Without an equation, how can we fit the model to data to find best-fit values of V_{max}, V_D and K_m? Actually, we don't need an *equation*. What we need is a computer *procedure* that calculates Y as a function of time, V_{max}, V_D and K_m. We can do this without calculus by doing brute force calculations. The idea is quite simple, and is best understood by an example.

Based on previous literature, here are some initial parameters we can use when fitting the curve. We found the parameters in a textbook of pharmacology. The parameters were tabulated as per kg, and we multiplied by 65 kg to get the values shown here.

Parameter	Initial value	Comment
V_D	42 liter	Volume of distribution. Parameter to be fit.
V_{max}	16 mg/hr	Maximal elimination rate when enzyme is fully saturated. Parameter to be fit.
K_m	5.7 mg/liter	Drug concentration that half saturates the metabolizing enzyme. Parameter to be fit.
Dose	300 mg	Dose injected intravenously at time zero. Experimental constant.

Beginning from these initial values, we want to use nonlinear regression to find best-fit values for V_d, V_{max} and K_m.

We can calculate the drug concentration (Y value) we expect to see at time=0. At time zero, the concentration of drug (Y_0) equals the dose you injected (D in mg) divided by the volume of distribution (V_0 in mL), and so is 7.14 mg/liter. The graph above shows that the actual concentration of drug at time zero was a bit higher than that. Not a problem. The values shown in the table above are just initial values. We'll use nonlinear regression to get the best-fit values.

Let's figure out the drug concentration an instant later. The trick is to pick a time interval that is small enough so drug concentration changes just a small amount, so the value Y on the right side of the equation can be treated as a constant. To make the calculations easier to follow, we'll start with a time interval of 0.25 hr. (Later we'll switch to a smaller interval.)

What is the value of Y at t=0.25 hr? We can rewrite the previous equation to substitute Δt (a value we chose to equal 0.25 hr) for dt and and ΔY for dY (dY is a mathematical ideal; ΔY is an actual value).

$$\frac{\Delta Y}{\Delta t} \approx -\frac{V_{max} \cdot Y}{V_D(Km+Y)}$$

Now rearrange to compute the change in Y.

$$\Delta Y \approx -\frac{V_{max} \cdot Y}{V_D(Km+Y)} \cdot \Delta t$$

We have values for all the variables on the right side of the equation, so we can calculate the value of ΔY, which will be negative (the drug concentration is going down over time). For this example, ΔY equals -0.0530 mg/liter. Add this negative value to the drug concentration at time 0 (7.14 mg/liter), and you've computed the concentration at time 1 minute (7.090 mg/liter). Plug this new value of Y into the right of the equation (the other variables don't change) and calculate a new value of ΔY for the change from time 0.25 hr to 0.50 hr. This time, ΔY= -0.0528 mg/liter. Add this negative value to the concentration at time 0.25 hr and you've computed the concentration at time 0.50 hr, which is 7.090 mg/liter.

Repeat this procedure many times (with a program or spreadsheet), and you can graph predicted drug concentration (Y) as a function of time (X). This procedure, which doesn't require any calculus, is called *numerical integration*.

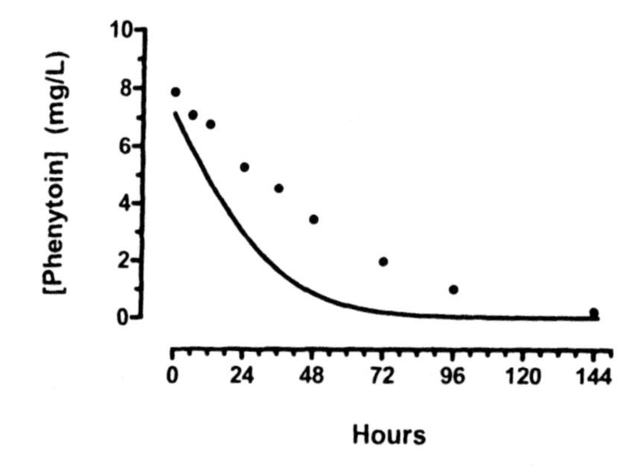

The curve doesn't go through our data. But that's ok. The curve was generated from some published (average) values. Now we want to fit the model to our data to find the kinetic parameters in this individual.

The only real trick to numerical integration is choosing a value for Δt. The method will be inaccurate if Δt is too large. In fact, we would have gotten slightly different (and more accurate) values if we had used a smaller Δt in our example. If Δt is small, the calculations will take longer. If Δt is really tiny, the calculations may lose accuracy. For details about choosing Δt and numerical integration, read *Numerical Recipes in C.*

To improve accuracy, we decreased a Δt to 0.01 hour, and fit the model to the data. Here are the best-fit values of the parameters, along with a graph of the resulting curve.

Parameter	Best-fit value	Units
V_d	38.2	liter
V_{max}	6.663	mg/hr
K_m	4.801	mg/liter

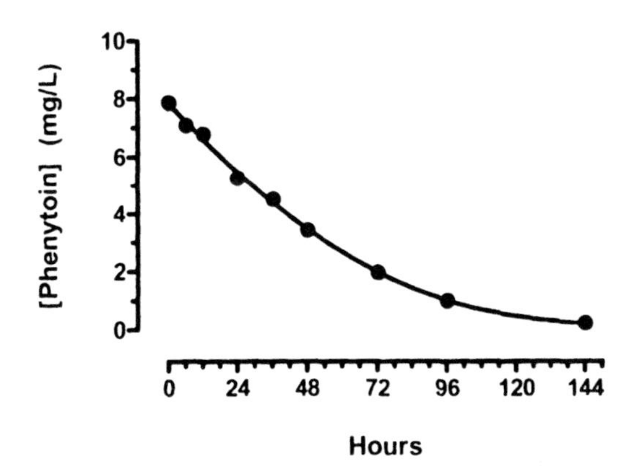

If your program lets you enter a model as a differential equation, you'll also need to define the Y value at time zero (for this example, you want to enter an equation that defines Y at time zero from the parameters). After that, curve fitting works as usual and there is nothing special about interpreting the results. Because numerical integration requires lots of calculations, and these need to be repeated over and over as nonlinear regression progresses, fitting a model defined by a differential equation takes longer than fitting a model defined by an ordinary equation.

> Note: Few nonlinear programs will let you express a model as a differential equation. GraphPad Prism 4.0 cannot fit models expressed this way, although this is a feature we are likely to add to a future version.

More complicated compartmental models

Here is an example of a more complicated compartmental model. It models the plasma concentration of estrogen.

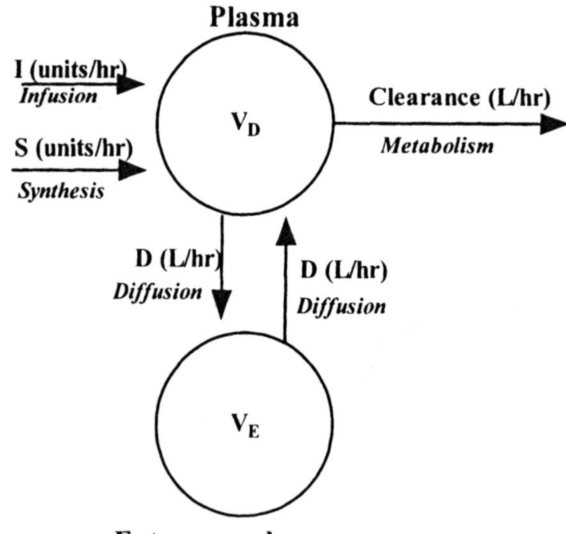

In addition to the plasma, estrogen also enters another extra-vascular compartment. In fact, this is probably the sum of several physiological spaces. But so long as they mix rapidly (compared to the time course of the experiment), we can consider it to be one compartment. This compartment is distinct from the plasma because it does not mix rapidly with the plasma. The diffusion constant D defines how rapidly the two compartments mix.

Starting from the compartmental diagram, we can construct the differential equations. Here is how to think it through:

- There are two methods that the plasma concentration increases, by infusion and by endogenous synthesis. So plasma concentration increases by the sum of S and I, both measured in units/hr.

- Liver metabolism equals the clearance (in liters/hr) divided by the volume of the plasma space (V_D in liters) and times the concentration of the drug in the plasma.

- Estrogen diffuses away from the plasma space. This flow equals the diffusion constant D (liters/hr) divided by the volume of the plasma space (V_D in liters) and times the concentration in the plasma space ($Conc_P$).

- Estrogen also diffuses into the extra-vascular space. This flow equals the diffusion constant D (liters/hr) divided by the volume of the extra-vascular space (V_E in liters) and times the concentration in the extra-vascular space ($Conc_E$).

- The two flows between the plasma and extra-vascular space need to enter into two differential equations, one for the change in plasma concentration, the other for the change in the extra-vascular concentration.

Turn those ideas into a set of differential equations. Each arrow in the compartmental diagram pointing in or out of the compartment we model becomes one term in the differential equation. The change in concentration equals the sum of all flows into that compartment minus the sum of all flows leaving the compartment.

$$\frac{dConc_{plasma}}{dt}=I+S-\frac{Clearance \cdot Conc_{Plasma}}{V_D}+\frac{D \cdot Conc_E}{V_E}-\frac{D \cdot Conc_{Plasma}}{V_D}$$

$$\frac{dConc_E}{dt}=\frac{D \cdot Conc_{Plasma}}{V_D}-\frac{D \cdot Conc_E}{V_E}$$

Note: Some programs (not from GraphPad) let you enter the model as a compartmental model. You draw boxes and arrows in the program, and it generates the differential equations automatically.

Now you can use the model to simulate data. Or use nonlinear regression (with an appropriate program, not from GraphPad) to fit experimental data and find best-fit parameters.

D. How nonlinear regression works

13. Modeling experimental error

Why the distribution of experimental error matters when fitting curves

The goal of regression is to find best-fit values for the parameters of the model (e.g., slope and intercept for linear regression, other parameters, such as rate constants, for nonlinear regression). More precisely, the goal is to find values for the parameters that are most likely to be correct. It turns out that you can't decide which parameter values are most likely to be correct without making an assumption about how the data are scattered around the line or curve.

> Note: Statisticians refer to the scatter of points around the line or curve as "error". This is a different use of the word than is used ordinarily. In statistics, the word "error" simply refers to deviation from the average. The deviation is usually assumed to be due to biological variability or experimental imprecision, rather than a mistake (the nontechnical meaning of "error").

Origin of the Gaussian distribution

Most linear and nonlinear regression assumes that the scatter follows a Gaussian (also called a "normal") distribution. This section explains what the Gaussian distribution is, and why it is so central to statistics.

Imagine a very simple "experiment". You pipette some water, and weigh it. Your pipette is supposed to deliver 10 mL of water, but in fact delivers randomly between 9.5 and 10.5 mL. If you pipette one thousand times, and create a frequency distribution histogram of the results, it will look like something like this.

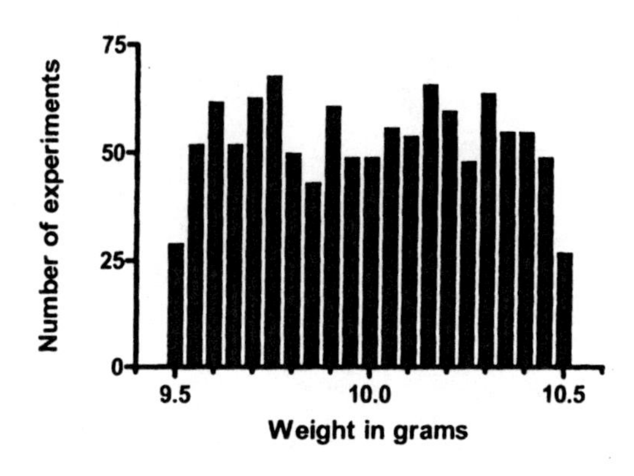

The average weight is 10 grams, the weight of 10 mL of water (at least on Earth). The distribution is flat, with random variation.

Now let's make the experiment slightly more complicated. We pipette two aliquots of 10 mL and weigh the result. On average, the weight will now be 20 grams. But the errors will cancel out much of the time. Values close to 20 grams will be more likely than values far from 20 grams. Here is the result of 1000 such (simulated) experiments.

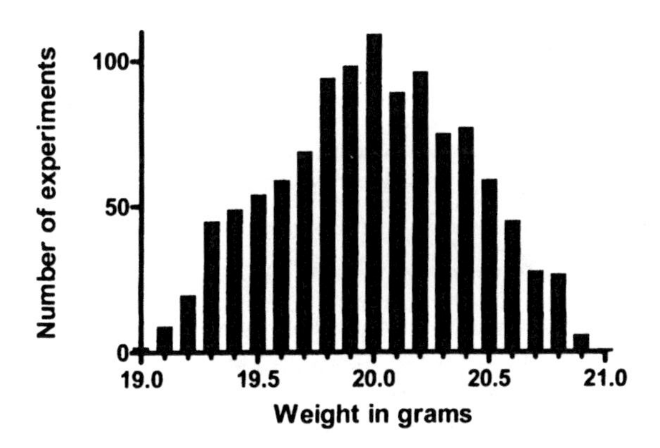

Each pipetting step has a flat random error. Add up the result of two pipetting steps, and the distribution is not flat. For example, you'll get weights near 21 grams only if both pipetting steps err substantially in the same direction, and that will happen rarely. In contrast, you can get a weight of 20 grams in lots of ways, so this happens more commonly.

Now let's make the experiment more realistic with ten steps. Here are the results of 15000 simulated experiments. Each result is the sum of ten steps.

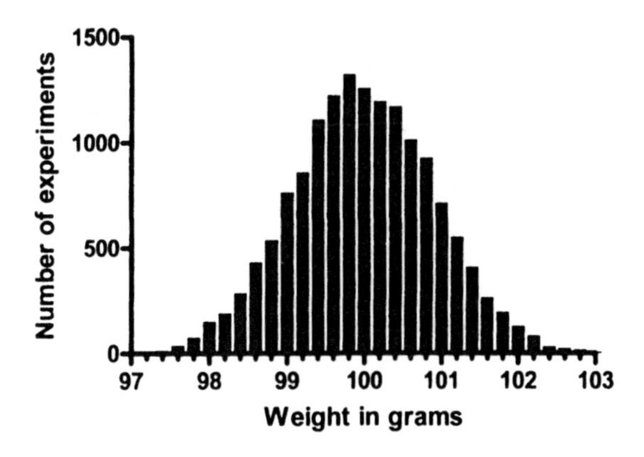

You'll recognize this distribution. It is the Gaussian bell-shaped distribution, which is central to much of statistics.

If your experimental scatter has numerous sources that are additive and of nearly equal weight, you expect the scatter of replicate values to approximate a Gaussian distribution. This also means that you'll expect the deviations of points around the curve to follow a Gaussian distribution.

The Gaussian distribution is a mathematical ideal, which extends from negative infinity to positive infinity. If the weights in the example above really followed a Gaussian distribution exactly, there would be some chance (albeit very small) of having a negative weight. Since weights can't be negative, the distribution cannot be exactly Gaussian. But it is close enough to Gaussian to make it ok to use statistical methods (like regression) that assume a Gaussian distribution. Few biological distributions, if any, completely follow the Gaussian distribution, but many are close enough.

> Tip: The Gaussian distribution is also called a *Normal* distribution. Don't confuse this use of the word "normal" with its other meanings of "ordinary", "common", or "disease free".

From Gaussian distributions to minimizing sums-of-squares

Given the assumption that scatter follows a Gaussian distribution, how do we decide which parameter values are most likely to be correct?

Mathematical statisticians approach this problem using a method called *maximum likelihood*. Start with a data set, a model, and an assumption about how the data are scattered around the model. Given a model, a set of parameters, and an assumption about scatter, it is possible to calculate the likelihood of generating your data. Some sets of parameters are very unlikely to have generated your data, because the curve those parameters define is far from your data. Other sets of parameters (those that define a curve near your data) are more likely. The maximum likelihood method finds the set of parameter values that are most likely to have generated your data.

The maximum likelihood method can be used with almost any assumption about the distribution of the data. However, standard nonlinear (and linear) regression is almost always based on the assumption that the scatter follows a Gaussian distribution. Given this specific assumption, it can be proven that you can find the most likely values of the parameters by minimizing the *sum of the squares* of the *vertical* distances ("residuals") of the points from the line or curve.

Unless you have studied statistics in depth, it isn't obvious how the assumption of a Gaussian distribution of residuals (distances of points from curve) leads to the rule that you find the most likely parameters by minimizing the sum-of-squares. But the two are linked. If you make some other assumption about the scatter of replicates, then you'll need to use a different rule (not least-squares) to find the values of the parameters that are most likely to be correct.

If you want to understand more about maximum likelihood methods, and how the least-squares rule was derived, read section 15.1 of *Numerical Recipes in C*, which you can find online at www.nr.com.

Here is an intuitive way to get a sense of why minimizing the sum of the squares of the distances makes sense. Two points from a set of data scattered randomly according to a Gaussian distribution are far more likely to have two medium size deviations (say 5 units each) than to have one small deviation (1 unit) and one large (9 units). A procedure that minimized the sum of the absolute value of the distances would have no preference over a line (or curve) that was 5 units away from two points and a line (or curve) that was 1 unit away from one point and 9 units from another. The sum of the distances (more precisely, the sum of the absolute value of the distances) is 10 units in each case. If, instead, you choose a procedure that minimizes the sum of the squares of the distances, then a fit that produces a line 5 units away from two points (sum-of-squares = 50) is a much better fit than one that produces a line 1 unit away from one point and 9 units away from another (sum-of-squares = 82). If the scatter is Gaussian (or nearly so), the line determined by minimizing the sum-of-squares is most likely to be correct.

Regression based on nongaussian scatter

The assumption of a Gaussian distribution is just that, an assumption. It is a useful assumption, because many kinds of data are scattered according to a Gaussian distribution (at least approximately) and because the mathematical methods based on the Gaussian distribution work so well.

Not all data are scattered according to a Gaussian distribution, however. Some kinds of data, for instance, follow a Poisson distribution, and regression methods have been developed that assume such a distribution for experimental errors.

It is also possible to assume a very wide distribution, such as the Lorentzian distribution (also called the Cauchy distribution), shown below. This distribution is much wider than a Gaussian distribution.

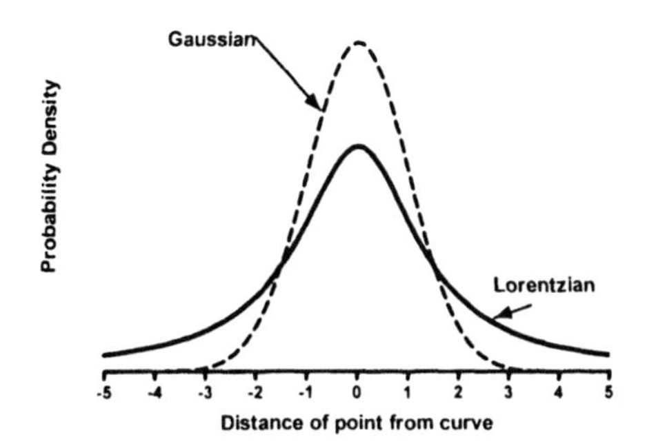

If you fit data to a model assuming this distribution, outliers (points far from the curve) will have less influence. The Lorentzian distribution is the same as a t distribution with 1 degree of freedom. The Gaussian distribution is the same as a t distribution with many degrees of freedom. By varying the number of degrees of freedom, the t distribution gradually changes from having a wide distribution to a compact one.

> GraphPad note: GraphPad Software, in collaboration with AISN Software, is working hard to develop a version of Prism that can fit assuming a Lorentzian distribution. Curve fitting done this way is much more robust to outliers. For details, go to www.graphpad.com and search for "robust".

14. Unequal weighting of data points

Standard weighting

As outlined in the previous chapter, regression is most often done by minimizing the sum-of-squares of the vertical distances of the data from the line or curve. Points farther from the curve contribute more to the sum-of-squares. Points close to the curve contribute little. This makes sense, when you expect experimental scatter to be the same, on average, in all parts of the curve.

The graph below is an example of data collected with twenty replicates at each drug dose. The scatter is Gaussian and the standard deviation among the replicates is about the same at each dose. With data like these, it makes sense to minimize the sum-of-squares of the absolute distances between the points and the curve.

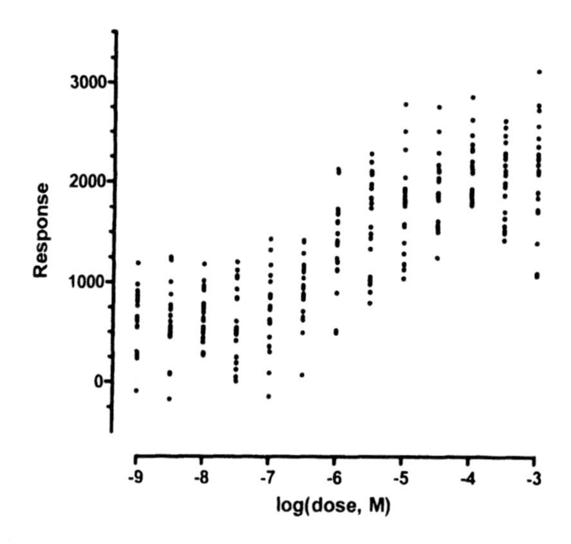

> Note: This example purposely exaggerates the scatter to make the example easier to follow.

Relative weighting (weighting by $1/Y^2$)

In many experimental situations, you expect the average distance (or rather the average absolute value of the distance) of the points from the curve to be higher when Y is higher. The relative distance (distance divided by Y), however, remains about the same all the way along the curve.

Here is another example dose-response curve. At each dose, twenty replicate values are plotted. The distribution of replicates is Gaussian, but the standard deviation varies. It is small at the left side (bottom) of the graph and large at the right side (top). While the standard deviation of the error is not consistent along the graph, the relative variability is consistent. In fact, these data were simulated by selecting random numbers from a Gaussian distribution with a standard deviation equal to 20% of the mean value.

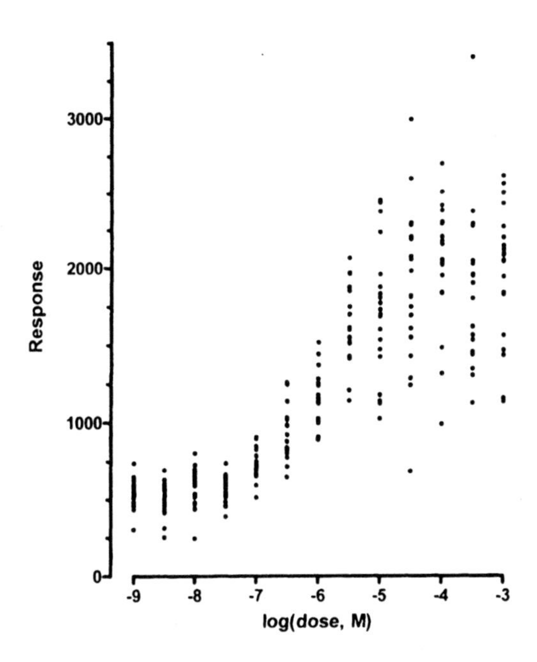

In this common situation, minimizing the sum-of-squares of the absolute vertical distances would be inappropriate. If you did that, points with high Y values would tend to have large deviations from the curve and so would have a large impact on the sum-of-squares value. In contrast, points with smaller Y values would have little influence. This is not desirable – you want all the points to have about equal influence on the goodness-of-fit. To do this, minimize the sum of the square of the *relative* distances.

The relative distance of a data point from the curve is given by

$$\frac{Y_{Data}-Y_{Curve}}{Y_{Data}}$$

If you choose relative weighting, nonlinear regression minimizes the sum of squares of relative distances:

$$\sum\left(\frac{Y_{Data}-Y_{Curve}}{Y_{Data}}\right)^2$$

Relative weighting is often called weighting by $1/Y^2$. Rearranging the expression shows why.

$$\sum\frac{1}{Y_{Data}^2}\left(Y_{Data}-Y_{Curve}\right)^2$$

Note: Some programs weight by the square of the Y value of the curve (the predicted Y value) rather than the Y value of the data points. This distinction (whether the denominator is the Y value of the data or curve) rarely will affect the results very much.

Poisson weighting (weighting by 1/Y)

Weighting by 1/Y is a compromise between minimizing the *actual* (absolute) distance squared and minimizing the *relative* distance squared. One situation where 1/Y weighting is appropriate is when the Y values follow a Poisson distribution. This would be the case when Y values are radioactive counts and most of the scatter is due to counting error. With the Poisson distribution, the standard deviation among replicate values is approximately equal to the square root of that mean. To fit a curve, therefore, you minimize the sum of squares of the distance between the data and the curve divided by the square root of the value. In other words, nonlinear regression minimizes this expression:

$$\sum\left(\frac{Y_{Data}-Y_{Curve}}{\sqrt{Y_{data}}}\right)^2$$

Rearrangement shows why it is sometimes called weighting by 1/Y.

$$\sum\frac{1}{Y_{Data}}\left(Y_{Data}-Y_{Curve}\right)^2$$

Note: Weighting by 1/Y is one way to cope with data distributed according to a Poisson distribution, but a better way is to use a regression method that is based upon an assumption of a Poisson regression. Look for a program that can perform “Poisson regression”.

Weighting by observed variability

Relative and Poisson weighting assume that you know how the scatter varies as the Y values vary. Often you don’t know. It sounds appealing, therefore, to base the weighting on the observed variation among replicates. If you have collected triplicate values, why not just base the weighting on the standard deviation of those values? Give points a high weight when the triplicates are tightly clustered with a small SD. Give points a low weight, when the triplicates are scattered with a large SD. In other words, minimize this quantity:

$$\sum\left(\frac{Y_{Data}-Y_{Curve}}{SD}\right)^2=\sum\frac{1}{SD^2}\left(Y_{Data}-Y_{Curve}\right)^2$$

Tip: We suggest that you rarely, if ever, use this method. Read this section carefully before weighting by the inverse of the standard deviation. The method sounds more useful than it is.

This method assumes that the mean of replicates with a large standard deviation is less accurate than the mean of replicates with a small standard deviation. This assumption sounds sensible, but it is often not true. When you only have a few replicates, the standard deviation will vary considerably just by chance. The figure below makes this point. It shows fifteen sets of triplicate values. Each value was chosen from a Gaussian distribution with a mean of 50 and a SD of 10. There is no difference between the fifteen data sets except for that due to random sampling. Compare the first (A) and last (O) data sets. The SD of the last data set is three times that of the first data set. Yet the mean of last data set is closer to the true mean (known to be 50 since these data were simulated) than is the mean of the first data set.

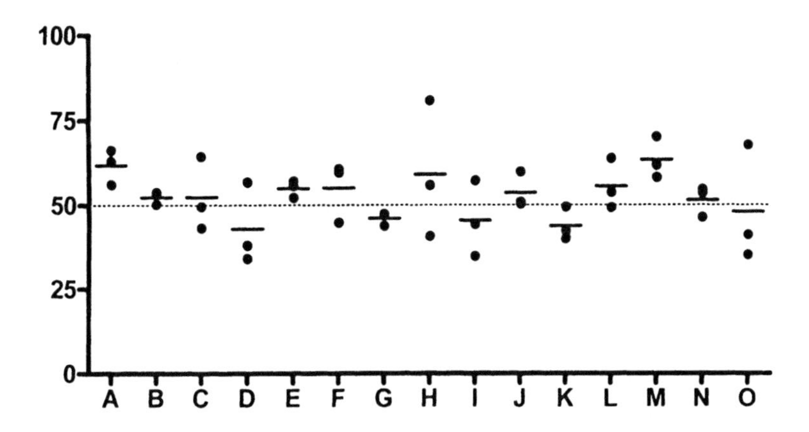

With a small number of replicates, the observed scatter jumps all over the place just by chance. The mean of a set of replicates with a small SD is not necessarily more accurate than the mean of a set of replicates with a larger SD. The small SD tells you that the replicates are clustered near each other, but this may not be close to the true value.

You want to choose a weighting scheme to account for systematic differences in the predicted amount of variability if you were to repeat the experiment many times. You should not choose weighting based on variability you happened to observe in one small experiment. That won't be helpful. Therefore it only makes sense to weight by the observed SD among replicates when you have lots (dozens) of replicates.

Error in both X and Y

Conventional linear and nonlinear regression assume that all the uncertainty is in the Y direction and that you know X precisely. A special kind of linear regression, called Model II or Deming regression (see page 50) allows for the possibility that there is error in both variables. Some methods have also been developed to account for error in both X and Y variables of nonlinear models, but these are not commonly used, and are beyond the scope of this book.

Weighting for unequal number of replicates

In most experiments, you collect replicate Y values at every value of X. How you analyze your replicates depends on your experimental design.

Independent replicates

In most experiments, it is fair to consider each replicate to be an independent data point. Each particular replicate is subject to random factors, which may increase or decrease its value. Each random factor affects individual replicates, and no random factor affects the replicates as a group. In any kind of biochemical experiment, where each value comes from a test tube or plate well, the replicates are almost certain to be independent.

When your replicates are independent, you should treat each replicate as a separate point. With some programs, this means you'll have to repeat your X values, as shown in the table shown on the left below. With other programs (such as Prism) you can put your replicates side by side, as shown on the table on the right below.

Dose	Response
1	3.4
1	2.9
1	3.1
2	4.5
2	5.1
2	4.7
3	5.9
3	6.2
3	6.3

Dose	Response		
X	Y1	Y2	Y3
1	3.4	2.9	3.1
2	4.5	5.1	4.7
3	5.9	6.2	6.3

Every replicate is treated by the nonlinear regression program as a separate data point. If there are four replicates at one X value and two at another, the four replicates will automatically get twice the weight, since the program considers them to be four separate data points.

An alternative is to average the replicate Y values and treat the mean as a single data point. If you have different number of replicates for different values, then you definitely should not average the replicates and fit to the means. The results will be wrong. If you do have the same number of replicates for every point (e.g., no missing values), fitting to means or fitting to individual replicates will give you the same best-fit values of the parameters and thus exactly the same best-fit curve. But the standard errors and confidence intervals won't be the same. Fitting to the means, rather than the replicates, can make the standard errors larger or smaller (the confidence intervals wider or narrower), depending on the data set.

Tip: If your replicates are independent, treat each one as a separate data point. Don't try to fit the averages.

Replicates that are not independent

In some experimental situations, the replicates are not independent. Random factors can affect all the replicates at once. Two examples:

- You performed a binding experiment with a single tube at each concentration, but measured the radioactivity in each tube three times. Those three values are not independent. Any experimental error while conducting the experiment would affect all the replicates.

- You performed a dose-response experiment, using a different animal at each dose with triplicate measurements. The three measurements are not independent. If one animal happens to respond more than the others, that will affect all the replicates. The replicates are not independent.

Treating each replicate as a separate data point would not be appropriate in these situations. Most of the random variation is between tubes (first example) or animals (second example). Collecting multiple replicates does not give you much additional information. Certainly, each replicate does not give independent information about the values of the parameters. Here is one way to look at this. Imagine that you have performed a dose-response experiment with a separate animal for each dose. You measure one animal in duplicate (for one dose) and another animal (another dose) ten times. It would be a mistake to enter those as individual values, because that would give five times more weight to the second dose compared to the first. The random factors tend to affect the animal, not the measurement, so measuring an animal ten times does not give you five times more information about the true value than measuring it two times.

Since each tube (first example, above) or animal (second example) is the experimental unit, you should enter each tube or animal once. If you measured several replicates, average these and enter the average. Don't enter individual values. Don't weight the means by sample size. Doing so would inflate the number of degrees of freedom inappropriately, and give you SE that are too small and CI that are too narrow. Doing so, when you have unequal number of replicates would give artificial, and undeserved, weight to the tubes or animals with more replicates, so would affect the best-fit curve and you would get less than optimal best fit parameter values.

Giving outliers less weight

If you are completely sure that all of the variation follows a Gaussian distribution, then minimizing the sum-of-squares gives you the best possible curve. But what if one of the values is far from the rest? If that deviation was generated because of an experimental mistake, it may not be part of the same Gaussian distribution as the rest. If you minimize the sum of the square of distances of points from the curve, an outlier like this can have enormous influence on the curve fitting. Remember, the distance of the point from the curve is large and you are squaring that.

One approach to dealing with outliers is to use a curve fitting method that does not assume a Gaussian distribution, but rather assumes a wider distribution of residuals. If the distribution is wider, large deviations are more common, and they don't influence the curve fitting so much. See page 83.

Another approach is to use a weighting function to reduce the influence of outlying points. One such scheme is called the Tukey Biweight. It works like this:

1. Determine the distance of each point from the curve (the residuals). Convert to absolute values, so only the distance from the curve matters, and it doesn't matter whether a point is above the curve (positive residual) or below the curve (negative residual).

2. Find the median (50th percentile) of the absolute values of the residuals. Call this M.

3. Calculate a cutoff value, C, that equals $6 \times M$. The value 6 is arbitrary, but this is how Tukey defined this method.

4. Give any point whose residual (distance from the curve) exceeds C a weight of zero. These points are so far from the curve that they can't possibly be right. By giving them a weight of zero, the points are ignored completely.

5. Give the other points a weight determined by the equation below, where R is the absolute value of the distance of that point from the curve, and C was defined in step 4 above. From this equation, you can see why this method is termed a *bisquare* method.

$$\text{Weight}=\left(1-\left(\frac{R}{C}\right)^2\right)^2$$

6. When computing the weighted sum-of-squares, the contribution of each point is the distance of the point from the curve (R) squared times the weighting factor above. In other words, after ignoring points whose distance from the curve exceeds C, this method asks the curve fitting program to minimize:

$$\sum R^2\left(1-\left(\frac{R}{C}\right)^2\right)^2$$

7. With each iteration of nonlinear regression, recompute the weights. As the curve fitting progresses, the curve will change, so the distribution of residuals will change as will the assignment of bisquare weights.

The graph below shows the contribution of a point to this adjusted sum-of-squares as a function of its distance from the curve. Half the points, by definition, are closer to the curve than the median value. For these points ($R<1$), the least-squares and Tukey biweight methods treat the points about the same. As the points get farther from the curve, up to $R=3$ (three times the median distance) their influence still increases with the Tukey biweight method, but not as much as they'd increase with standard least-squares regression. Points that are between 3 and 6 times the median distance from the curve still contribute to the curve fitting, but their influence decreases as they get farther from the curve. Points that are more than six times the median distance are ignored completely by the bisquare method.

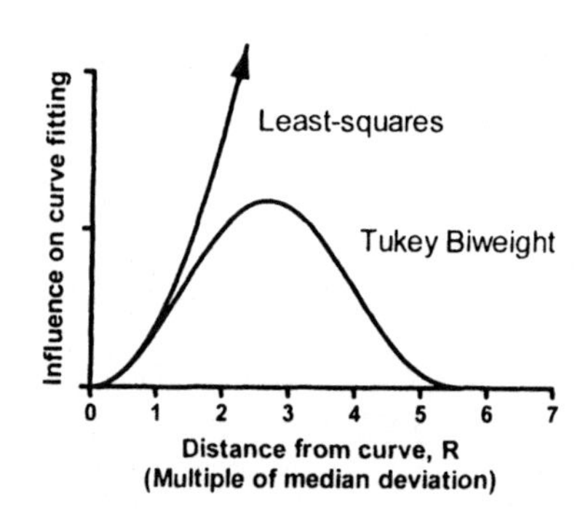

GraphPad note: No GraphPad program performs regression with the Tukey biweight method.

15. How nonlinear regression minimizes the sum-of-squares

Nonlinear regression requires an iterative approach

Nonlinear regression is more complicated than linear regression. You can't compute the best-fit values of the parameters directly from the X and Y values of the data points. Instead, nonlinear regression requires an iterative approach.

You won't be able to understand the mathematical details of nonlinear regression unless you first master matrix algebra. But the basic idea is pretty easy to understand. Every nonlinear regression method follows these steps:

1. Start with an initial estimated value for each parameter in the equation.
2. Generate the curve defined by the initial values.
3. Calculate the sum-of-squares (the sum of the squares of the vertical distances of the points from the curve).
4. Adjust the parameters to make the curve come closer to the data points. There are several algorithms for adjusting the variables, as explained in the next section.
5. Adjust the parameters again so that the curve comes even closer to the points. Repeat.
6. Stop the calculations when the adjustments make virtually no difference in the sum-of-squares.
7. Report the best-fit results. The precise values you obtain will depend in part on the initial values chosen in step 1 and the stopping criteria of step 6. This means that different programs will not always give *exactly* the same results.

Step 4 is the only difficult one. Most nonlinear regression programs use the method of Marquardt and Levenberg, which blends two other methods, the method of steepest descent and the method of Gauss-Newton. The next section explains how these methods work.

How the nonlinear regression method works

Fitting one parameter

The example below shows a signal that decreases over time. The results are normalized to run from 100% at time zero to 0% at infinite times.

Minutes	Percent
0.5	83.9
1.0	72.9
1.5	56.5
2.0	39.4
2.5	33.3
3.0	38.1
3.5	25.0
4.0	23.8
4.5	14.3
5.0	12.0
5.5	19.0
6.0	6.0

We wish to fit this to an exponential decay model, fixing the top to 100 and the bottom plateau to 0, leaving only one parameter, the rate constant (k), to fit to this model:

$$Y=100\cdot e^{-k\cdot X}$$

Since we are only finding the best-fit value of one parameter, we can do so in a brute-force manner. By looking at the data, and knowing the meaning of the rate constant, we can be quite sure the best-fit value is between 0.2 and 0.7. From the goals of the experiment, we can say that we only need to know the best-fit value to two decimal places. So there are only 50 possible values of k (0.20, 0.21, 0.22, ..., 0.69, 0.70). For each of these values, construct a curve, and compute the sum-of-squares of the distances of the points from the curve. Below is a graph of sum-of-squares as a function of k. Note that this graph was constructed specifically for this example. With different data, the graph would look different.

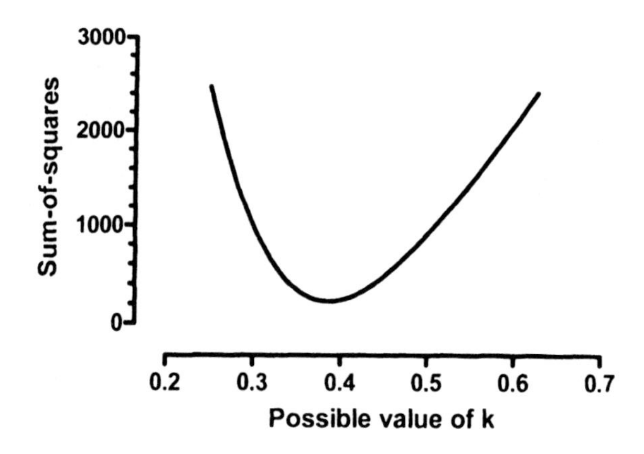

The sum-of-squares was smallest when k equals 0.39. That is our best-fit value. We're done.

Since we are only fitting one parameter, that brute force method worked fine. But if you fit several parameters, the number of possible combinations of values becomes astronomical. We need a more efficient way to find the best-fit values.

Nonlinear regression begins with an initial value of k that you give the program (or the program generates from the range of the X values). For this example, the initial value is

0.3. Mathematicians have developed several methods to move from the initial values to find the values of the parameters that minimize the sum of squares.

The **method of steepest descent** follows a very simple strategy. You start with the initial estimate of the parameters (k in this example). This initial value generates a curve, and you can compute the sum-of-squares of the points from this curve. This corresponds to a point on the graph shown above. From that point, figure out which way is downhill. Is the value of k too high or too low? Our initial value of 0.3 is too low. So we increase the k a small amount. The step size is somewhat arbitrary. We'll take a step of 0.1, to reach the value of k=0.4. Generate the curve defined by this value of k, and compute the sum-of-squares. Indeed it is lower. Repeat many times. Each step will usually reduce the sum-of-squares. If the sum-of-squares goes up instead, the step must have been so large that you went past the bottom and back up the other side. This means your step size is too large, so go back and take a smaller step. After repeating these steps many times, gradually decreasing the step size, you'll reach the bottom. You'll know you are on the bottom, because every step you take (every small change in the value of k), in either direction, leads up hill (increases the sum-of-squares)

The **Gauss-Newton method** is a bit harder to understand. As with the method of steepest descent, start at the spot defined by the initial estimated values. Then compute how much the sum-of-squares changes when you make a small change in the value of k. This tells you the steepness of the sum-of-squares curve at the point defined by the initial values. If the model really were linear, the sum-of-squares curve would have a regular shape. Knowing one point, and the steepness at that point completely defines that curve, and so defines its bottom. With a linear equation, you go from initial estimate values to best-fit values in one step using the Gauss-Newton method. With nonlinear equations, the sum-of-squares curve has an irregular shape so the Gauss-Newton method won't find the best-fit value in one step. But in most cases, the spot that the Gauss-Newton method predicts is the bottom of the surface is closer to the bottom. Starting from that spot, figure out the steepness again, and use the Gauss-Newton method to predict where the bottom of the valley is. Change the value of k again. After repeating many iterations, you will reach the bottom.

This method of steepest descent tends to work well for early iterations, but works slowly when it gets close to the best-fit values. In contrast, the Gauss-Newton method tends to work badly in early iterations, but works very well in later iterations (since the bottom of the sum-of-squares curve tends to approximate the standard shape, even for nonlinear models). The two methods are blended in the method of **Marquardt** (also called the **Levenberg-Marquardt method**). It uses the method of steepest descent in early iterations and then gradually switches to the Gauss-Newton approach. Most commercially-available computer programs use the Marquardt method for performing nonlinear regression.

Any of these methods finds that the best-fit value of k is 0.3874 min^{-1}). Here is the best-fit curve.

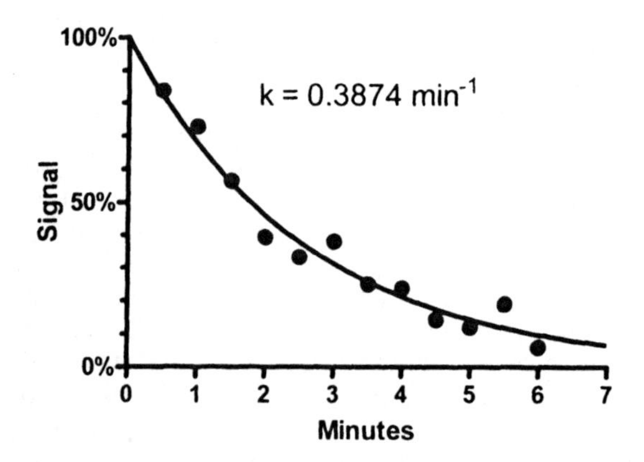

Two parameters

Here are some data to be fit to a typical binding curve (rectangular hyperbola).

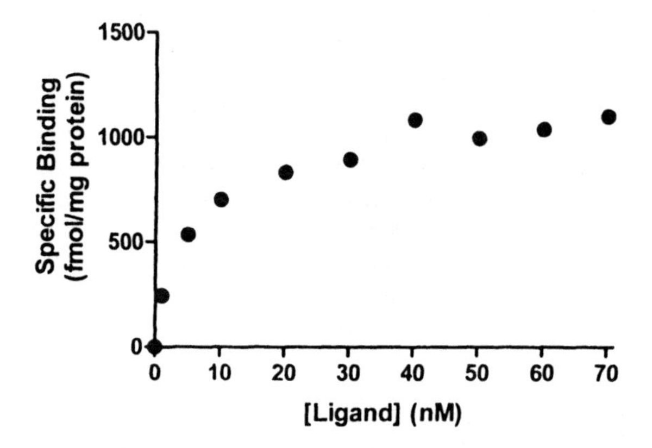

You want to fit a model defining equilibrium binding. This equation has two parameters, B_{max} and K_d.

$$Y=\frac{B_{max}\cdot X}{K_d+X}$$

How can you find the values of B_{max} and K_d that fit the data best? You can generate an infinite number of curves by varying B_{max} and K_d. Each pair of B_{max} and K_d values generates a curve, and you can compute the sum-of-squares to assess how well that curve fits the data. The following graph illustrates the situation.

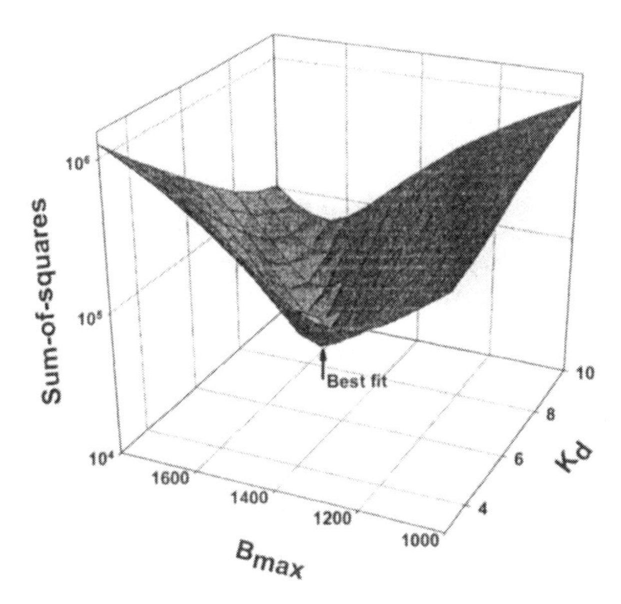

The X and Y axes correspond to two parameters to be fit by nonlinear regression (B_{max} and K_d in this example). Every pair of values for K_d and B_{max} define a curve, and the Z-axis plots the sum-of-squares of the vertical distances of the data points from that curve. Note that this graph was created specifically for the data set shown previously. The data are not graphed. Instead the graph shows how those particular data fit curves defined by various values of B_{max} and K_d.

The goal of nonlinear regression is to find the values of B_{max} and K_d that make the sum-of-squares as small as possible. In other words, the goal is to find the point on that surface that is lowest (has the smallest sum-of-squares). More simply, the goal is to find the bottom of the valley. The three methods described in the previous section all can be used, and find the bottom of the surface. Rather than adjust one parameter during each iteration, both are adjusted at once, working our way down the surface to find its low point.

More than two parameters

It was pretty easy to visualize the optimization of one parameter on a two-dimensional graph. And it wasn't too hard to visualize optimization of two parameters on a three dimensional graph. Once you have three or more parameters, it is not possible to visualize the search for optimum parameter values graphically. But the sum-of-squares graphs shown in the previous two sections were just shown to explain the idea behind the method. The method actually works via matrix algebra, and it works for any number of parameters. Nonlinear regression can fit any number of parameters.

The Simplex method

While the Marquardt method is used most often, some programs use an alternative method called the **Simplex** method. While the Marquardt method starts with one set of initial estimated values (one for each parameter), the Simplex method requires K+1 sets of initial values, where K is the number of parameters you are fitting. The example shown above fits two parameters, so the Simplex method requires three sets of initial values (one more than the number of parameters). These three sets of initial values become three points on the 3D surface shown in the figure above. For each set of initial values, the Simplex method generates the curve and calculates the sum-of-squares. One set of initial values has the highest sum-of-squares (the worst fit). The Simplex method discards that set of values and finds a new set that fits better (using a complicated set of rules). This process is repeated until the sets of parameters become indistinguishable. With two parameters (as in our example), this can be visualized as a triangular amoeba crawling down the surface, gradually becoming smaller, until the points of the triangle are almost superimposed. The advantage of the Simplex method is that it can sometimes find a good solution in difficult situations where the other methods give up with an error message. The disadvantage, and it is a major one, is that the Simplex method does not compute a standard error or confidence interval for the parameters. You get best-fit values, but you don't know how precise they are.

> Note: While the Simplex method does not automatically create confidence intervals, it is possible to create confidence intervals using Monte Carlo simulations (Chapter 17) or via method comparison (see Chapter 18).

Independent scatter

Nonlinear regression, as usually implemented, assumes not only that the scatter is Gaussian, but also that each data point has independent scatter. We'll discuss the meaning of independence in terms of replicates on page 87. If X is time, it is also important that adjacent data points do not influence each other.

In the diagram below, the horizontal line is the ideal curve, which happens to be horizontal for this simplified example. The data point (circle) on the left is the measurement at time 1. By chance, experimental error made that value fall below the ideal curve. If the measurements are entirely independent, as is assumed by linear and nonlinear regression, then the value at time 2 should not be affected by the value at time 1. This means that:

- C is the most likely value.
- D and B are equally likely.
- A is quite unlikely.

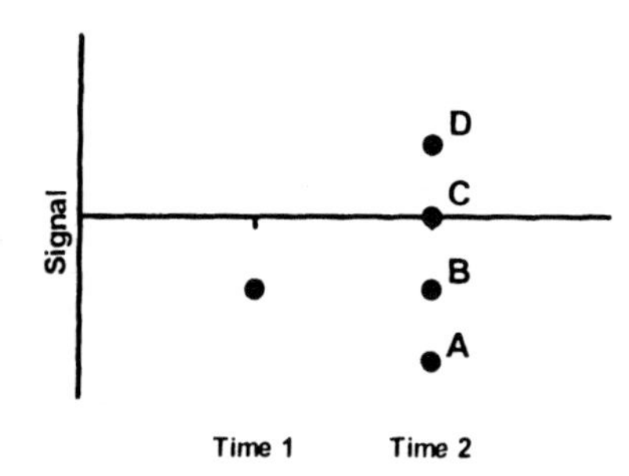

In some experimental systems, there is a carryover from one time point to the next. In these systems, the most likely value at time 2 is B, with A and C equally likely, and D very unlikely. If your data are like this, standard linear and nonlinear regression are not appropriate. Extensions to nonlinear regression are being developed to handle this situation (J. Celantano, personal communication, 2002).

E. Confidence intervals of the parameters

16. Asymptotic standard errors and confidence intervals

Interpreting standard errors and confidence intervals

In most situations, the whole point of regression is to find the best-fit values of the parameters. Your goal is to understand your system, or perhaps to look at how a parameter changes with a treatment. If this is your goal, then you will certainly care about how precisely you know the best-fit values of the parameters. It isn't enough to know, for example, that the best fit value of a rate constant is 0.10 min^{-1}. Your conclusion will be very crisp if the program tells you that you can be 95% sure the true rate constant lies between 0.09 and 0.11 min^{-1}. On the other hand, you'll know the results are pretty sloppy if the program tells you that you can be 95% sure the true rate constant lies between 0.0 and 0.5 min^{-1}. This tells you very little except that you need to redesign your experiment. Your entire interpretation of the results, and your decision on what to do in future experiments, depends on knowing the confidence intervals (or standard errors) of the parameters.

You can only interpret nonlinear regression results if the assumptions of nonlinear regression are true or at least not badly violated (see page 30). If you accept these assumptions, the 95% CI is supposed to be an interval that has a 95% chance of containing the true value of the parameter. More precisely, if you perform nonlinear regression many times (on different data sets) you expect the confidence interval to include the true value 95% of the time, but to exclude the true value the other 5% of the time (but you won't know when this happens).

The standard errors reported by most nonlinear regression programs (including Prism) are "approximate" or "asymptotic". Accordingly, the confidence intervals computed using these errors should also be considered approximate.

Does it matter that the confidence intervals are only approximately correct? That depends on how you want to use those intervals. In most cases, you'll use the confidence intervals to get a sense of whether your results are any good. If the confidence intervals are narrow, you know the parameters precisely. If the confidence intervals are very wide, you know that you have not determined the parameters very precisely. In that case, you either need to collect more data (perhaps over a wider range of X) or run the nonlinear regression again, fixing one or more parameters to constant values.

If you use the confidence intervals in this qualitative way, you won't really care if the intervals are exactly correct or just an approximation. You can accept the values reported by your nonlinear regression program and skip the next two chapters.

In a few cases, you may really want to know the value of a parameter with exactly 95% confidence. Or you may want to use the standard error of a parameter in further statistical calculations. In these cases, you should read the next two chapters to learn more about why the standard errors and confidence intervals are only approximately correct, and to learn about alternative methods to compute them.

How asymptotic standard errors are computed

Calculating asymptotic standard errors is not easy. If your program doesn't report standard errors, you should get a different program rather than trying to compute them by hand. But it is easy to understand what factors affect the calculation:

- The scatter of data around the curve, quantified by the sum-of-squares. Everything else being equal, you'll get smaller standard errors if your data are less scattered.
- The number of data points. Everything else being equal, you'll get smaller standard errors if you have more data.
- The X values you chose for your experiment. The standard errors are affected not only by the number of data points, but also by which X values you choose. Some parameters are defined mostly by the first few points (lowest X values), others are defined mostly by the last few points (highest X values), and others are defined equally by all the data.
- Whether or not you chose to fix parameters to constant values. When you fit a curve, you need to decide which parameters (if any) to fix to constant values. In most cases, fixing one parameter to a constant value will reduce the standard error for the remaining parameters.
- Whether you chose a global model. If the model is global (you share one or more parameters between several data sets) then the number of data points and sum-of-squares is for the entire set of data sets, not just one, so the standard errors will be smaller.

To understand the method a nonlinear regression program uses to compute the standard errors (and thus the confidence intervals) of the parameters, you need to master the matrix algebra and calculus that make nonlinear regression work. That is beyond the mathematical depth of this book, but here is a taste of how it works. Each standard error is computed from three terms:

- The first term is the hardest to calculate. At each value of X, compute how much the predicted value of Y changes if you change the value of each parameter a little bit from its best fit value (compute dY/dA, where A is a parameter and Y is computed from your model holding each parameter fixed to its best-fit value). Combine all these together into a matrix and do some heavy-duty matrix manipulations to obtain a value that depends on the model, the parameters, the number of data points, and the values of X. Note that this value is not influenced at all by the scatter of your data.
- The sum-of-squares of the vertical distances of the data points from the curves. This quantifies how scattered the data are.
- The number of degrees of freedom, computed as the number of data points minus the number of parameters.

The value of the first term is different for each parameter. The other two terms are the same for each parameter.

> Note: The calculations only work if nonlinear regression has converged on a sensible fit. If the regression converged on a false minimum, then the sum-of-squares as well as the parameter values will be wrong, so the reported standard error and confidence intervals won't be helpful.

An example

This example shows the decrease in the plasma concentration of a drug as a function of time.

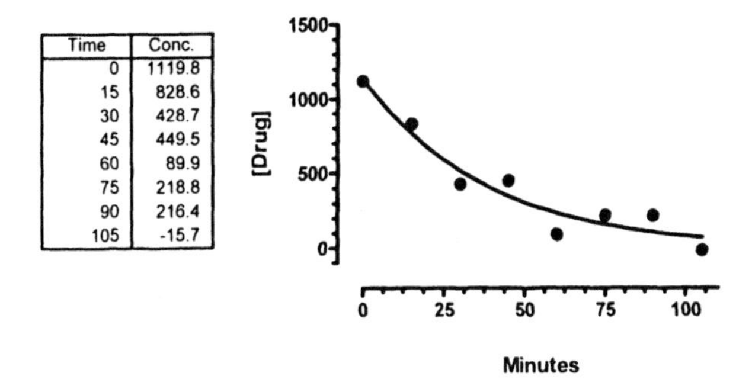

Time	Conc.
0	1119.8
15	828.6
30	428.7
45	449.5
60	89.9
75	218.8
90	216.4
105	-15.7

We fit the data to this model:

$$Drug_t = Drug_0 \cdot e^{-k \times t}$$

$$Y = Y_0 \cdot e^{-k \cdot X} = Y_0 \cdot \exp(-k \cdot X)$$

At time zero, drug concentration equals Y_0 and it then decays, following an exponential decay (see page 60) to a plateau of zero. We don't fit the bottom plateau – we know it has to plateau at zero eventually. We do fit the concentration at time zero even though we have a data point at time zero. That point is subject to experimental error, so we want to fit the value at time zero from the overall shape of the curve.

Nonlinear regression presents the best-fit value for each parameter, along with a standard error and 95% confidence interval.

Parameter	Best-fit value	Standard Error	95% CI
Y_0	1136	94.50	904.8 to 1367
k	0.02626	0.003842	0.01686 to 0.03567

The standard errors assess the precision of the best-fit values and are used to compute the confidence intervals. We find that confidence intervals are much easier to interpret, and recommend that you ignore the standard error values and focus instead on the confidence intervals.

We are 95% sure that the true value for the starting drug concentration (Y_0) is between 905 and 1367, and that the true value for the rate constant is between 0.017 and 0.036 min^{-1}. It can be hard to interpret rate constants unless you work with them a lot. We prefer to think about the half-life, the time it takes for the concentration to get cut in half. The half-life equals 0.693/k, so the 95% confidence interval for the half-life ranges from 19.4 to 41.1 minutes. More simply, the true half-life is pretty certain to be between 20 and 40, a factor of two.

How you interpret these intervals depends on the context of the experiment. We know with 95% confidence that the half-life ranges over a factor of about two. In some contexts, this will be very satisfactory. In other contexts, that is way too much uncertainty, and means the experiment is worthless. It all depends on why you did the experiment.

Because asymptotic confidence intervals are always symmetrical, it matters how you express your model

The asymptotic method described above determines a single SE value for each parameter. It then multiplies that SE by a value determined from the t distribution, with a value close to 2.0 in most cases. It then adds and subtracts that value from the best-fit value to obtain the 95% confidence interval. Since it is calculated this way, the confidence interval is symmetrical around the best-fit value. For details on computing the confidence interval, see page 103.

This symmetry means that expressing a parameter differently can result in a different confidence interval. The next two chapters discuss alternative methods to obtain confidence intervals where this is not true.

The example presented earlier defined Y at any time via an exponential equation that included the rate constant k. This rate constant is expressed in units of inverse time, $minutes^{-1}$. The example used nonlinear regression to find the best-fit value of this rate constant along with its 95% confidence interval.

The equation can be presented in an alternative form, where the exponential decay is defined by a time constant τ (tau), expressed in units of time, minutes.

$$Drug_t = Drug_0 \cdot e^{t/\tau}$$

$$Y = Y_0 \cdot e^{X/\tau} = Y_0 \cdot \exp(X/\tau)$$

The two models are equivalent. Some fields of science tend to use rate constants and others tend to use time constants.

If we fit the example data with a model expressed with a time constant, the best-fit value of Y_0 is still 1136, and the SE and 95% CI are the same. Changing from rate constant to time constant does not change the best-fit value at time zero.

The best-fit value of the time constant is 38.07 minutes, with a SE of 5.569 and a 95% confidence interval ranging from 24.45 to 51.70.

A rate constant is the reciprocal of the time constant. So to compare the best-fit time constant with the best-fit value of the rate constant determined earlier, we take the reciprocal of the best-fit time constant, which is 0.026267. That value is identical to the best-fit rate constant. No surprise here. Expressing the parameter in a different way won't change which curve fits the data the best. The best-fit value of the parameter is equivalent no matter how you express the model.

Taking the reciprocal of each end of the confidence interval of the time constant, we get a confidence interval of the rate constant ranging from 0.0193 to 0.0409. As you can see more easily on the graph below, the two confidence intervals are not the same. Our decision to express the model using a rate constant or a time constant affected the confidence interval.

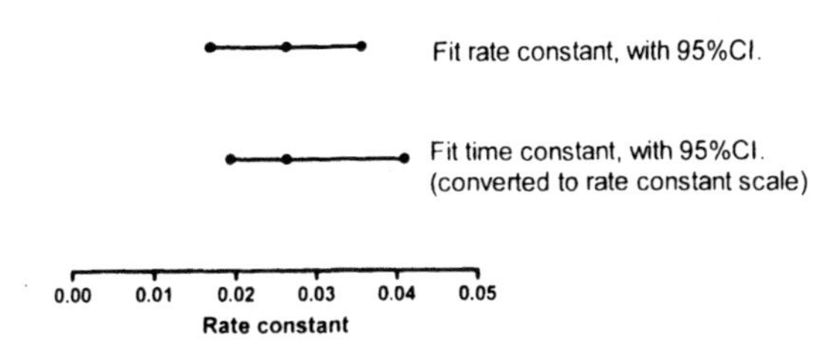

The graph above showed the results as rate constants. The graph below shows the confidence intervals as time constants. If you fit the time constant, the confidence interval (expressed as a time constant) is symmetrical. If you fit the rate constant, and then transform the results to display as time constants, the 95% confidence interval is not symmetrical.

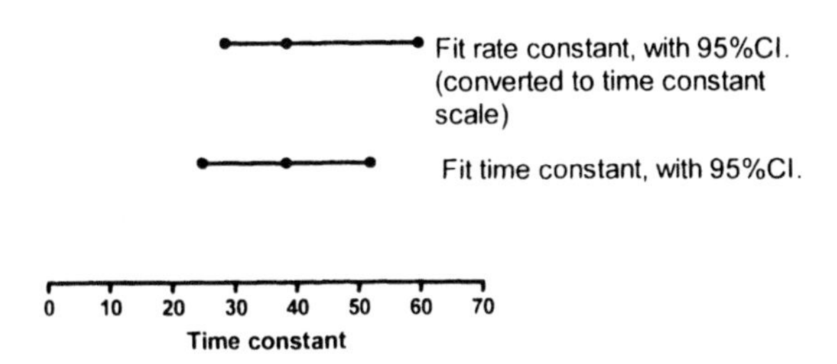

It is a bit disconcerting that our choice of how to express the parameter also determines, to some extent, the confidence interval. This is a fundamental problem with asymptotic confidence intervals. But in this example, note that the difference between the two confidence intervals is not huge, and won't greatly change how you interpret the results.

Christopoulos (Trends Pharmacol. Sci, 19:351-357, 1998) has used simulations to compare various forms of commonly used pharmacological model parameters to figure out which form gives more valid confidence intervals. He found that certain commonly-used parameters, such as drug affinity (e.g., K_d, K_b, K_i, etc) or drug potency (e.g., EC_{50}), are only associated with valid asymptotic confidence intervals if non-linear regression is performed using models where these parameters are expressed as logarithms (e.g., $LogEC_{50}$). So, with respect to the reliability of the error associated with model parameters, it really does matter how you express your parameters.

Problems with asymptotic standard errors and confidence intervals

Asymptotic confidence intervals are only approximately correct. In addition to the problem discussed in the previous section (the intervals are always symmetrical), there are two other issues.

- They are based on a mathematical simplification. When applied to nonlinear equations, the method is an approximation. If you have plenty of data that clearly define the curve without a huge amount of scatter, the approximation is a good one. The confidence intervals give you a good sense of how precisely your data define the parameters. With marginal data, the approximation doesn't always work so well.

- By looking at the standard error and confidence interval for each parameter, you won't learn about the relationships between parameters. The calculations don't completely take into account the degree to which the parameters are intertwined. As you'll see in the next two chapters, best-fit parameters are often not independent.

Despite these problems, the asymptotic standard error and confidence intervals reported by most nonlinear regression programs have proven to be very useful in giving you a good sense of how precisely you know the parameters.

The next two chapters explain alternative methods to compute confidence intervals, and Chapter 19 compares the methods.

What if your program reports "standard deviations" instead of "standard errors"?

Some programs report a standard deviation for each parameter rather than a standard error. In this context (uncertainty of best-fit values determined by regression), there is no distinction between the standard error and standard deviation.

The term *standard error* can be used to quantify how precisely you know a computed value. For example, the standard error of the mean tells you how uncertain that mean is. Its value depends on the number of data points and their scatter (as quantified by the standard deviation). The standard error of a mean is your best estimate for the value you would obtain if you repeated the experiment many times, obtaining a different mean for each experiment, and then computed the standard deviation of those means. So the standard error of a mean can also be called the standard deviation of the mean (but it is very different from the standard deviation of the data).

Similarly, the standard error of a slope is an assessment of how precisely you have determined that slope. It is your best estimate for the value you would obtain if you repeated the experiment many times, obtained a different slope for each experiment, and then computed the standard deviation of those slopes. You could also call that value the standard deviation of the slope.

It is more conventional to refer to the standard error of a parameter rather than the standard deviation of a parameter.

How to compute confidence intervals from standard errors

If your program doesn't report the 95% confidence intervals of the parameters, it is easy to compute them by hand. The confidence interval of each parameter is computed from the best-fit value and the SE of those best-fit values using this equation:

$$\text{BestFit} - t^{*} \cdot \text{SE} \;\text{ to }\; \text{BestFit} + t^{*} \cdot \text{SE}$$

If you didn't constrain the parameter, the confidence interval is always centered at the best fit value and extends the same distance above and below it. This distance equals the standard error of the parameter (discussed in the previous section) times the constant t^*. This value comes from the t distribution, and depends on the amount of confidence you want (usually 95%, but you could choose some other value) and the number of degrees of freedom (df). For nonlinear regression, df equals the number of data points minus the number of parameters fit by nonlinear regression.

You can find the value of t* in tables in most statistics books. If you have plenty of degrees of freedom (more than a dozen or so), t* will have a value near 2.0. You can find the value of t* for 95% confidence intervals using this Excel formula (substitute 0.01 for 0.05 to get the value of t* to compute 99% confidence intervals).

```
=TINV(0.05,df)
```

Our example had 8 data points and 2 parameters, and so has 6 degrees of freedom. The value of t* (for 95% confidence) is 2.4469. If you had more data (more degrees of freedom) this value would be lower. As degrees of freedom increase, t* (for 95% confidence) approaches 1.96.

17. Generating confidence intervals by Monte Carlo simulations

An overview of confidence intervals via Monte Carlo simulations

As discussed in the previous chapter, the asymptotic method built in to most nonlinear regression programs is only an approximate method to determine the confidence interval of a best-fit parameter. This chapter presents one alternative method, and the next chapter presents still another alternative method.

The idea here is to simulate a bunch of data sets, each with different random scatter. Then fit each simulated data set to determine the best-fit values of the parameters. Finally, we'll use the distribution of best-fit parameters among the simulated data sets to create confidence intervals.

This approach for determining confidence intervals is called Monte Carlo simulations.

Monte Carlo confidence intervals

General idea

From the curve fitting results in the previous chapter, we know the best-fit values of Y_0 and k. The curve fitting program also reports $S_{y.x}$, which is the best-fit estimate of the standard deviation of the residuals. Its value is computed from the sum-of-squares, the number of data points and number of parameters (see page 33). It quantifies how much the data are scattered around the curve.

Follow these steps to determine the confidence intervals by Monte Carlo simulation:

1. Generate an ideal data set. Use the same X values as used in the actual data. Generate the Y values using the same model you used to fit the experimental data (an exponential decay model in this case). Use the best-fit values from nonlinear regression of the experimental data, as the ideal parameters (k=0.02626 and Y0=1136 for this example) to generate the ideal data.

2. Add random scatter. To each ideal point, add random scatter drawn from a Gaussian distribution with a mean of zero and an SD equal to the value of $S_{y.x}$ reported by nonlinear regression of our experimental data. For this example, we want to choose random numbers with a standard deviation of 105.3 (the reported $S_{y.x}$ from the fit of our experimental data).

3. Fit the simulated data with nonlinear regression, and record the best-fit value of each parameter. Here are the first two such simulated data sets, along with the best-fit curves.

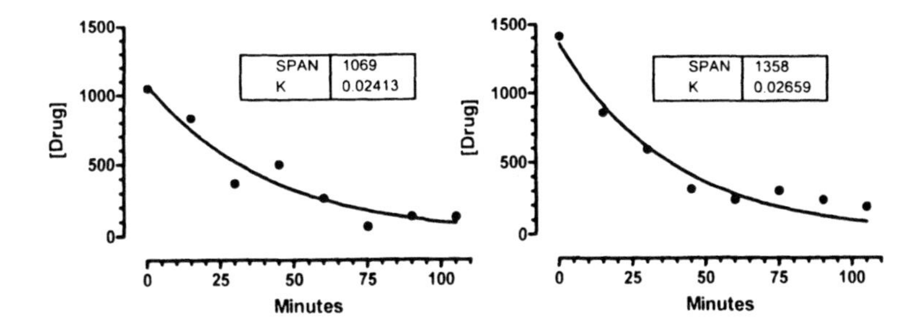

4. Repeat steps 2 and 3 many (perhaps 1000) times. Each time, compute new random numbers so each of the simulated experiments has different data. The idea here is that these simulated data sets are similar to what you would have observed if you repeated the experiment many times.

5. Find the 2.5 and 97.5 percentile values for each parameter. The range between these values is the confidence interval.

Confidence interval of the rate constant

Here is a graph of the frequency distribution for the 1000 rate constants determined from 1000 randomly generated data sets.

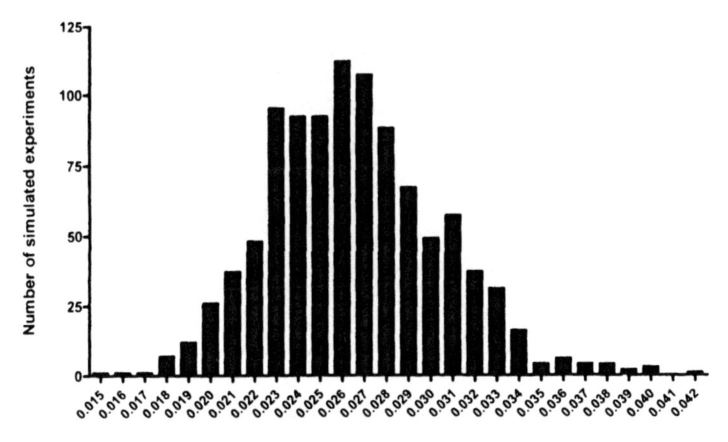

We chose to simulate 1000 experiments. If we had simulated more experiments, the frequency distribution might have been smoother. To obtain the 95% confidence interval, we sort the 1000 best-fit values, and remove the 25 (2.5%) at each end. The range of the rest is the 95% confidence interval for the rate constant, which extends from .01970 to .03448.

We repeated the simulations with 10,000 (rather than 1000) simulated data sets, and the 95% confidence interval was about the same. It went from 0.01948 to 0.03520.

Note that this method of obtaining the confidence interval creates intervals that are not necessarily symmetrical around the best-fit value. These intervals also don't depend on how we express the model. We could have fit each of the simulated data sets to a model expressed in terms of time constant instead of rate constant. Each simulated data set would have a best-fit value of the time constant that is the reciprocal of the best-fit value of the rate constant. Rank the one thousand best-fit values of the time constant, remove the 25 highest and lowest values, and you'll get the confidence interval of the time constant. This is completely equivalent to taking the inverse of each end of the confidence interval of the rate constant.

Our decision to express the model in terms of rate constant or time constant affected only how the confidence intervals are presented. With Monte Carlo simulations (unlike the asymptotic method built-in to most nonlinear regression programs), you'll get equivalent intervals no matter how you choose to express the parameter.

Confidence interval of Y_0

Here is a graph of the frequency distribution for the 1000 values of Y_0 determined from 1000 randomly generated data sets.

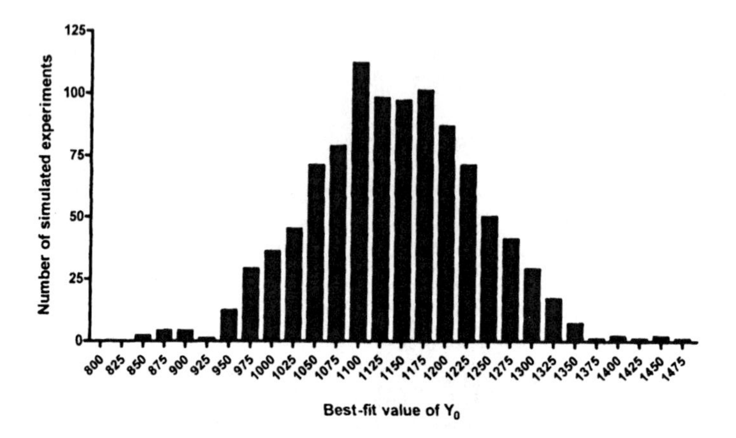

To obtain the 95% confidence interval, we sort the 1000 best-fit values, and remove the 25 at each end. The range of the rest is the 95% confidence interval, which extends from 964 to 1317.

When we repeated the simulations with 10,000 (rather than 1000) simulated data sets, the 95% confidence interval extends from 956 to 1328 (about the same).

Monte Carlo simulations shows how the two parameters are related

So far, we have determined a confidence interval for each parameter separately. But the two parameters are related. Here is a graph of our one thousand simulated experiments, showing the best-fit value of rate constant and Y_0. Each dot is the result of analyzing one simulated data set, plotting the best-fit value of the rate constant on the X axis and the best-fit value of Y_0 on the Y axis.

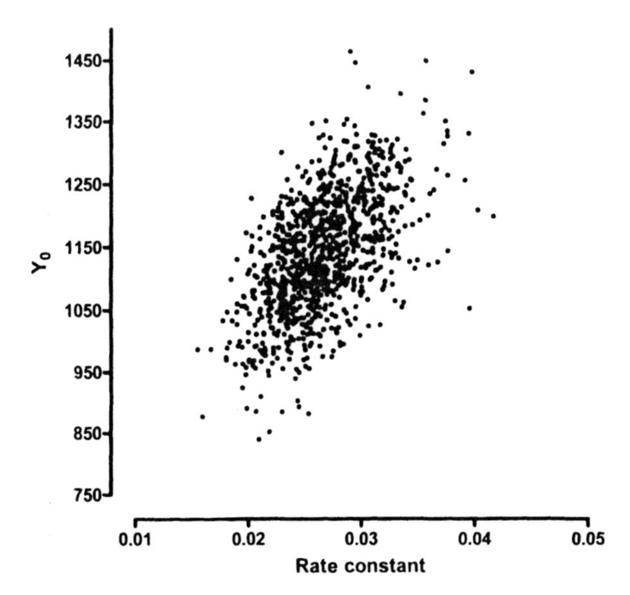

The relation between parameters makes sense. If the true Y_0 is larger, that means there is more drug in the plasma at time zero. Therefore, the rate constant is likely to be larger (faster drug elimination). If the true Y_0 is smaller (less drug), there is less drug present so the rate constant is likely to be lower (slower elimination).

Perspective on Monte Carlo methods

If you are encountering Monte Carlo methods for the first time, you may think that it is essentially magic. You start with one data set, simulate lots of "data sets" that aren't really data, and then use the analyses of those simulated data sets to learn about the uncertainty in the parameters you estimated from the actual data.

This method has a solid mathematical foundation (plenty of theorems have been proven) and has been shown to be useful in practice. Although this method seems strange when you first hear about it, it really does work.

How to perform Monte Carlo simulations with Prism

It isn't hard to do the Monte Carlo simulations with Prism. Use the "Simulate data" analysis to generate a simulated data set. You'll need to choose the X values, the model, the parameters, and the SD of random scatter. For parameters, enter the best-fit values from fitting your data. For SD of the random scatter, enter the $S_{y.x}$ reported by nonlinear regression. Choose Gaussian error.

This "analysis" will create a results table with the simulated data. From that table, click Analyze, choose nonlinear regression, and pick the model and choices just as if you analyzed the original data.

To simulate 1000 data sets, you need to run a script (a choice on the File menu). Here is the Prism script we used for the first example:

```
shortlog
setpath "c:\temp"
Openoutput "Monte Carlo CI.txt" Clear
wtext "Top" ;
wtext "K"
Foreach 1000
  GoTo R 1
  regenerate
  GoTo R 2
  Wcell 3,1;
  wcell 4,1
Next
```

The first line says not to create a detailed log. The next two lines define the folder and file that will contain the results. The next two lines write a header on top of the file. Then the loop starts, to create 1000 data sets. Of course, you can increase that number to get more precise confidence intervals. For each loop, the script goes to the first results page (the simulation) and regenerates it (new random numbers). Then it goes to the second results page (the nonlinear regression results) and records the values in the first column, third and fourth row (values of K and Y_0 for our example; you may need to adjust the row numbers).

Tip: Read about Prism scripts in the last chapter of the Prism 4 User's Guide.

After the script is finished (a few minutes), import the resulting 1000 row text file into Excel. To determine the 2.5 and 95.7 percentiles from a list of values in column B, use these Excel equations.

```
=Percentile(B:B, 0.025)
=Percentile(B:B, 0.975)
```

Variations of the Monte Carlo method

The method presented in this chapter is not the only way to do Monte Carlo simulations to generate confidence intervals of best-fit parameters. Here are some choices you have when programming a Monte Carlo simulation:

- Add random error to what? With each simulation, you can add random error to the actual experimental data, or you can add random data to ideal data predicted from the model.

- How do you generate the random error? In our discussion, we generated the random error from a Gaussian distribution. This makes a lot of sense, but requires assuming that the scatter of data around the curve does indeed follow a Gaussian distribution. One alternative is to bootstrap the errors. Create a list of residuals – the vertical distances of the data from the best-fit curve. Now for each simulation, for each data point, randomly choose one of those residuals. You don't generate new random error. Instead, you randomly reapply the actual residuals observed in the experiment.

- How do you generate the confidence interval? We used the 2.5 and 97.5 percentiles. Other methods are possible.

18. Generating confidence intervals via model comparison

Overview on using model comparison to generate confidence intervals

In this chapter, we'll develop a third method to obtain the confidence interval of a best-fit parameter (to complement the asymptotic and Monte Carlo methods described in the previous two chapters). It is based on the method to be discussed below in Chapter 22. That chapter explains how to compare two related fits to a set of data. This method generates an F ratio, from which you can determine a P value. If the P value is less than 0.05, we say that the fit of the two models differs significantly.

Conceptually, it isn't hard to use that method to develop confidence intervals for the best-fit parameters. If we change the value of the parameters from their best-fit values, we know the fit will get worse. But how much worse? We'll compare this new fit with the best-fit curve determined by nonlinear regression. If the P value is greater than 0.05, than this new fit is not significantly worse, so this set of parameter values must be within our confidence interval. If the P value is less than 0.05, then the new parameter values lead to a curve that fits the data significantly worse than the best-fit values, so these parameters are outside the confidence interval.

A simple example with one parameter

Later in this chapter we'll return to the example used in the previous two chapters. But first, let's go back to a simpler example (see page 91) where there is only one parameter to fit.

We measured a signal over time and normalized the results to run from 100% at time zero to 0 at infinite times. We fit this to an exponential decay model, fixing the top to 100 and the bottom plateau to 0, leaving only one parameter, the rate constant, to fit. The best-fit value of k is 0.3874 min^{-1}. How precisely do we know this value of k? Let's see how the sum-of-squares varies as we change the value of k.

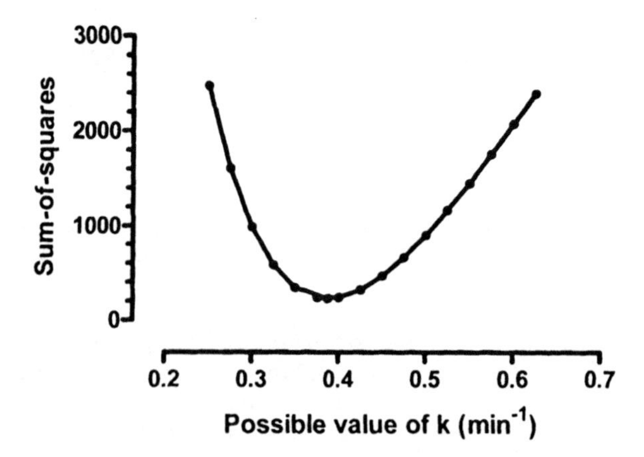

The graph above plots sum-of-squares (a measure of how well the curve fits the data) as we vary the value of the rate constant k. The best-fit value of k, 0.3874 min-1) is the value where the sum-of-squares is lowest.

The model comparison method for generating a confidence interval is based on the graph above. Chapter 22 shows how to compare two curves and calculate F using this equation.

$$F = \frac{(SS_{null} - SS_{alt})/SS_{alt}}{(DF_{null} - DF_{alt})/DF_{alt}}$$

In this context, substitute $SS_{best\text{-}fit}$ for SS_{alt}, This is the sum-of-squares of the curve determined by nonlinear regression. Also substitute $SS_{all\text{-}fixed}$ for SS_{null}. This is the sum-of-squares when we fix all the parameters (in this example, there is only one parameter) to a fixed value we choose. Also substitute N-P for DF_{alt} and N for DF_{null}, where N is the number of data points and P is the number of parameters (don't mix this up with the P value). Now the equation looks like this:

$$F = \frac{(SS_{all\text{-}fixed} - SS_{best\text{-}fit})/SS_{best\text{-}fit}}{P/(N\text{-}P)} = \frac{\left(\frac{SS_{all\text{-}fixed}}{SS_{best\text{-}fit}}\right) - 1}{P/(N\text{-}P)}$$

This equation is usually used to compare two fits -- to compute F from which we obtain a P value. Here, we are want to work backwards to get a confidence interval. We are striving for a 95% confidence interval, so we want the P value to equal 0.05. To get this P value, we know (can look up) the value of F we need. So F is a constant we set, not a variable we compute. Similarly $SS_{best\text{-}fit}$ is always the same, as determined by our data and model. We got its value by using nonlinear regression. We also know the values of N (number of data points) and P (number of parameters). What we need to do is find values of the rate constant that create a curve where the sum-of-squares equals $SS_{all\text{-}fixed}$. So let's rewrite the equation, putting $SS_{all\text{-}fixed}$ on the left.

$$SS_{all\text{-}fixed} = SS_{best\text{-}fit}\left(F\frac{P}{N\text{-}P} + 1\right)$$

Now we know the values of all the variables on the right.

Variable	Comment	Value
$SS_{best\text{-}fit}$	Sum-of-squares when nonlinear regression fits the curve, adjusting both parameters.	SS=228.97
P	Number of parameters. We are fitting the rate constant (k) and the value of Y at time zero (Y0) so P=2. Don't confuse this use of the variable P with a P value.	P=1
N	Number of data points. We had twelve X values and only collected one Y value at each X, so N=12.	N=12
F	The critical value of the F distribution for a P value of 0.05 (for 95% confidence) with P df in the numerator (1 in our example) and N-P (12-1=11 in our example) degrees of freedom in the denominator. You can look this up in a statistics table, use the free calculator at www.graphpad.com, or use this Excel formula: =FINV(0.05,1,11)	F= 4.8483
$SS_{all\text{-}fixed}$	Computed from the other parameters in the equation. This is our target value for SS. We will vary the rate constant to achieve a sum-of-squares equal to this target.	SS=329.89

What this means is that if we change the value of k a small amount, so that the resulting curve has a sum-of-squares less than 329.89, the fit will not be significantly different than the fit determined by nonlinear regression. These values of k define the 95% confidence interval. If k is outside this range, then the resulting curve will have a sum-of-squares greater than 329.89 and so fit the data significantly worse.

So what range of values for k satisfies this criterion? As you can see from the graph below, a range of values of k from about 0.35 to 0.43 give sum-of-square values less than 329.9 so therefore generate curves that are not significantly different (at the 5% significance level) than the best fit curve.

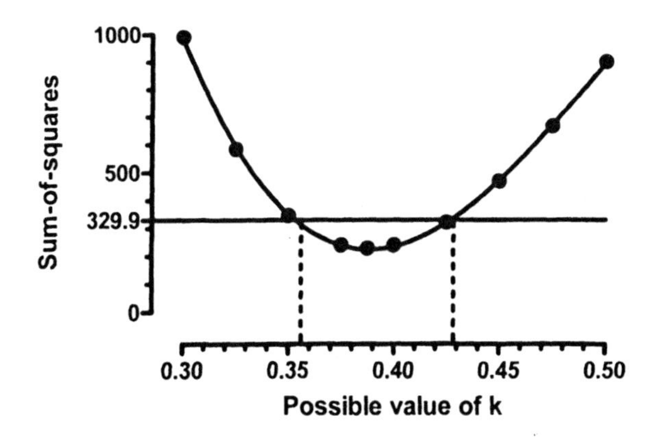

To compute the confidence intervals more exactly, we can alter k over a finer range of values. Or, as explained below on page 116, we can use Excel's solver to find the two values of k that lead to a sum-of-squares of 329.89. The confidence interval determined this way ranges from 0.3525 to 0.4266. Any value of k outside that range would result in a curve whose fit is significantly worse than the best-fit curve. Note that this confidence interval is not symmetrical around the best-fit value.

Confidence interval for the sample data with two parameters

The rest of this chapter shows how to apply this method to the example data used in the previous two chapters. Since the two parameters are related, you'll get the most information by finding a confidence region – a set of values for the two parameters. We'll go through the calculations in the next two sections. But first, the answer. The graph below plots various possible values of the rate constant on the X axis vs. possible values of Y_0 on the Y axis. The two parameters are correlated. If the value of k is high, then the value of Y_0 is also very likely to be high. If the value of k is low, then the value of Y_0 is also very likely to be low.

The plus in the middle denotes the best-fit value of k and Y_0 determined by nonlinear regression. The egg-shaped contour shows values of k and Y_0 that generate curves that are just barely significantly different than the best-fit curve. So any values of k and Y_0 that lie within the contour (the gray area) lead to curves that are not significantly different (at the 5% significance level) than the best-fit curve. Therefore, this contour defines the confidence interval.

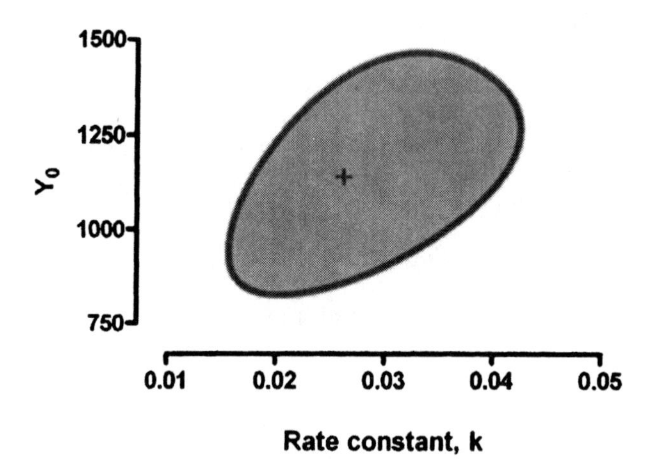

The next section explains how this graph was constructed.

Using model comparison to generate a confidence contour for the example data

The method to be discussed below in Chapter 22 lets you compare two fits (to one set of data) by looking at the difference between sum-of-squares. This method computes an F ratio, which leads to a P value.

We will repeatedly compare the sum-of-squares from curves defined by two sets of parameters.

One curve is always defined by the best-fit results obtained by nonlinear regression. Call the number of parameters you fit to be P. Call the sum-of-squares from this fit to be $SS_{best\text{-}fit}$.

The other curve will change as we repeatedly apply the method. To create this curve, we will fix all of the parameters to constant values. You'll see soon how we pick these values. Since we set all the parameter to fixed (constant) values, we don't ask the program to fit anything. Instead, we only ask a program to compute the sum-of-squares. Call this sum-of-squares $SS_{all\text{-}fixed}$.

Compare the fits using an equation explained earlier in this chapter.

$$SS_{all\text{-}fixed} = SS_{best\text{-}fit}\left(F\frac{P}{N\text{-}P}+1\right)$$

For this example, here are the values.

Variable	Comment	Value
$SS_{best\text{-}fit}$	Sum-of-squares when nonlinear regression fits the curve, adjusting both parameters.	SS= 66489.16
P	Number of parameters. We are fitting the rate constant (k) and the value of Y at time zero (Y0) so P=2. Don't confuse this use of the variable P with a P value.	P=2
N	Number of data points. We had eight X values and only collected one Y value at each X, so N=8.	N=8
F	The critical value of the F distribution for a P value of 0.05 (for 95% confidence) with P df in the numerator (2 in our example) and N-P (8-2=6 in our example) degrees of freedom in the denominator. You can look this up in a statistics table, use the free calculator at www.graphpad.com, or use this Excel formula: =FINV(0.05,2,6)	F= 5.1432
$SS_{all\text{-}fixed}$	Computed from the other parameters in the equation. This is our target value for SS. We will vary parameters to achieve a sum-of-squares equal to this target.	SS=180479.3

We want to vary one parameter, holding the rest constant, increasing SS until it gets to 180479.3. When it reaches that value, we know that F will equal 5.1432, so the P value will equal 0.05. This target value for $SS_{all\text{-}fixed}$ was computed from the equation above.

Let's start by holding Y_0 constant at its best-fit value of 1136, and gradually decrease k. As you reduce the value of k from the best-fit value, the curve will get further from the data points, so the sum-of-squares will go up. Adjust the value of k until the sum-of-squares equals 180479.3. This happens when k= 0.01820. We plot on the graph below the point k=0.01820, Y_0=1136.

Now let's increase the value of k from its best-fit value until the sum-of-squares again hits that target value. This happens when k= 0.04002. Add to the point k=0.04002, Y_0=1136 to the graph we are building (below).

You might think we have determined the confidence interval of k as ranging from 0.01820 to 0.04002. But this isn't really the confidence interval we care about. It is the confidence interval of k, assuming that Y_0 equals exactly 1136 (its best-fit value). But Y_0 is itself subject to variation. What you want to know is the lowest and highest values of k that generate the target value of SS at any value of Y_0.

Now let's fix the rate constant at its best-fit value, and increase and decrease Y_0 until the sum-of-squares is 180479.3. That happens when Y_0 equals 886.5 and 1385.5. We'll add two more points to the graph, at k=0.02626, Y_0=886.5 and k=0.02626, Y_0=1385.5. The graph below now shows four pairs of values of k and Y_0 where the sum-of-squares is 180479.3., along with lines showing the best-fit values of Y_0 and k.

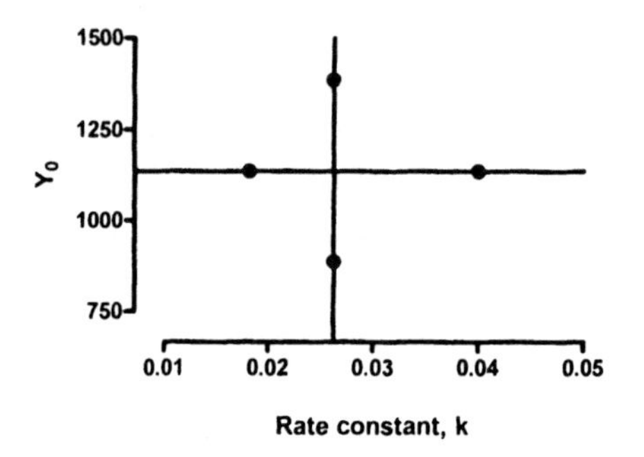

Let's continue this process. We hold one parameter constant (to various values) and find the two values of the other parameter that make the sum-of-squares equal 180479.3. Then hold the first parameter constant to a different value, and again find two values of the other parameter that makes the sum-of-squares meet the target value. It doesn't really matter if you hold k constant and vary Y_0 or hold Y_0 constant and vary k. The graph below was made using a combination of the two methods. Repeat with many sets of values, and here is the result. (The next section explains how we used Excel to facilitate these calculations.)

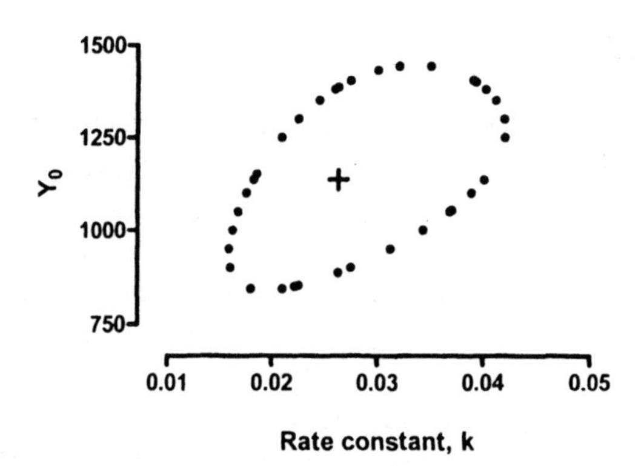

The plus in the middle of the graph represents the best-fit values of k and Y_0. Each point represents a pair of values for k and Y_0 that define a curve that is far enough from our sample data so the sum-of-squares equals 180479.3. If you compare the fit of any curve defined by the parameter values graphed here with the best-fit nonlinear regression curve, the F ratio equals 5.143 so the P value equals 0.05. Even though the shape (in this example) looks more like an egg than an ellipse, it is often called a *confidence ellipse*. A better term is *confidence contour*.

Any pair of values of k and Y_0 that is outside the contour define a curve that is farther from the points, and generates a higher sum-of-squares. Therefore the value of F is larger than 5.143 and so the P value is less than 0.05. These values of k and Y_0 define curves that fit the data significantly worse than the best-fit curve. Any pair of k and Y_0 that is inside the contour defines a curve that is further from the points than the best-fit curve. But if we compare the two curves, the F ratio will be less than 5.143 so the P value is greater than 0.05. The curves generated by parameter values inside the egg do not fit the data significantly worse than the best-fit values. Therefore we are 95% sure that the true value of k and Y_0 lies inside the confidence contour.

Converting the confidence contour into confidence intervals for the parameters

To find the 95% CI for k, we find the lowest and highest values of k on the contour, as shown in the figure below. The 95% CI of the rate constant determined this way ranges from 0.0159 to 0.0420.

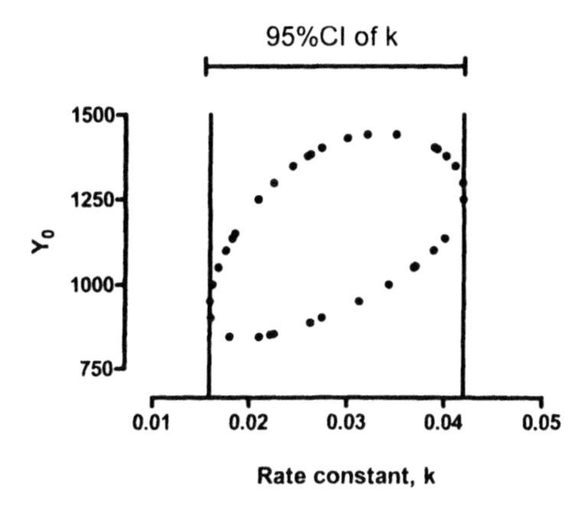

To find the 95% confidence interval for Y_0, we find the lowest and highest values of Y_0 on the contour, as shown below. The 95% confidence interval for Y_0 ranges from 844.00 to 1442.75.

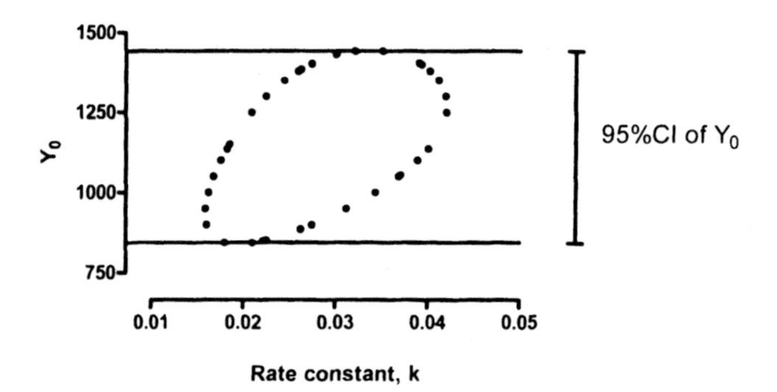

How to use Excel's solver to adjust the value of a parameter to get the desired sum-of-squares

The description above left out an important step. How exactly do we adjust the value of a parameter to get the desired value of the sum-of-squares? This procedure could be included in computer programs, but it is not part of commonly used nonlinear regression programs (it is not part of Prism 4).

We used Excel's solver. The solver comes with Excel, but it is not installed by default. You may need to install again from a CD and choose to install the Solver. Then you need to go to the Add-in manager and check the option to enable the solver.

Here is a screen shot of our Excel worksheet.

	A	B	C	D
1			Y0	1200.00
2			k	0.0200
3				
4	X	Y	ypred	diff sq
5	0	1119.8	1200.00	6428.19
6	15	828.6	888.98	3651.83
7	30	428.7	658.57	52844.38
8	45	449.5	487.88	1471.53
9	60	89.9	361.43	73717.37
10	75	218.8	267.76	2401.02
11	90	216.4	198.36	324.62
12	105	-15.7	146.95	26454.30
13				
14			Sum-of-squares	167293.2

The data were entered in columns A and B from row 5 down, and were never changed. Values of Y_0 and k were entered in rows 1 and 2 of column D. We set one of these, and ask the Solver to find a value for the other, as you'll see. Column C shows the predicted value of Y, determined from our model and from the values of Y_0 and k shown at the top of the worksheet. Each cell had a formula, so its value automatically changed when the value of k or Y_0 was changed. Column D shows the square of the difference between column B (the Y value of the data) and column C (the Y value of the curve). Cell D14 shows the sum-of-squares. When we change Y_0 or k, the values in columns C and D automatically update, and so does the sum-of-squares.

Launch the Solver from Excel's tools menu. If you don't see a Solver command, make sure you have installed the Solver and selected it in the Add-in dialog. Here is the Solver dialog:

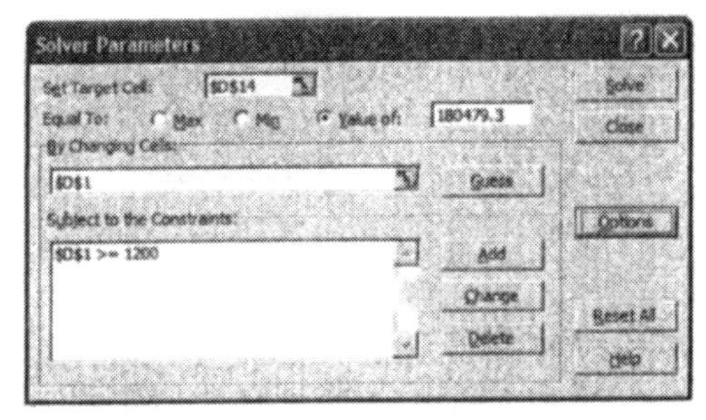

Here we ask it to adjust the value of cell D1 (the value of Y_0) in order to get the value of cell D14 (the sum-of-squares) to equal 180479.3 (the target we computed above). There will be two values of Y_0 that work, so we give the Solver the constraint that Y_0 (cell D1) must be greater than 1200. Click Solve. Excel tells you that the Solver found a solution.

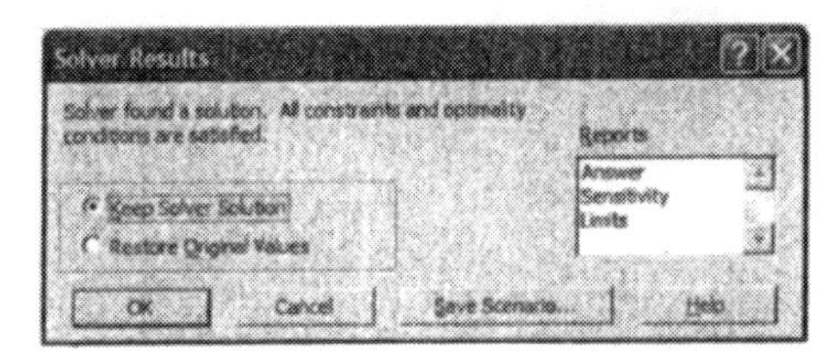

Click OK, and Prism changes the value of Y_0 to 1216.7, which leads to a sum-of-squares (as requested) of 180479.3.

	A	B	C	D
1			Y0	1216.70
2			k	0.0200
3				
4	X	Y	ypred	diff sq
5	0	1119.8	1216.70	9385.65
6	15	828.6	901.36	5300.52
7	30	428.7	667.74	57143.07
8	45	449.5	494.67	2038.68
9	60	89.9	366.46	76474.62
10	75	218.8	271.48	2780.16
11	90	216.4	201.12	232.75
12	105	-15.7	148.99	27123.86
13				
14			Sum-of-squares	180479.3

Now we have determined one set of values for Y_0 and k (1216.7 and 0.0200) that defines one point on our graph.

Repeat this process for many values of k, and then repeat again with the same values of k but with the opposite constraint (Y_0<1200) and you'll create enough points to define the egg shown earlier.

Instead of constraining the value of Y_0, you can simply change the starting value (the value you enter into cell D1). There are two solutions to the problem, and the Solver will almost always find the solution that is closest to the starting value. So run the Solver twice, once starting with a high value of Y_0 and once starting with a low value of Y_0.

More than two parameters

If you are fitting more than two parameters, it becomes much harder to apply this method. And it even becomes hard to figure out how to display the results. But the idea is still the same. You want to find all sets of parameters that define curves whose sum-of-squares is just enough worse than the sum-of-squares of the best-fit curve so that an F test gives a P value of 0.05.

19. Comparing the three methods for creating confidence intervals

Comparing the three methods for our first example

The last three chapters presented three distinct methods for obtaining confidence intervals from our sample data.

The three methods gave similar, but not identical, results as shown in this graph.

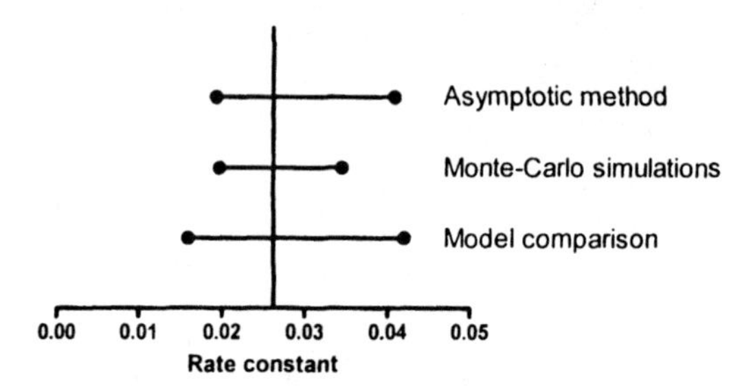

The simple method always gives a confidence interval that is symmetrical and tends to be a bit optimistic (too narrow). Monte Carlo simulations also tend to be a bit optimistic – the interval is too narrow. The model comparison method (via F test) is probably the most accurate method, but is also most cumbersome.

Here is the comparison, showing the time constant rather than the rate constant.

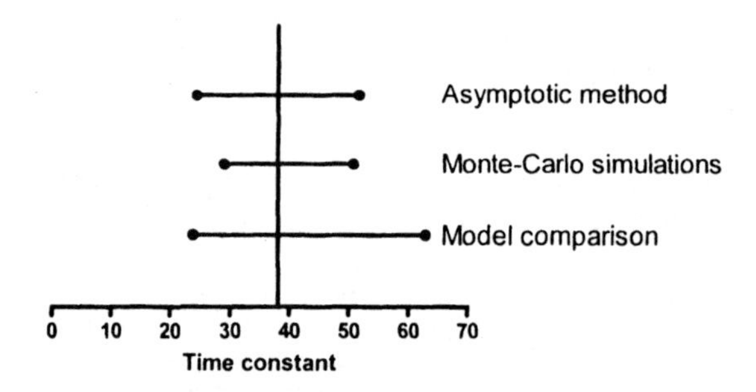

Here is the comparison of the three different methods to compute the confidence intervals of Y_0.

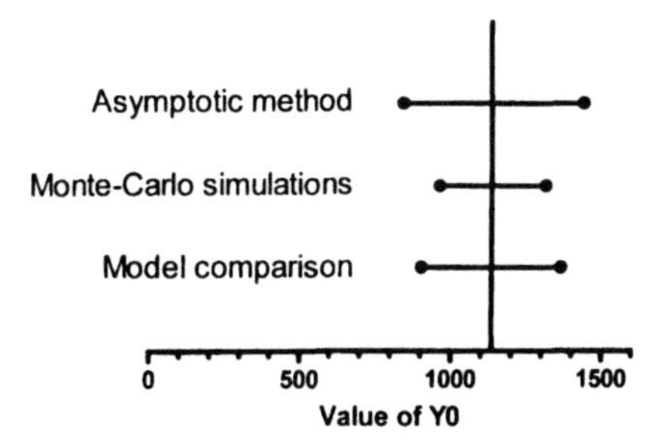

Here we superimpose the egg shape determined by model comparison (circles) with the results determined by Monte Carlo simulations (plusses). The two have a very similar shape, but the Monte Carlo simulations are a bit more compact.

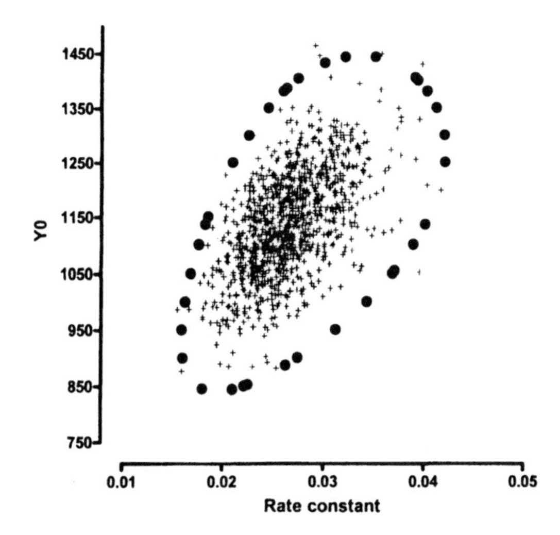

A second example. Enzyme kinetics.

This example shows enzyme activity as a function of substrate concentration. There are only seven data points, and the experiment does not use high enough substrate concentrations to properly define the maximum velocity.

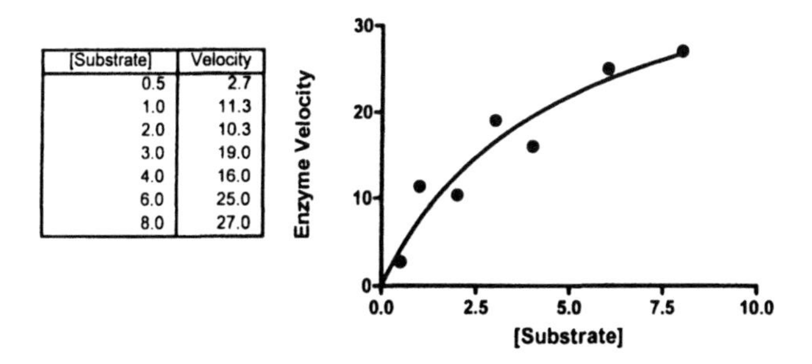

[Substrate]	Velocity
0.5	2.7
1.0	11.3
2.0	10.3
3.0	19.0
4.0	16.0
6.0	25.0
8.0	27.0

We fit the data to this model:

$$\text{Velocity}=\frac{\text{Velocity}_{max}\cdot[\text{Substrate}]}{K_M+[\text{Substrate}]}$$

$$Y=\frac{V_{max}\cdot X}{K_M+X}$$

Asymptotic (approximate) standard errors and confidence intervals

The best-fit parameter values, with their standard errors and confidence intervals are shown in the table below. The sum-of-squares is 42.3, and $S_{y.x}$ is 2.909.

Parameter	Best-fit value	Standard Error	95% CI
V_{max}	42.40	10.81	14.61 to 70.19
K_m	4.687	2.374	-1.42 to 10.79

The confidence interval for V_{max} is pretty wide, telling you what you already knew -- that this experiment wasn't designed very well so doesn't define V_{max} very well. The asymptotic confidence interval is symmetrical around the best-fit value, as it must be (given how it is computed). But the lower limit doesn't make much sense. It is too low -- lower than half of the Y values in the data.

The K_m is the concentration of substrate that leads to half-maximal enzyme velocity. The confidence interval for K_m extends down to negative values. Since this is nonsense, we'll truncate it to 0.0 to 10.79. This means that the K_m could be anywhere in the range of concentrations we used, which means this experiment gave us no real information on the value of K_m.

Monte Carlo confidence intervals

Using the Monte Carlo method, the 95% confidence interval for V_{max} ranged from 25.78 to 92.88. The 95% confidence interval for Km ranged from 1.376 to 16.13.

The two parameters are extremely related, as this graph shows. Each point represents the best-fit value of V_{max} (X axis) and K_m (Y axis) for one simulated data set. The left panel shows the results of all thousand simulations. The right panel drops the results of a few dozen simulations to emphasize the bulk of the results.

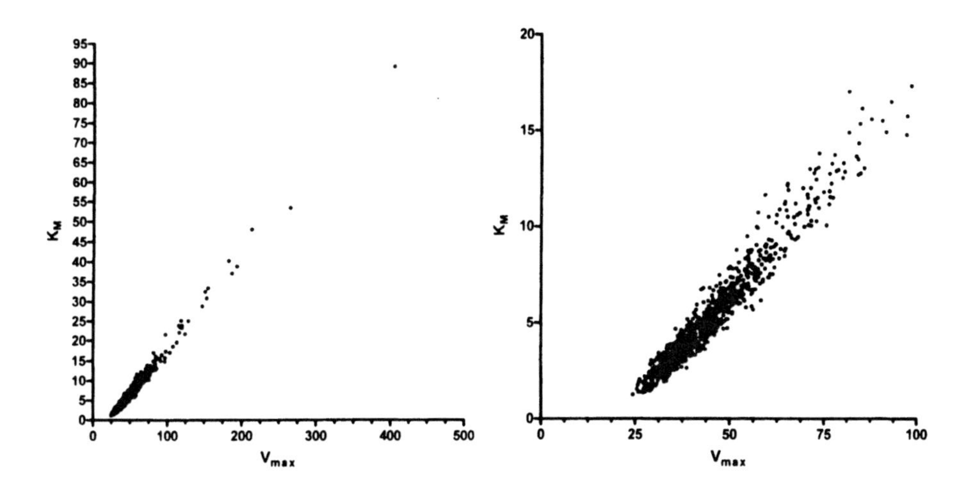

Confidence intervals via model comparison

Again, we'll compare the best-fit curve generated by nonlinear regression with the fit of a curve where both parameters are fixed to values we select. Again, we'll use the equation below to figure out the sum-of-squares needed to make a fit be different enough that its F value is high enough to yield a P value of exactly 0.05.

$$SS_{all\text{-}fixed}=SS_{best\text{-}fit}\left(F\frac{P}{N\text{-}P}+1\right)$$

Variable	Comment	Value
$SS_{best\text{-}fit}$	Sum-of-squares from nonlinear regression.	SS= 42.303
P	Number of parameters.	P=2
N	Number of data points.	N=7
F	The critical value of the F distribution for a P value of 0.05 (for 95% confidence) with P df in the numerator (2 in our example) and N-P (7-2=5 in our example) degrees of freedom in the denominator.	F= 5.7861
$SS_{all\text{-}fixed}$	Computed from the other parameters in the equation. This is our target value for SS.	SS=142.211

We set K_m to various values, and used Excel's Solver (see page 116) to find two values of V_{max} (for each K_m) that lead to the target sum-of-squares. The graph below shows the best fit values of V_{max} and K_m as the horizontal and vertical line. Each point on the graph represents a pair of values for V_{max} and K_m that cause the sum-of-squares to equal the target value and so define the 95% confidence interval.

For this example, the interval is not a closed loop. There is no upper limit for K_m or V_{max}. For any K_m you propose, there are a range of values for V_{max} that lead to curves that are not significantly different from the best-fit curve.

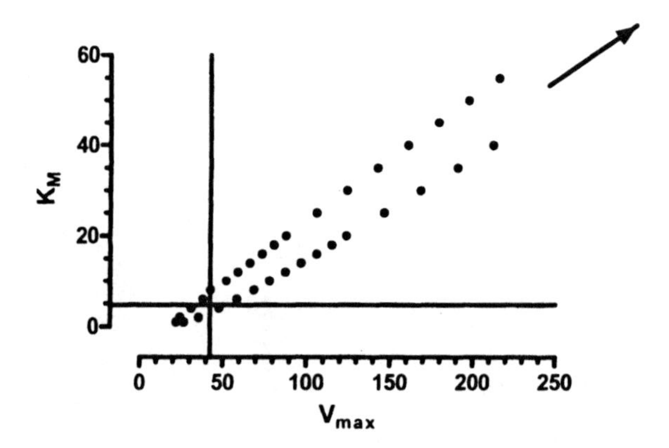

As the graph below shows, this is because with large values of K_m and V_{max}, the data are all in the early portion of the curve that is essentially a straight line. This portion of the curve is almost the same with any large values for the parameters. The fit of any of these curves is not significantly worse than the best-fit curve. That means that the data simply don't define (with 95% certainty) an upper limit for K_m and V_{max}.

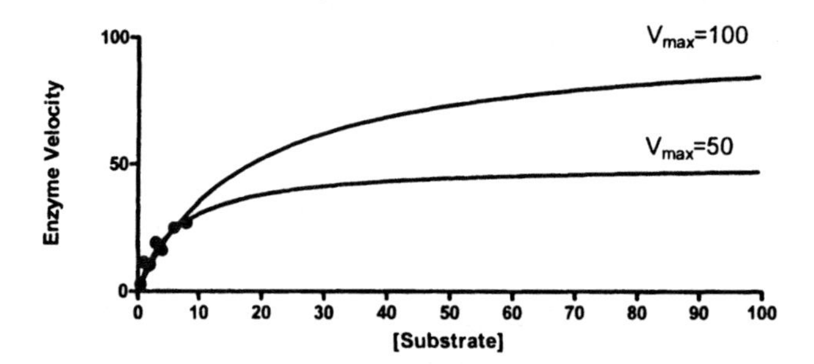

This graph superimposes the results of Monte Carlo simulations (circles) with the results of the F ratio comparison method (pluses). The two give very similar results.

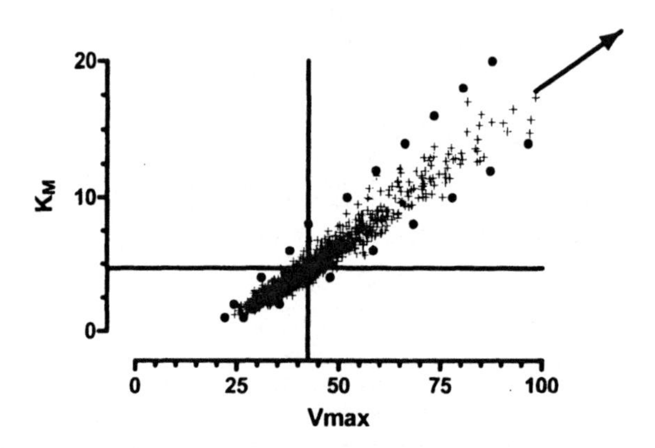

Comparing the confidence interval of the three methods

This graph compares the three methods used to determine the 95% confidence interval for V_{max}. Note that the third method (compare fits with F ratio) has no upper limit.

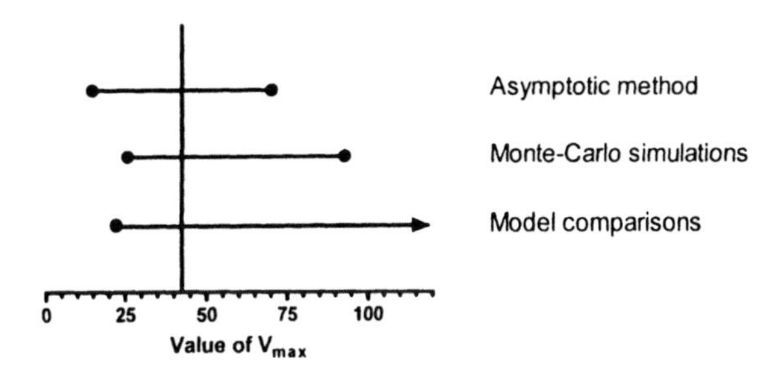

This graph compares the three methods used to determine the 95% confidence interval of the K_m. Again, note that the third method (model comparison with F ratio) has no upper limit.

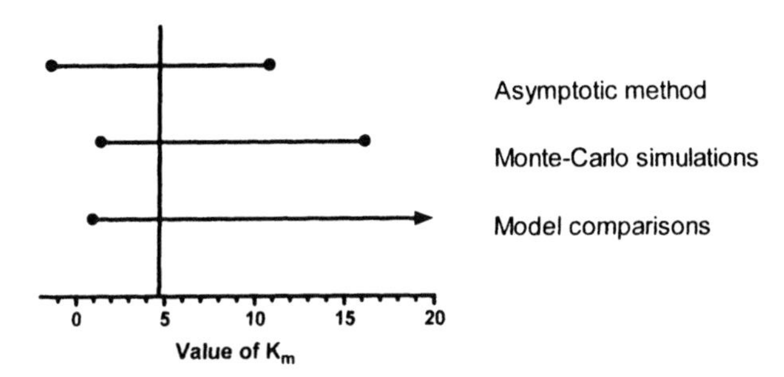

In this example, our data simply don't define the curve. The most accurate method, comparison with F ratio, simply can't find an upper limit for either parameter.

For this example, the standard method built-in to nonlinear regression programs gives you a confidence interval that is somewhat misleading. But an alert scientist would notice two things:

- The lower confidence limit for V_{max} is obviously too low, since it is lower than half the data points. Since the confidence intervals must by symmetrical, if one limit is too low, the other limit must not be high enough.
- The confidence interval for K_m includes all the substrate concentrations used in the experiment (the confidence interval goes up to 12, but the highest concentration used in the experiment is 8).

Even though we can't interpret the asymptotic confidence intervals literally, they tell us what we need to know – that our data don't define K_m or V_{max} very well.

A third example

Here is a third example. This is a dose response curve without too much scatter.

log(conc)	Percent Respone		
	Y1	Y2	Y3
-9.0	8.7	11.4	-2.0
-8.5	9.4	2.7	1.5
-8.0	3.1	4.9	-1.9
-7.5	24.4	23.7	9.0
-7.0	30.2	36.1	39.1
-6.5	60.6	65.0	48.4
-6.0	72.8	67.7	76.3
-5.5	77.6	76.2	76.9
-5.0	100.7	88.5	91.0
-4.5	97.2	98.0	94.6
-4.0	102.4	107.2	106.6
-3.5	114.1	108.8	100.5
-3.0	98.6	93.3	104.3

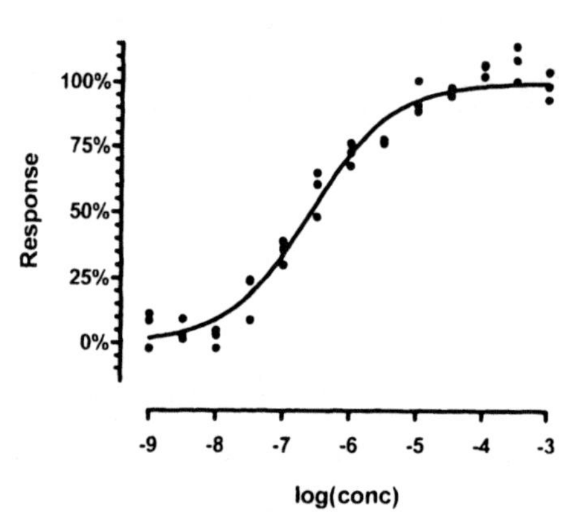

The results were normalized to run from 0 to 100. We fit a sigmoidal dose-response curve to find the best-fit values of the $\log EC_{50}$ (the concentration that gives a 50% response; the middle of the curve) and the Hill Slope (a measure of how steep the curve is). We fixed the bottom plateau to 0.0 and the top to 100.0, so only fit two parameters (the $\log EC_{50}$ and the Hill Slope). The triplicate values were treated as separate data points, so there a total of 39 points, so 37 degrees of freedom (39 data points minus two parameters to fit).

Note: We normalized the data to control values so the data run from 0 to 100. But there is random variation, so some data points have normalized values less than zero and others have normalized values that greater than one hundred. We left those in to avoid biasing the analyses.

Asymptotic standard errors and confidence intervals

Here are the standard errors and confidence intervals, as computed by the standard method built-in to all nonlinear regression programs. The sum-of-squares is 1644.2 and the $S_{y.x}$ is 6.6661.

Parameter	Best-fit value	Standard Error	95% CI
$\log EC_{50}$	-6.590	0.05264	-6.697 to -6.484
Hill Slope	0.6996	0.05237	0.5934 to 0.8057

Monte Carlo simulations

The results of Monte Carlo simulations are graphed below. The two parameters are not related. The 95% confidence interval for the $\log EC_{50}$ ranges from -6.687 to -6.484. The 95% confidence interval for the Hill Slope ranges from 0.6077 to 0.8231. These confidence intervals are very close to those determined by the simple method built in to nonlinear regression programs.

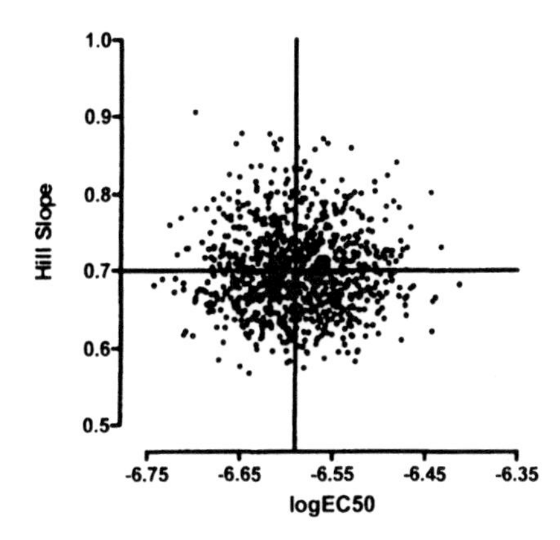

Method comparison via F test

Here are the values we used to find confidence interval via model comparison.

$$SS_{all\text{-}fixed}=SS_{best\text{-}fit}\left(F\frac{P}{N\text{-}P}+1\right)$$

Variable	Comment	Value
$SS_{best\text{-}fit}$	Sum-of-squares from nonlinear regression.	SS= 1644.2
P	Number of parameters.	P=2
N	Number of data points.	N=39
F	The critical value of the F distribution for a P value of 0.05 (for 95% confidence) with P df in the numerator (2 in our example) and N-P (39-2=37) df in the denominator.	F= 3.252
$SS_{all\text{-}fixed}$	Computed from the other parameters in the equation. This is our target value for SS.	SS=1933.2

Here are the results. Each point represents a pair of values for $logEC_{50}$ and Slope that generate a curve whose sum-of-squares is just enough different from the best-fit curve to give a F ratio of 3.252 and thus a P value of 0.05. Any points inside the ellipse define a curve that is not significantly different from the best-fit curve. Any points outside the ellipse define a curve that is significantly different from the best fit curve. The ellipse, therefore, defines our 95% confidence interval.

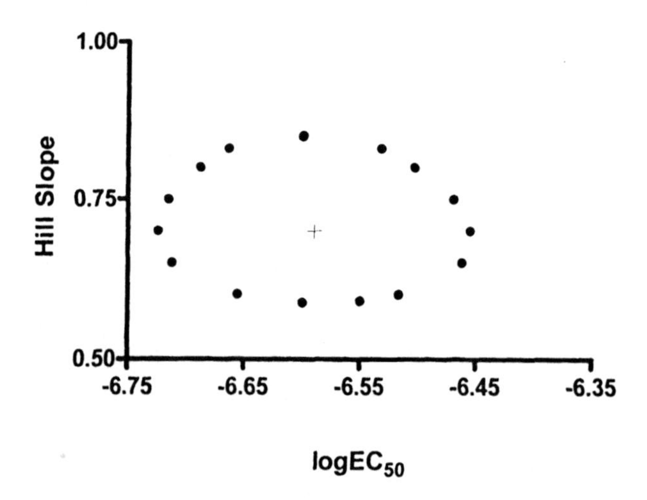

The 95% confidence intervals are taken from the highest and lowest value of each parameter on the ellipse. The 95% confidence interval for the logEC50 ranges from -6.72 to -6.45. The 95% confidence interval for the Hill Slope ranges from 0.59 to 0.85. These confidence intervals are very close to those determined by the simple method built in to nonlinear regression programs.

Comparing the three methods

The graph below compares the confidence intervals for each parameter as generated by the three methods. The three methods are essentially the same for all practical purposes.

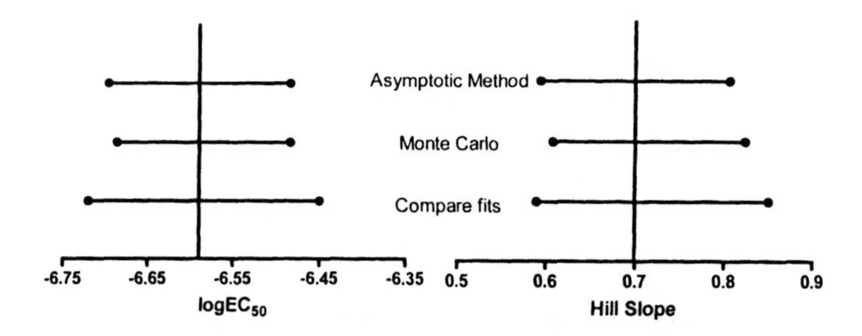

The graph below compares the Monte Carlo method with the method that compares fits with the F ratio. The circles show the border of the 95% confidence region as defined by the model comparison method. The pluses show the individual results of Monte Carlo simulations. If the methods were equivalent, you'd expect 95% of the simulated fits to lie within the ellipse defined by the model comparison method, and that is about what you see.

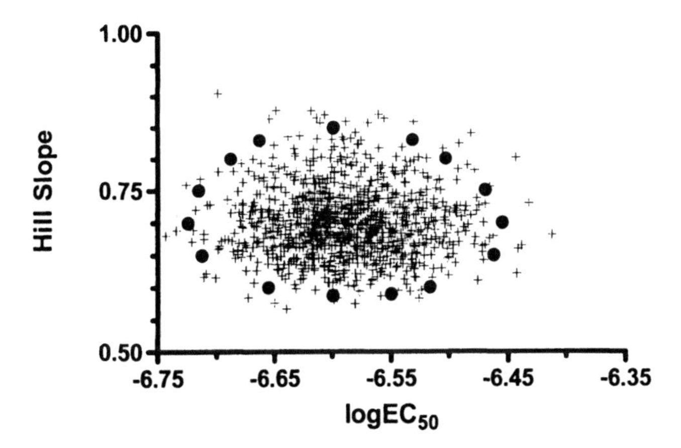

Conclusions

When you evaluate the results of nonlinear regression, you don't only care about the best-fit values of the parameters. You also want to know how precisely you have determined them. The easiest way to do so is to look at the asymptotic confidence intervals reported by your nonlinear regression program.

Ideally you have collected plenty of data points without too much scatter and have spaced these sensibly over appropriate range of X. If so, the asymptotic confidence intervals will be quite useful. As we saw in the third example above, all three methods for computing confidence intervals give almost identical results in cases like this.

In a less ideal situation (you have collected few data points, they have a lot of experimental scatter, and you have chosen the X values poorly), the reported confidence interval will be much wider. In these cases, the asymptotic interval reported by nonlinear regression may be too narrow, too optimistic. The uncertainty may even be worse than what the program reports. But the wide confidence intervals tell you what you need to know – that you haven't determined the parameters very well and should either collect more data or run nonlinear regression constraining a parameter to a constant value.

It would be a mistake to assume that the "95% confidence intervals" reported by nonlinear regression have exactly a 95% chance of enclosing the true parameter values. The chance that the true value of the parameter is within the reported confidence interval may not be exactly 95%. Even so, the asymptotic confidence intervals will give you a good sense of how precisely you have determined the value of the parameter.

When you need more accurate confidence intervals, use Monte Carlo simulations or compute confidence intervals by model comparison.

20. Using simulations to understand confidence intervals and plan experiments

Chapter 17 explained how to use Monte Carlo simulations to create confidence intervals. You start with best-fit results from fitting one particular data set, and use the simulations to create confidence intervals for that data set. Monte Carlo simulations can also be used more generally to help you decide how to design experiments and to help you figure out how to express the model.

Example 1. Should we express the middle of a dose-response curve as EC_{50} or $\log(EC_{50})$?

The simulated data sets below each had ten data points, equally spaced on a log scale from 1 nM to 10 μM. The true curve had a bottom plateau at 0.0, a top plateau at 100, and an EC_{50} of 0.1 μM. (The EC_{50} is the concentration required to get a response half-way between the minimum and maximum.) Random noise was added to each point, using a Gaussian distribution with a mean of 0 and a SD of 15, which simulates data in a system with lots of scatter. Three typical data sets are superimposed in the graph below.

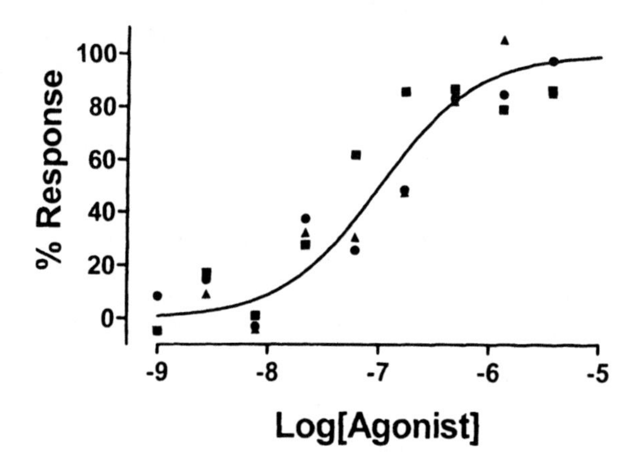

We simulated five thousand data sets and fit each one twice. First we fit to a dose-response curve written either in terms of EC_{50} and then to a dose-response curve written in terms of the $\log(EC_{50})$. The distribution of these 5000 EC_{50} and $\log(EC_{50})$ values are shown below along with a superimposed Gaussian distributions.

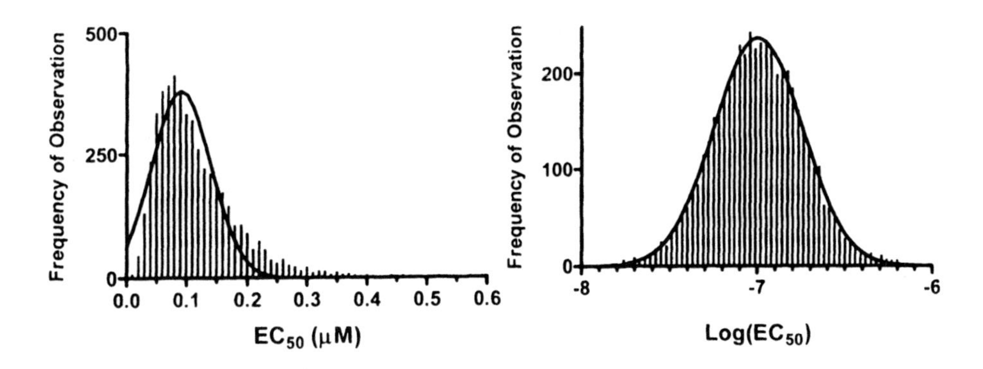

Clearly, the distribution of log(EC_{50}) values is much closer to Gaussian. The normality test confirms this impression. The distribution of log(EC_{50}) passes the normality test. The distribution of EC_{50} values fails the normality test with $P<0.0001$.

If a parameter follows a Gaussian distribution, the asymptotic confidence interval should be fairly accurate. To confirm this, we recorded the asymptotic confidence intervals for EC50 and LogEC50 for each of the simulated data sets.

When the data were fit to a dose-response equation written in terms of log(EC50), 94.20% of the "95%" confidence intervals of log(EC_{50}) contained the true value. In contrast, when we fit the model written in terms of EC_{50}, only 91.20% of the 5000 "95%" confidence intervals contained the true value. While that isn't terribly far from 95%, a bigger problem is that 78% of these confidence intervals began with a negative number. Of course negative concentrations are impossible, so we would call the lower limit zero in these cases. This means that in majority of the simulated data sets, the 95% confidence interval of the EC_{50} gave us no information at all about the lower limit of the EC_{50}.

> Note: When you are fitting real data, you'll never know if the confidence interval contains the true parameter value or not. You only know the value determined from your data. But in the discussion above, we were analyzing simulated data sets, so knew the true value of the parameter. This allowed us to tabulate how often the confidence interval included the true parameter value.

These simulations show a clear advantage to expressing the dose-response equation in terms of log(EC_{50}), at least for data similar to our simulations where the concentrations were equally spaced on a logarithmic axis. This makes sense. Since the concentrations of drug are equally spaced on a log scale, it makes sense that the uncertainty of the log(EC_{50}) will be symmetrical, but the uncertainty of the EC_{50} will not be. If the true uncertainty is not symmetrical, then the asymptotic confidence intervals will not be very useful.

Example simulation 2. Exponential decay.

Many biological and chemical events follow an exponential decay model. We'll compare three ways to express this model. In all cases, one of the parameters is the starting point, which we will call Y_0. The second parameter quantifies how rapidly the curve decays. We will compare three ways to express this value, as a rate constant in units of inverse time, as a time constant in units of time, or as a log(rate constant). The three equations are:

$$Y = Y_0 \cdot e^{-k_{rate} \times t}$$

$$Y = Y_0 \cdot e^{-(1/k_{time}) \times t}$$

$$Y = Y_0 \cdot e^{-10^{\log(k_{rate})} \times t}$$

Is it better to express the exponential decay equation in terms of rate constant, time constant, or log(rate constant)? If we want to rely on the asymptotic confidence intervals provided by nonlinear regression programs, we want to write the model so the parameter distribution is close to Gaussian.

Which distribution is closer to Gaussian?

To find out which of the three equations uses a parameter that is closer to Gaussian, we'll simulate data.

First, we need to choose some parameters. We chose a curve that starts at Y=100 and decays exponentially toward 0 with a rate constant (k_{off}) of 0.3 min^{-1} and a half-life of a bit more than 2 minutes ($\ln(2)/k_{off}$). Our simulations generated 10 data points equally spaced between 0 and 20 minutes, adding Gaussian random error with a standard deviation of 10.The graph below shows three sample simulations.

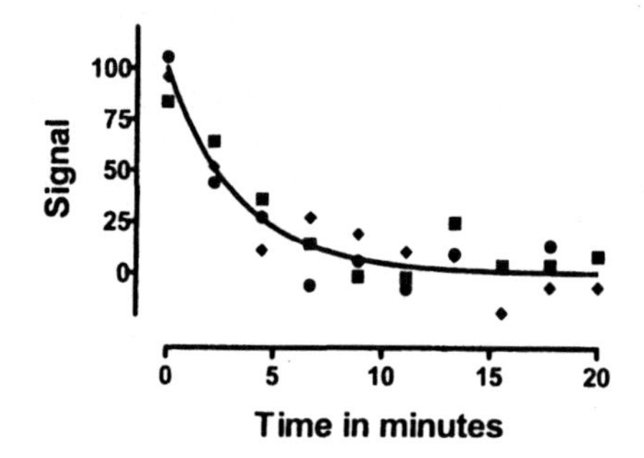

We simulated 5000 sets of data and fit each data set three times, using the three different expressions of the exponential decay model. The distribution of the rate constant, time constant, and log(rate constant) are shown in the following figures, which also superimpose ideal Gaussian distributions.

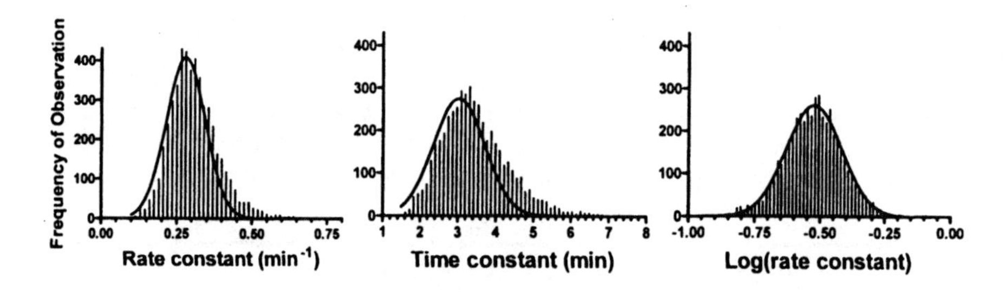

At first glance, all three distributions look roughly Gaussian. Looking more carefully, you can see that the distribution of time constants is skewed to the right. Careful scrutiny reveals that the rate constant distribution is also a bit skewed. These impressions can be confirmed by a normality test. The results are shown in the following table.

Model	Rate constant	Time constant	log(Rate constant)
KS	0.06359	0.07169	0.01339
P value	P<0.0001	P<0.0001	P > 0.10

The KS value is the largest discrepancy between the actual cumulative distribution and an ideal cumulative Gaussian distribution (expressed as a fraction). None of the distributions are far from Gaussian. The distribution of time constants is the furthest from Gaussian; the distribution of log(rate constant) is closest. The P value answers the following question: If the true distribution is Gaussian, what is the chance of obtaining a KS value as large as, or larger than, we observed? The distribution of both rate constants and time constants deviated significantly from the Gaussian ideal, while the distribution of log(rate constants) is indistinguishable from Gaussian.

Because we simulated so many data sets, the KS test has the power to detect even modest deviations from a Gaussian distribution. For this example, the distribution of log(rate constant) comes the closest to a Gaussian distribution, but the other two are not very far off.

How accurate are the confidence intervals of time constants and rate constants?

Using simulated data, we can ask how often the asymptotic 95% confidence interval contains the true value. Since these are simulated data sets, we know the true value (not true when you analyze experimental data). So for each fit of each of the 5000 simulated data sets, we can ask whether the 95% confidence interval includes the true value. The results are:

Model	% of "95% CIs" containing true value
Rate constant	93.72%
Time constant	94.44%
log(rate constant)	94.94%

No matter how we expressed the model, the confidence intervals contained the true value almost 95% of the time. The difference between 95% confidence and 93% confidence is unlikely to alter your interpretation of experimental results.

These simulations showed us that writing the model in terms of the logarithm of rate constant will give us a parameter that is closest to a Gaussian distribution and thus gives a asymptotic 95% confidence interval that is closest to correct. But the simulations also showed that the other two forms of the model (time constant and rate constant) work quite well.

How to generate a parameter distribution with Prism

Prism provides an "analysis" that simulates data. You need to choose the range and spacing of the X values, enter or choose an equation, enter values for the parameter, and choose the kind of random scatter (Gaussian with SD=10 in this example).

To get a sense of how the random numbers work, go to the graph, drop the Change menu and choose Simulate again. Repeat a few times to see how the data changes.

The next step is to analyze the simulated data to determine the best-fit values. From the results page, click Analyze, choose Nonlinear regression, and choose the equation.

Create a script to instruct Prism to generate and analyze 5000 sets of data, and record the best-fit values. To learn details about scripting, read the chapter on scripts in the Prism User's Guide. Using Notepad or some other text editor, create a file with the following script. Enter the lines shown in the left column. The right column explains what each command does, and should not be typed. Give the file a name with the extension pzc.

Script line	**Explanation**
`Shortlog`	Don't add a line to the log for each iteration.
`Setpath c:\sims`	Use this folder.
`Open disskin.pzm`	Open a Prism file.
`OpenOutput kinfit.txt`	Create a text file to hold the best-fit values.
`ForEach 5000`	Loop 5000 times.
`Goto R 1`	Go to the first results page.
`Regenerate`	Create new random numbers.
`Goto R 2`	Go to the second results page.
`WCell 4,1`	Write the value in the fourth row of the first column into the output file. This is the best-fit value of the rate constant.
`Next`	Loop again.

Run this script from Prism by choosing the Run Script command from the File menu. Selecting the script you just created, and click Run to execute it. Depending on the speed of your computer, this task should take a few minutes to execute.

Import the resulting file (kinfit.txt for this example) into a fresh Prism table, and analyze to create a frequency distribution and/or column statistics.

The example above created a list of parameter values, which could be imported into Prism for further analyses. In some cases, you want to record the list of 95% confidence intervals that Prism reports. But then you'll encounter a problem. In the nonlinear regression results, Prism reports the confidence interval of each parameter. There is no way to instruct Prism to report the two individual values separately, rather than as one text string. You can't import these back into Prism as two separate values. But you can get Excel to do it.

Assume that you simulated data with a time constant of 3.0 minutes. You run a script that saves the confidence interval of the time constant for each simulated data set. Import this list into column A of Excel, where each confidence interval becomes a single string stored in a single Excel cell. For example, cell A2 might contain the text "1.86 to 4.14". We'll use the Excel formulae shown below to extract the lower and upper confidence limit from the text and then determine whether this interval contains the true parameter value (3.0 for this example).

Cell	Comment	Put in Excel
A2	Confidence interval	1.86 3 to 4.137
B2	Lower confidence limit	=Value(LEFT(A2,FIND(" to",A2,1)))
C2	Upper confidence limit	=Value(RIGHT(A2,LEN(A2) - (FIND(" to",A2) +3)))
D2	Does interval start below the true value?	=(B2<3.0)
E2	Does the interval end above the true value?	=(C2>3.0)
F2	Does the interval contain the true value?	=IF(AND(D2,E2),1,0)

Cell A2 is the confidence interval as imported from the file created by the Prism script. Cell A3 will have the next interval, and A4 the next, and so on. Cell B2 extracts the lower confidence limit from the left part of the interval and C2 extracts the upper limit. D2 is True if the lower confidence limit (B2) is less than the true parameter value (3.0) and is otherwise false. E2 is True if the upper confidence limit (C2) is greater than the true parameter value (3.0). Cell F2 combines those two Booleans. If both are true, F2 has a value of 1.0. This happens when the interval begins below the true value and ends above it. In other words, F2 will equal 1.0 when the confidence interval contains the true value, and otherwise it will equal 0.

Copy these formula down the table, then add up column F. The sum is the number of the intervals (each computed from a separate simulated data set) that contain the true value. Divide by the total simulated data sets to compute the fraction of the intervals that contain the true value.

F. Comparing models

21. Approach to comparing models

Why compare models?

When fitting biological data with regression, your main objective is often to *discriminate* between different models. Perhaps you wish to test whether your data are more consistent with one possible mechanism relative to another.

You have only a limited number of data points, and these include experimental scatter. Given those limitations, there is no way to know for sure which model is best. All you can do is address the question statistically and get answers in terms of probabilities. This chapter explains how.

Before you use a statistical approach to comparing models

Sometimes the more complicated model just doesn't fit

Sometimes you may not even get the opportunity to compare a simple model to a more complicated model because the nonlinear regression program cannot fit the more complicated model. You may receive an error message such as "cannot converge", "floating point error", "can't invert matrix", or "division by zero". In general, the exact wording of the error message is not helpful, but the presence of an error message tells you that the program is not able to fit that model.

If you get an error message, don't assume there is something wrong with the program or that you have made incorrect choices. Rather, follow these steps:

1. Check to make sure the model is sensible. Would it make more sense to fix one of the parameters to a constant value (for example, fixing the bottom plateau to zero)? Would it make more sense to constrain a parameter to a range of values (for example, constrain a fraction to be between zero and one)?

2. Check that the initial values create a curve that goes near your data points.

3. Try other initial values in hopes that the program can converge on a solution.

4. If all your attempts end in an error message, then you've learned something important – that it is not possible to fit this model to your data. Your data simply don't define the chosen model.

5. If you really want to pursue the idea that the more complicated model is correct, consider repeating the experiment with more data points and perhaps a wider range of X values.

Don't compare models statistically if one model fits with nonsensical best-fit parameter values

Before relying upon statistical comparisons between models, use common sense. Reject any model if the best-fit values make no sense. For example, reject a model if the best-fit value of a rate constant is negative (which is impossible). Also reject a fit if the confidence intervals of the best-fit values are very wide. You don't need any statistical test to tell you that you shouldn't accept such a model. It might be possible to salvage the situation by telling the nonlinear regression program to constrain the value to be within a certain range.

> Tip: Only use statistical calculations to compare the fits of two models if the results of both fits are sensible and the more complicated model fits better.

If both models fit the data with sensible best-fit values, your next step is to compare goodness-of-fit as quantified by the sum-of-squares (or weighted sum-of-squares). If the simpler model fits better (has lower sum-of-squares) than the more complicated model, then you are done. This model is both simpler and fits the data better. What more could you want? Accept this model with no statistical calculations. This will happen rarely, however, as the curve generated by the more complicated equation (the one with more parameters) will nearly always have a lower sum-of-squares, simply because it has more inflection points (it "wiggles" more). In most cases, therefore, you will need to use a more powerful method to discriminate one model from the other. You need to use a method that looks at the tradeoff between lower sum-of-squares and more parameters.

The model with the lower sum-of-squares may not be preferred

Nonlinear regression quantifies goodness-of-fit as the sum of squares of the vertical distances of the data points from the curve. You might assume, therefore, that when comparing two models, you always choose the one that fits your data with the smallest sum-of-squares. In fact, it is not that simple to compare one model to another.

The problem is that a more complicated model (more parameters) gives the curve more flexibility than the curve defined by the simpler model (fewer parameters). This means that the curve defined by more parameters can "wiggle" more and thus fit the data better. For instance, a two-site binding model almost always fits competitive binding data better than a one-site binding model. A three-site model fits the data even better, and a four-site model better still!

To compare models, therefore, you can't just ask which model generates a curve that fits your data with the smallest sum-of-squares. You need to use a statistical approach.

Statistical approaches to comparing models

Choose an approach to compare nested models

When you compare models, most often the two models are related. The term "related" has a specific meaning when applied to models. Two models are said to be related when one is a simpler case of the other. For example, a one-site competitive binding model is a simple case of a two-site competitive binding model. Similarly, a kinetic model with one component or phase is a simple case of the same model with multiple components or phases. When one model is a simpler case of the other, mathematicians say that the models are *nested*.

If your models are related (nested), you can use two distinct approaches to comparing models.

- The first method is based on statistical *hypothesis testing* and ANOVA (analysis of variance). It is based on analyzing the difference between the sum-of-squares of the two models. This is why it is sometimes called the *extra sum-of-squares test*. These ANOVA calculations compute an F ratio and a P value. If the P value is low, you conclude there are sufficient data to convince you to reject the simpler model (the null hypothesis) in favor of the alternative hypothesis that the more complicated model is correct. This F test approach is explained in Chapter 22.

- The second method for comparing models is not based on hypothesis testing at all, but rather on information theory. This method calculates Akaike's Information Criterion (AIC) which answers the questions you care about: "Which model is more likely to have generated the data?", and "How much more likely?". This approach is discussed in Chapter 23.

Don't use the F test if your models are not nested

The F test (extra sum-of-squares) should only be used to compare nested models. If your models are not nested, use the AIC approach.

Compare the fit of two models to the same data set

The methods used to compare models only work when both models are fit to *exactly* the same data. Some examples of comparisons that *cannot* be made using methods designed to compare models are listed below.

- It is not appropriate to use model comparison methods to compare two ways of expressing the data (say Y values expressed as logarithms vs. plain numbers). The Y values must be the same in both fits.

- It is not appropriate to use model comparison methods to compare fits that use different weighting methods. The weights must be the same in both fits.

- It is not appropriate to use model comparison methods to compare a fit to a full data set vs. the fit to just part of the data set (say leaving off the early time points). The data must be the same in both fits.

- It is not appropriate to use model comparison methods to compare a fit to a data set with the fit to the same data set minus an outlying value.

When you compare two models, both models must fit the same data using the same weighting scheme.

Using global fitting to turn questions about comparing curves into questions about comparing models

So far, we have only discussed comparing the fit of two models to one data set.

Often you wish to ask a different question. You want to compare two different curves. In some situations you want to ask whether a particular parameter (say the $\log EC_{50}$) differs between two data sets. In other situations, you may want to ask a more general question – are two curves distinguishable at all?

It turns out that you can answer these questions using the same methods used to compare models. The trick is to turn a question about comparing curves into a question about comparing models. This is done via global or shared fitting. The details are explained in Chapter 27, so we'll just give an overview here.

Assume you measure a decaying signal over time to determine a rate constant. You want to know if the rate constant differs significantly between two treatments. How can you turn this question into a question about comparing two models? The first model is that the two curves are distinct, so you fit two separate curves and define the overall sum-of-squares as the total of the sum-of-squares for each separate curve. The second model uses global fitting. You fit all the data at once, finding one shared value of the rate constant. Now that we have turned the question about comparing curves into a question about comparing models, we can use methods developed to compare models. The next two chapters explain these methods.

22. Comparing models using the extra sum-of-squares F test

Introducing the extra sum-of-squares F test

When you compare two nested models, the model with more parameters will almost always fit the data better (have a lower sum-of-squares) than the model with fewer parameters. It is not enough to compare sum-of-squares. We need to use a statistical approach to decide which model to accept.

As its name suggests, the extra sum-of-squares F test is based on the difference between the sum-of-squares of the two models. It also takes into account the number of data points and the number of parameters of each model. It uses this information to compute an F ratio, from which it calculates a P value. The P value answers this question: If the simpler model (the one with fewer parameters) were really correct, what is the chance that you'd happen to obtain data where the difference between sum-of-squares is as large (or larger) than obtained in this experiment. ?

If the P value is small, there are two possibilities:

- The more complicated model is correct.
- The simpler model is correct, but random scatter led the more complicated model to fit better. The P value tells you how frequently you'd expect this to happen.

So which model is correct? Statistics can't answer that question.

Which model fits better? The more complicated model does, but you'd expect this by chance, so that is the wrong question to ask.

Which model should we accept? This is answered by statistical hypothesis testing. First set a significance level (also called alpha), which is usually set to 0.05. If the P value is less than this significance level, conclude that the alternative model fits the data significantly better than the null hypothesis model. Otherwise, conclude that there is no compelling evidence supporting the alternative model, so accept the simpler (null) model.

Note that the F test isn't really helping you decide which model is correct. What it does is help you decide whether you have sufficient evidence to reject the simpler null hypothesis model.

The F test is for comparing nested models only

The F test is *only* valid for comparing nested models. It *cannot* be used for non-nested models. In this latter circumstance, you will need to use an alternative method based on information theory rather than hypothesis testing. This is explained in the following chapter.

GraphPad note: Prism does not enforce this restriction. It lets you compare any two models with the F test. The results will only be meaningful, however, if one model is a simple case of the other.

How the extra sum-of-squares F test works

One-way ANOVA as a comparison of models

Using the F test to compare fits to nonlinear models is an adaptation of analysis of variance (ANOVA). We'll first review one-way ANOVA, and then show how it can be viewed as a comparison of two models. We'll then extend this to comparing the fits of two models.

One-way ANOVA quantifies how sure you can be that the groups (treatments) really have different means. In these sample data below, a response was measured from six subjects in each of three groups.

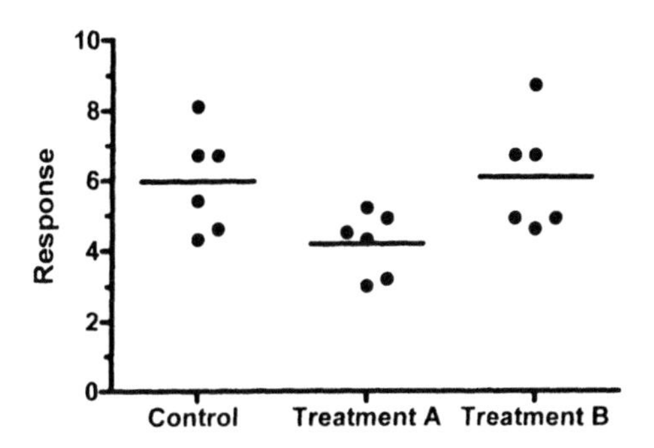

Conventional ANOVA table

ANOVA helps you decide whether the differences between groups could be due to chance. It does this by partitioning the variability into variation between groups and variation within groups.

Here is a conventional ANOVA table. Depending on the program that you use, some of the terms may be different, for example, "error" instead of "residual", and "groups" instead of "columns".

Source of variation	SS	df	MS
Treatment (between columns)	13.61	2	6.804
Residual (within columns)	27.23	15	1.815
Total	40.84	17	

The first row quantifies how far the three (in this case) group means are from each other. The second row quantifies how much variability there is within the groups. The third row (total) quantifies the total variability among all the values (ignoring group altogether).

The first column is the sum-of-squares. The second column is degrees of freedom (df), which accounts for number of values and number of groups. The final column divides the sum-of-squares by the df to create something called mean-square. These mean-square values are comparable. The ratio of mean squares treatment divided by the mean square residual is called the F ratio. For this example, F= 3.75. If all the variability were random (the treatments were really all the same), you'd expect F to be near 1.0. Our F was much higher. This could be caused by a real difference among column means, or it could be a coincidence. A P value helps you consider these possibilities. For this example, the P value is 0.0479. This means that if the three treatments were really identical, there is a 4.79% chance of happening to find as much variability among group means as we observed here.

An alternative ANOVA table emphasizing model comparison

Below is an alternative view of the same table, which emphasizes that ANOVA compares two models.

Model	SS	df
Null hypothesis	40.84	17
Alternative hypothesis	27.23	15
Difference	13.61	2
Relative difference	0.50	0.13
Ratio (F)	3.75	

The ANOVA compares two models, or hypotheses:

- The null hypothesis is that all the data are sampled from populations with the same mean, so the differences we observe between the sample means are due to chance. We just happened to sample larger values from one group and smaller values from another. The fit of the data to this model is determined by summing the square of the difference between each value and the grand mean. The result is 40.84. There are 18 values altogether, and we use the grand mean in the calculations, so there are 17 degrees of freedom.

- The alternative hypothesis is that the groups really do have distinct means, and that the variability among group means is not due to chance. The fit of the data to this model is quantified by summing the square of the difference between each value and its own group (column) mean. The result is 27.23. Each of the three groups has six values, but we use the group mean of each in the calculations. So there are 15 degrees of freedom (18 values minus 3 group mean, or equivalently, three groups with 5 degrees of freedom each).

The third row compares the models by computing the difference between the two models. You always expect the alternative model to fit the data better (have a smaller sum-of-squares). The group means are computed from the data, so of course the values are closer, on average, to the group means than to the grand mean. The question is how much better the alternative model works, and how likely is that difference to be due to chance.

To make sense of these values, we need to express the difference as a ratio (so it doesn't matter what units we used) and to account for degrees of freedom. This is done by expressing the difference in sum-of-squares and df as a percentage of the value for the alternative hypothesis. The difference in sum-of-squares is 50.0% of the sum-of-squares for the alternative hypothesis. The difference in degrees of freedom is 13.3% of the degrees of freedom for the alternative hypothesis.

We won't attempt to present a mathematical proof here, but statisticians have shown that if the null hypothesis were true, you'd expect the relative difference in sum-of-squares to be approximately equal to the relative difference in degrees of freedom. The ratio of the two percentages is called the F ratio, 3.75 for this example.

If the null hypothesis were true – that really all three samples are drawn from populations with the same mean, and all differences are due to chance – then we expect the F ratio to be near 1. Statisticians know the distribution of the F ratio under the null hypothesis, so we can find out how unlikely it would be to have an F ratio greater than or equal to 3.75. The answer depends on both degrees of freedom values (for the alternative hypothesis and for the difference). For this example, the answer – the P value – is 0.0479. This means that if the three treatments were really identical, there is a 4.79% chance of happening to find as much variability among group means as we observed here.

Why rearrange the ANOVA table?

The alternative presentation of the ANOVA table in the previous section emphasizes that ANOVA compares two models. We find this an easier way to understand ANOVA results than the conventional presentation. But both approaches show the same results – they just arrange the values differently.

We present this alternative approach to one-way ANOVA here simply as a way to introduce the use of an F test to compare models.

The F test to compare models

The following table generalizes the concepts presented in the preceding section.

Model	**SS**	**df**
Null hypothesis	SS_{null}	DF_{null}
Alternative hypothesis	SS_{alt}	DF_{alt}
Difference	$SS_{null} - SS_{alt}$	$DF_{null} - DF_{alt}$
Relative difference	$(SS_{null} - SS_{alt}) / SS_{alt}$	$(DF_{null} - DF_{alt}) / DF_{alt}$
Ratio (F)	$\frac{(SS_{null}-SS_{alt})/SS_{alt}}{(DF_{null}-DF_{alt})/DF_{alt}}$	

The idea is simple. You fit your data to two models, and quantify goodness-of-fit as the sum of squares of deviations of the data points from the model. You quantify the complexity of the models with the degrees of freedom (df), which equal the number of data points minus the number of parameters fit by regression.

If the simpler model (the null hypothesis) is correct, you expect the relative increase in the sum of squares to be approximately equal the relative increase in degrees of freedom. If the more complicated (alternative) model is correct, then you expect the relative increase in sum-of-squares (going from complicated to simple model) to be greater than the relative increase in degrees of freedom:

The F ratio equals the relative difference in sum-of-squares divided by the relative difference in degrees of freedom.

$$F=\frac{(SS_{null}-SS_{alt})/SS_{alt}}{(DF_{null}-DF_{alt})/DF_{alt}}$$

That equation is more commonly shown in this equivalent form:

$$F=\frac{(SS_{null}-SS_{alt})/(DF_{null}-DF_{alt})}{SS_{alt}/DF_{alt}}$$

or

$$F=\frac{(SS1-SS2)/(DF1-DF2)}{SS2/DF2}$$

where the numbers 1 and 2 refer to the null (simpler) and alternative (more complex) models, respectively. F ratios are always associated with a certain number of degrees of freedom for the numerator and a certain number of degrees of freedom for the denominator. This F ratio has DF1-DF2 degrees of freedom for the numerator, and DF2 degrees of freedom for the denominator.

How to determine a P value from F

The F distribution is known, so a P value can be calculated from the F ratio and the two df values.

> Tip: When you use a program or table to find a P value that corresponds with the F ratio, take extra care to be sure that you don't mix up the two df values. You'll get the wrong P value if you accidentally reverse the two df values.

GraphPad offers a free web calculator to calculate P from F (go to www.graphpad.com, then to QuickCalcs). You can also use this Microsoft Excel function:

```
=FDIST(F,dfn, dfd)
```

23. Comparing models using Akaike's Information Criterion (AIC)

The previous chapter explained how to compare nested models using an F test and choose a model using statistical hypothesis testing. This chapter presents an alternative approach that can be applied to both nested and non-nested models, and which does not rely on P values or the concept of statistical significance.

Introducing Akaike's Information Criterion (AIC)

Akaike developed an alternative method for comparing models, based on information theory. The method is called Akaike's Information Criterion, abbreviated AIC.

The logic is not one of hypothesis testing, so you don't state a null hypothesis, don't compute a P value, and don't need to decide on a threshold P value that you deem statistically significant. Rather, the method lets you determine which model is more likely to be correct and quantify how much more likely.

Unlike the F test, which can only be used to compare nested models, Akaike's method can be used to compare either nested or nonnested models.

The theoretical basis of Akaike's method is difficult to follow. It combines maximum likelihood theory, information theory, and the concept of the entropy of information (really!). If you wish to learn more, read *Model Selection and Multimodel Inference -- A practical Information-theoretic approach* by KP Burnham and DR Anderson, second edition, Springer, 2002. It presents the principles of model selection in a way that can be understood by scientists. While it has some mathematical proofs, these are segregated in special chapters and you can follow most of the book without much mathematical background.

How AIC compares models

While the theoretical basis of Akaike's method is difficult to follow, it is easy to do the computations and make sense of the results.

The fit of any model to a data set can be summarized by an information criterion developed by Akaike.

If you accept the usual assumptions of nonlinear regression (that the scatter of points around the curve follows a Gaussian distribution), the AIC is defined by the equation below, where N is the number of data points, K is the number of parameters fit by the regression plus one (because regression is "estimating" the sum-of-squares as well as the values of the parameters), and SS is the sum of the square of the vertical distances of the points from the curve.

$$\mathrm{AIC}=\mathrm{N}\cdot\ln\left(\frac{\mathrm{SS}}{\mathrm{N}}\right)+2\mathrm{K}$$

When you see a new equation, it is often helpful to think about units. N and K are unitless, but SS is in the square of the units you choose to express your data in. This means that you can't really interpret a single AIC value. An AIC value can be positive or negative, and the sign of the AIC doesn't really tell you anything (it can be changed by using different units to express your data). The value of the AIC is in comparing models, so it is only the difference between AIC values that you care about.

Define A to be a simpler model and B to be a more complicated model (with more parameters). The difference in AIC is defined by:

$$\begin{aligned}\Delta AIC &= AIC_B - AIC_A \\ &= N\left[\ln\left(\frac{SS_B}{N}\right) - \ln\left(\frac{SS_A}{N}\right)\right] + 2(K_B - K_A) \\ &= N \cdot \ln\left(\frac{SS_B}{SS_A}\right) + 2 \cdot (K_B - K_A)\end{aligned}$$

The units of the data no longer matter, because the units cancel when you compute the ratio of the sum-of-squares.

The equation now makes intuitive sense. Like the F test, it balances the change in goodness-of-fit as assessed by sum-of-squares with the change in the number of parameters to be fit. Since model A is the simpler model, it will almost always fit worse, so SS_A will be greater than SS_B. Since the logarithm of a fraction is always negative, the first term will be negative. Model B has more parameters, so K_B is larger than K_A, making the last term positive. If the net result is negative, that means that the difference in sum-of-squares is more than expected based on the difference in number of parameters, so you conclude Model B (the more complicated model) is more likely. If the difference in AIC is positive, then the change in sum-of-squares is not as large as expected from the change in number of parameters, so the data are more likely to have come from Model A (the simpler model).

The equation above helps you get a sense of how AIC works – balancing change in goodness-of-fit vs. the difference in number of parameters. But you don't have to use that equation. Just look at the individual AIC values, and choose the model with the smallest AIC value. That model is most likely to be correct.

A second-order (corrected) AIC

When N is small compared to K, mathematicians have shown that AIC is too small. The corrected AIC value (AIC_C) is more accurate. AIC_C is calculated with the equation below (where N is number of data points, and K is the number of parameters plus 1):

$$AIC_C = AIC + \frac{2K(K+1)}{N-K-1}$$

If your samples are large, with at least a few dozen times more data points than parameters, this correction will be trivial. N will be much larger than K, so the numerator is small compared to the denominator, so the correction is tiny. With smaller samples, the correction will matter and help you choose the best model.

> Note: The AIC_c can only be computed when the number of data points is at least two greater than the number of parameters.

We recommend that you always use the AIC_c rather than the AIC. With small sample sizes commonly encountered in biological data analysis, the AIC_c is more accurate. With large sample sizes, the two values are very similar.

The change in AICc tells you the likelihood that a model is correct

The model with the lower AIC_c score is the model more likely to be correct. But how much more likely?

If the AIC_c scores are very close, there isn't much evidence to choose one model over the other. If the AICc scores are far apart, then the evidence is overwhelming. The probability that you have chosen the correct model is computed by the following equation, where Δ is the difference between AIC_c scores.

$$\text{probability} = \frac{e^{-0.5\Delta}}{1+e^{-0.5\Delta}}$$

> Note: The probabilities are based on the difference between AIC_c scores. The probabilities are the same if the AIC_c scores are 620,000 and 620,010 as they are if the AIC_c scores are 1 and 11. Only the absolute difference matters, not the relative difference.

This graph shows the relationship between the difference in AIC (or AIC_c) scores and the probability that each model is true.

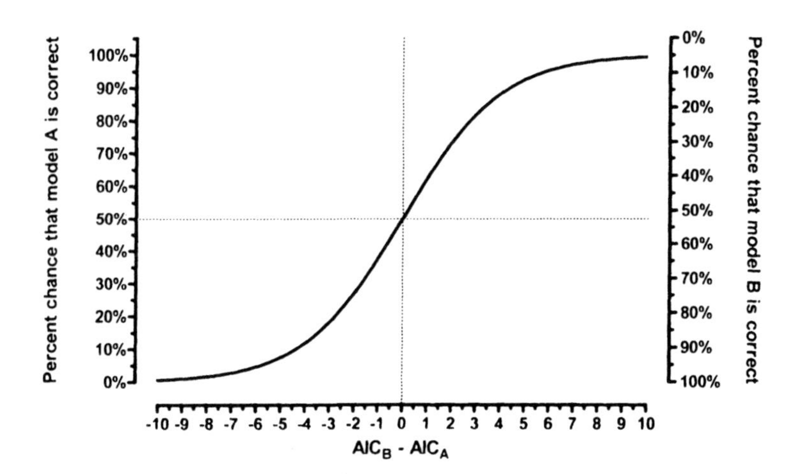

> Note: These probabilities are also called Akaike's weights.

If the two AIC_c values are the same, then the difference is zero, and each model is equally likely to be correct. The graph shows that there is a 50% chance that model A is correct, and a 50% chance that model B is correct. If the difference is 2.0, with model A having the lower score, there is a 73% probability that model A is correct, and a 27% chance that model B is correct. Another way to look at this is that model A is 73/27 or 2.7 times more likely to be correct than model B. If the difference between AIC_c scores is 6.0, then model

A has a 95% chance of being correct so is 20 (95/5) times more likely to be correct than model B.

> Note: Akaike's weights compute the relative probability of two models. It can be extended to compute the relative probabilities of a family of three or more models. But it is always possible that another model is even more likely. The AIC_c method only compares the models you choose, and can't tell you if a different model is more likely still.

The relative likelihood or evidence ratio

When comparing two models, you can divide the probability that one model is correct by the probability the other model is correct to obtain the *evidence ratio*, which is defined by this equation.

$$\text{Evidence Ratio} = \frac{\text{Probability that model 1 is correct}}{\text{Probability that model 2 is correct}} = \frac{1}{e^{-0.5 \cdot \Delta AIC_c}}$$

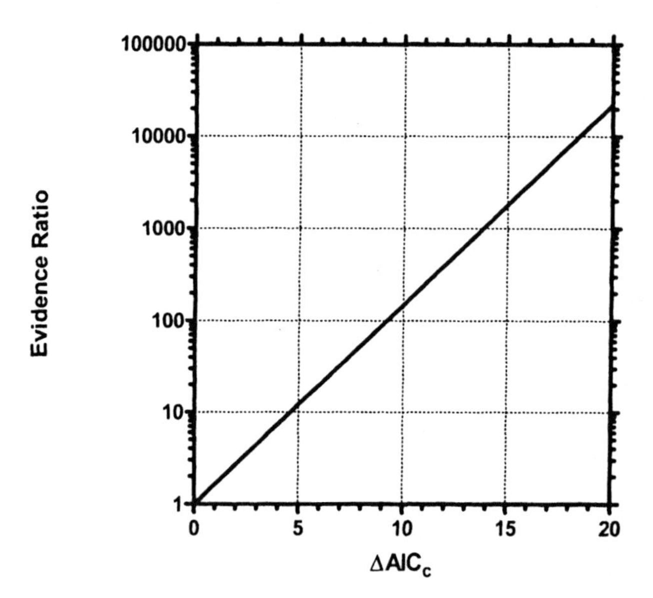

> Note: The evidence ratio is based on the absolute difference between AIC_c scores, not the relative difference. You'll get the same evidence ratio with AIC_c scores of 4567 and 4577 and AIC_c scores of 1 and 11. In both cases, the difference is 10.

For example, if the AIC_c scores differ by 5.0, then the evidence ratio equals 12.18. The model with the lower AIC_c score is a bit more than twelve times more likely to be correct than the other model. Most people don't consider this to be completely persuasive. If the difference in AIC_c scores equals 10, then the evidence ratio is 148, so the evidence is overwhelmingly in favor of the model with the lower AIC_c.

Don't overinterpret the evidence ratio. The ratio tells you the relative likelihood, *given your experimental design*, of the two models being correct. If you have few data points, the simpler model might fit best with a very high evidence ratio. This tells you that you can be quite sure that the simple model is adequate to explain your data. With so few points, you don't have any real evidence that the more complicated model is right. This doesn't mean that a more complicated model might not explain your system better. If you had lots of data, you might find that a more complicated model is more likely to be correct.

Terminology to avoid when using AIC_c

The AIC is derived from information theory, which is different from statistical hypothesis testing. You can use the AIC_c method to determine the relative likelihood of two (or more) models. You can't use AIC_c to decide whether the data fit "significantly" better to one model so you should "reject" the other. Therefore you should never use the terms "significant" or "reject" when presenting the conclusions of model comparison by AIC_c. These terms have very definite meanings in statistics that only apply when you compute a P value and use the construct of statistical hypothesis testing. Those terms carry too much baggage, so should only be used in the context of statistical hypothesis testing.

How to compare models with AIC_C by hand

Even if your nonlinear regression program does not compare models with the Akaike's Information Criteria, you can do so fairly simply. These steps review the equations presented in this chapter, so you can compute AIC_c by hand.

1. Fit the first model using nonlinear regression.

2. Look at the results of nonlinear regression and write down the sum-of-squares; call this value SS. If you used any weighting factors, then record the weighted sum-of-squares.

3. Define N to be the number of data points. Be sure to account for replicates properly. If you have 13 X values with duplicate determinations of Y, and you asked the program to treat each replicate as a separate value, then N is 26.

4. Define K to be the number of parameters fit by nonlinear regression plus 1. Don't count parameters that are constrained to constant values. If in doubt, count the number of distinct SE values reported by nonlinear regression, then add 1 to get the value of K. (Why add one? Because nonlinear regression is also "estimating" the value of the sum-of-squares.)

5. Compute AIC_C:

$$AIC_c = N \cdot \ln\left(\frac{SS}{N}\right) + 2K + \frac{2K(K+1)}{N-K-1}$$

6. Repeat steps 1-5 with the other model.

7. The model with the lower AIC_c score is more likely to be correct.

8. Calculate the evidence ratio from the difference in AIC_c scores:

$$\text{Evidence Ratio} = \frac{1}{e^{-0.5 \cdot \Delta AIC_c}}$$

One-way ANOVA by AICc

The previous chapter showed how to turn one-way ANOVA into a model comparison problem, and did the comparison by the extra sum-of-squares F test. Here are the same data, comparing models by AIC_c.

Model	SS	N	Pars	AIC_c	Probability	Evidence ratio
Null hypothesis	40.84	18	1	19.55	37.53%	1.66
Alternative hypothesis	27.23	18	3	18.53	62.47%	

The alternative hypothesis (that the group means are not all identical) has the lower AIC_c. Thus, it is more likely to be correct than the null hypothesis that all the data come from populations with the same mean. But the evidence ratio is only 1.66. So the alternative hypothesis is more likely to be correct than the null hypothesis, but only 1.66 times more likely. With so few data and so much scatter, you really don't have enough evidence to decide between the two models.

24. How should you compare models -- AIC_c or F test?

A review of the approaches to comparing models

Chapter 21 discussed the general approach to use when comparing models. Most important:

Before comparing models statistically, use common sense. Only use statistical comparisons to compare two models that make scientific sense, and where each parameter has a sensible best-fit value and a reasonably narrow confidence interval. Many scientists rush to look at statistical comparisons of models too soon. Take the time to ask whether each fit is sensible, and only rely on statistical comparisons when both models fit the data well.

If you wish to compare two models that are not related ("nested"), then your only choice is to compare the fits with the AIC_c method. The F test should not be used to compare nonnested models. Since you'll rarely find it helpful to compare fits of biological models that are not nested, you'll almost always compare nested models.

If the two models are related, and both fit the data with sensible parameter values, you can choose between the F test and AIC method. The rest of this chapter explains the advantages and disadvantages of the two approaches.

Pros and cons of using the F test to compare models

The F test is based on traditional statistical hypothesis testing.

The null hypothesis is that the simpler model (the one with fewer parameters) is correct. The improvement of the more complicated model is quantified as the difference in sum-of-squares. You expect some improvement just by chance, and the amount you expect by chance is determined by the number of degrees of freedom in each model. The F test compares the difference in sum-of-squares with the difference you'd expect by chance. The result is expressed as the F ratio, from which a P value is calculated.

The P value answers this question: If the null hypothesis is really correct, in what fraction of experiments (the size of yours) will the difference in sum-of-squares be as large as you observed or larger? If the P value is small, you'll conclude that the simple model (the null hypothesis) is wrong, and accept the more complicated model. Usually the threshold P value is set at its traditional value of 0.05. If the P value is less than 0.05, then you reject the simpler (null) model and conclude that the more complicated model fits significantly better.

> Reminder: The F test should only be used to compare two related ("nested") models. If the two models are not nested, use the AIC_c method to compare them.

The main advantage of the F test is familiarity. It uses the statistical hypothesis testing paradigm that will be familiar to the people who read your papers or attend your presentations. Many will even be familiar with the use of the F test to compare models.

The F test also has some disadvantages:

- You have to set an arbitrary value of alpha, the threshold P value below which you deem the more complicated model to be "significantly" better so "reject" the simpler model. Traditionally, this value of alpha is set to 0.05, but this is arbitrary.

- The value of alpha is generally set to the same value regardless of sample size. This means that even if you did an experiment with many thousands of data points, there is a 5% chance of rejecting the simpler model even when it is correct. If you think about it, this doesn't make much sense. As you collect more and more data, you should be increasingly sure about which model is correct. It doesn't really make sense to always allow a 5% chance of rejecting the simple model falsely.

- The F test cannot be readily extended to compare three or more models. The problem is one of interpreting multiple P values.

Pros and cons of using AIC_c to compare models

The AIC_c method is based on information theory, and does not use the traditional "hypothesis testing" statistical paradigm. Therefore it does not generate a P value, does not reach conclusions about "statistical significance", and does not "reject" any model.

The AIC_c model determines how well the data supports each model. The model with the lowest AIC_c score is most likely to be correct. The difference in AIC_c score tells you how much more likely. Two models can be compared with a likelihood ratio, which tells you how many times more likely one model is compared to the other.

The main advantage of the AIC approach is that it doesn't tell you just which model is more likely, it tells you how much more likely. This makes it much easier to interpret your results. If one model is hundreds or thousands of times more likely than another, you can make a firm conclusion. If one model is only a few times more likely than the other, you can only make a very tentative conclusion.

Another feature of the AIC approach (which we consider to be an advantage) is that the method doesn't make any decisions for you. It tells you which model is more likely to be correct, and how much more likely. It is up to you to make a decision about whether to conclude that one model is clearly correct, or to conclude that the results are ambiguous and the experiment needs to be repeated. You can simply accept the more likely model. You can accept the simpler model unless the more complicated model is much more likely. Or you can conclude that the data are ambiguous and make no decision about which model is best until you collect more data. This seems like a much more sensible approach to us than the rigid approach of statistical hypothesis testing.

A final advantage of the AIC_c approach is that it is easily extended to compare more than two models. Since you don't set arbitrary values of alpha and declare "statistical significance", you don't get trapped in the logical quandaries that accompany multiple statistical comparisons.

The only real disadvantage of the AIC_c approach is that it is unfamiliar to many scientists. If you use this approach, you may have to explain the method as well as your data.

Which method should you use?

When comparing models, by far the most important step comes before using either the F test or the AIC_c method. You should look at the graph of the fits and at the curve fitting results. You should reject a model if one of the fits has best-fit values that make no scientific sense, if the confidence intervals for the parameters are extremely wide, or if a two-phase (or two component) model fits with one phase having a tiny magnitude. Only when both fits are sensible should you go on to compare the models statistically.

If you compare nonnested models, then the F test is not appropriate. Use the AIC_c approach.

If you compare three or more models, then the whole approach of significance testing gets tricky. Use the AIC_c approach.

In most cases, you'll compare two nested models, so there is no clear advantage of one approach over the other. We prefer the AIC_c method for the reasons listed in the previous section, mostly because AIC_c quantifies the strength of the evidence much better than does a P value. However, this is not a strong preference, and the F test method works well and has the advantage of familiarity.

> Tip: Pick a method and stick with it. It is not appropriate to compare with both methods and pick the results that you like the best.

25. Examples of comparing the fit of two models to one data set

Example 1. Two-site competitive binding model clearly better.

The data

The F test can compare two nested models. In this example, we compare a one-site to a two-site competitive binding model.

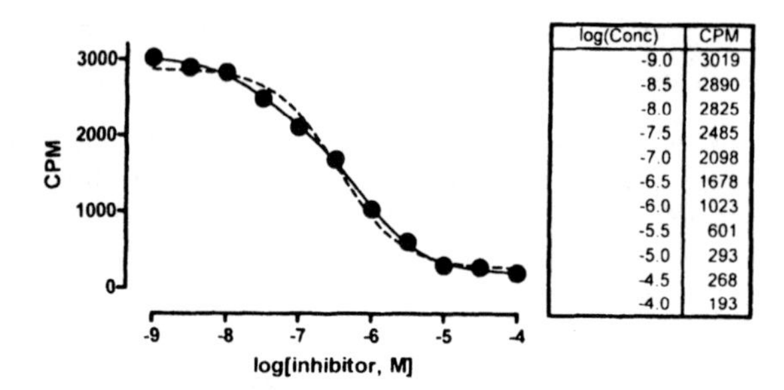

log(Conc)	CPM
-9.0	3019
-8.5	2890
-8.0	2825
-7.5	2485
-7.0	2098
-6.5	1678
-6.0	1023
-5.5	601
-5.0	293
-4.5	268
-4.0	193

Do both fits make sense?

The graph above shows the results of fitting both a one-site (dashed curve) and two-site (solid curve) competitive binding curve to some data. Here are the tabular results.

One site competition	
Best-fit values	
BOTTOM	257.9
TOP	2873
LOGEC50	-6.471
EC50	3.3814e-007
Std. Error	
BOTTOM	67.27
TOP	66.12
LOGEC50	0.07250
95% Confidence Intervals	
BOTTOM	102.8 to 413.0
TOP	2721 to 3026
LOGEC50	-6.638 to -6.304
EC50	2.3009e-007 to 4.9693e-007
Goodness of Fit	
Degrees of Freedom	8
R^2	0.9910
Absolute Sum of Squares	116882
Sy.x	120.9

Two site competition	
Best-fit values	
BOTTOM	188.8700
TOP	3030.036
FRACTION1	0.3130884
LOGEC50_1	-7.520541
LOGEC50_2	-6.121639
EC50_1	3.016191e-008
EC50_2	7.557197e-007
Std. Error	
BOTTOM	27.88553
TOP	39.85059
FRACTION1	0.05738087
LOGEC50_1	0.1907365
LOGEC50_2	0.07960846
95% Confidence Intervals	
BOTTOM	120.6341 to 257.1059
TOP	2932.522 to 3127.551
FRACTION1	0.1726775 to 0.4534995
LOGEC50_1	-7.987273 to -7.053809
LOGEC50_2	-6.316442 to -5.926837
EC50_1	1.029738e-008 to 8.834680e-008
EC50_2	4.825679e-007 to 1.183484e-006
Goodness of Fit	
Degrees of Freedom	6
R^2	0.9991944
Absolute Sum of Squares	10475.93
Sy.x	41.78503
Constraints	
FRACTION1	0 < FRACTION1 < 1.000000

Look at the graph of the one site fit model (dashed curve). It seems to deviate systematically from the data points, but not so badly that we will reject the model completely. The graph of the two-site fit looks better. Do the best-fit values of the two-site model make sense? The best-fit two-site curve has a high affinity site with a $logEC_{50}$ of -7.5 comprising 31% of the sites, and a low affinity site with a $logEC_{50}$ of -6.1. These results are certainly scientifically plausible. Next we look at the confidence intervals, which are reasonably narrow (given so few data points). Because the results are sensible, it makes sense to compare the sum-of-squares statistically.

Compare with the F test

The F test compares the goodness-of-fit of the two models (assessed by the sum-of-squares, SS) adjusting for difference in the number of degrees of freedom (df).

Here are the results.

Model	SS	df
Null hypothesis (1 site)	116882	8
Alternative hypothesis (2 sites)	10476	6
Difference	106406	2
Fractional difference	10.16	0.333
Ratio (F)	30.48	

The two sum-of-squares values come from the nonlinear regression results.

The degrees of freedom equal the number of data points (11 in this example) minus the number of parameters fit by the program (three parameters for the one-site model, five for the two-site model).

Going from the two-site model to the one site model, the degrees of freedom increased 33.3%. If the one-site model were correct, you'd expect the sum-of-squares to also increase about 33.3% just by chance. In fact, the fit to the one-site model had a sum-of-squares 1016% higher than the fit to the two-site model. The relative increase in sum-of-squares was 30.5 times higher than the relative increase in degrees of freedom, so the F ratio is 30.5 with 2 (numerator) and 6 (denominator) degrees of freedom.

The P value is 0.0007. The P value is a probability, and it is easy to interpret. Start by assuming that the null hypothesis, the simpler one-site model, is really correct. Now perform the experiment many times with the same number of data points, the same X values, and the same amount of scatter (on average). Fit each experiment to both one- and two-site models. The two-site model will almost always fit the data better than the one-site model. But how much better? The P value is the fraction of those hypothetical experiments that you'd expect to find as a large a difference between the sum-of-squares of the two-site model and one-site model as you actually observed.

You can't know for sure which model is more appropriate for your data. All you can say is that if the simpler (one-site) model were correct, it would be a very rare coincidence – one that happens 0.07% of the time, or one time in 1428 experiments – to find data that fit the two-site model so well. Since that P value is lower than the traditional (but arbitrary) threshold of 0.05, we'll conclude that the two-site model fits the data significantly better than the one-site model.

Compare with AIC_c

Let's compare the models using AIC_c. The sum-of-squares values come from the curve fit. There are 12 data points. The one-site model fits 3 parameters; the two-site model fits 5. Put these values into the equations presented in the previous chapter to calculate the AIC_c for each model.

Model	SS	N	Pars	AIC_c	Probability	Evidence Ratio
One site	116882	11	3	116.65	1.63%	60.33
Two site	10476	11	5	108.45	98.37%	

The two-site model has the lower AIC_c so is more likely to be correct. In fact, it is 98.37% certain to be correct, which means it is 60.33 times more likely to be correct than the one-site model.

Example 2: Two-site binding model doesn't fit better.

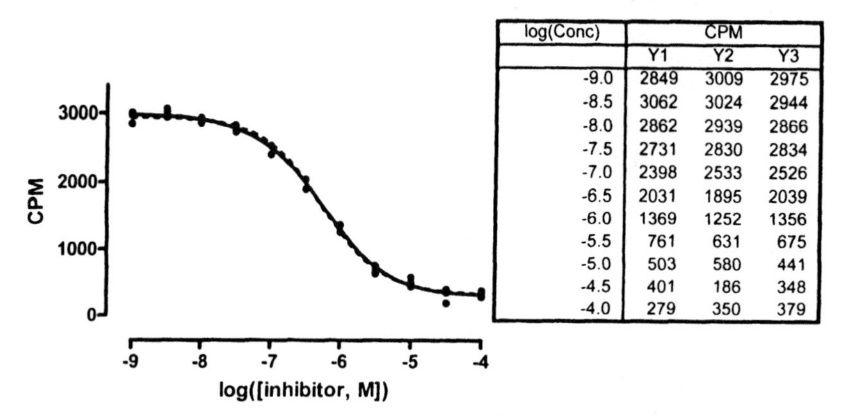

log(Conc)	CPM		
	Y1	Y2	Y3
-9.0	2849	3009	2975
-8.5	3062	3024	2944
-8.0	2862	2939	2866
-7.5	2731	2830	2834
-7.0	2398	2533	2526
-6.5	2031	1895	2039
-6.0	1369	1252	1356
-5.5	761	631	675
-5.0	503	580	441
-4.5	401	186	348
-4.0	279	350	379

When you look at the graph of the results in this example (above), the two curves are almost identical. You may even have difficulty seeing that the graph includes two curves.

Here are the tabular results of the two-site fit:

Best-fit values	
BOTTOM	299.0
TOP	2986
FRACTION1	0.1271
LOGEC50_1	-7.147
LOGEC50_2	-6.145
EC50_1	7.1281e-008
EC50_2	7.1579e-007
Std. Error	
BOTTOM	28.97
TOP	33.22
FRACTION1	0.1658
LOGEC50_1	0.7740
LOGEC50_2	0.1125
95% Confidence Intervals	
BOTTOM	239.6 to 358.3
TOP	2918 to 3054
FRACTION1	0.0 to 0.4667
LOGEC50_1	-8.732 to -5.562
LOGEC50_2	-6.376 to -5.915
EC50_1	1.8530e-009 to 2.7420e-006
EC50_2	4.2118e-007 to 1.2165e-006
Goodness of Fit	
Degrees of Freedom	28
R^2	0.9961
Absolute Sum of Squares	150257
Sy.x	73.26

The two EC_{50} values are about a log apart. But the best-fit value for the fraction of high affinity sites is only 13%, with a 95% confidence interval extending from 0 to 47%.

With such a wide confidence interval for the fraction of high-affinity sites, and with the two-site curve being barely distinguishable from the one-site curve, you should be dubious about the validity of the two-site fit. But those results are not nonsense, so we'll push ahead and do the formal comparison with both the F test and the AIC_c method.

Here are the results of the comparison using the extra sum-of-squares F test:

Model	**SS**	**df**
Null hypothesis (1 site)	174334	30
Alternative hypothesis (2 sites)	150257	28
Difference	24077	2
Fractional difference	0.1602	0.0714
Ratio (F)	2.24	

The P value is 0.1248. If the one-site model were correct, and you performed many experiments like this one, you'd find that data fit a two-site model this much better than a one-site model in about 12% of the experiments (about 1 in 8). Since the P value is higher than the traditional threshold of 0.05, we conclude that we do not reject the null hypothesis. The data do not fit a two-site model significantly better than they fit a one-site model.

Here are the AIC_c calculations.

Model	SS	N	Pars	AIC_c	Probability	Evidence Ratio
One site	174334	33	3	290.88	61.03 %	1.57
Two site	150257	33	5	289.98	38.97 %	

The one-site model is more likely to be correct. But the evidence ratio is only 1.57. With such a small evidence ratio, you can't be very sure that the one-site model is right. The data are simply ambiguous.

Example 3. Can't get a two-site binding model to fit at all.

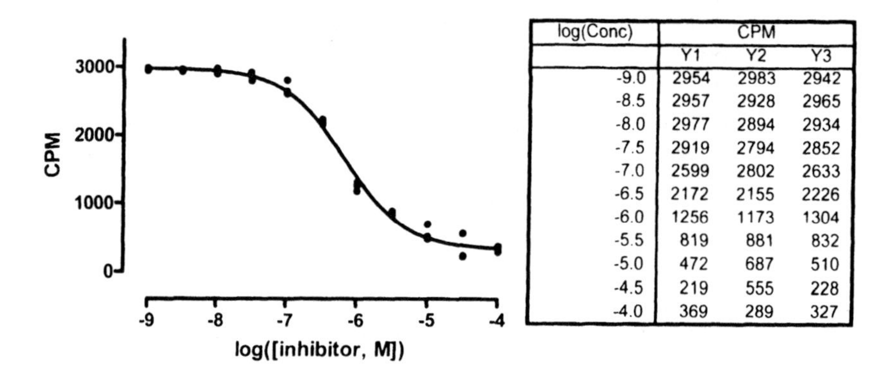

log(Conc)	CPM		
	Y1	Y2	Y3
-9.0	2954	2983	2942
-8.5	2957	2928	2965
-8.0	2977	2894	2934
-7.5	2919	2794	2852
-7.0	2599	2802	2633
-6.5	2172	2155	2226
-6.0	1256	1173	1304
-5.5	819	881	832
-5.0	472	687	510
-4.5	219	555	228
-4.0	369	289	327

Here the data fit the one-site model just fine, resulting in the curve shown in the graph.

When we attempted to fit the data to a two-site model, the nonlinear regression program said "cannot converge". We tried several sets of initial values for the parameters, but kept seeing that error message. The two-site model that we used was constrained such that the parameter *Fraction*, which determines the fraction of the high-affinity site, must have a value between 0.0 and 1.0. When we lifted the constraint, the program was able to fit the model, but with a best-fit value of fraction equal to -0.3. This makes no sense, so that fit is nonsense.

Since we can't fit the data to a sensible two-site model, we accept the one-site model. The problems of fitting to a two-site model are not due to program bugs or invalid choices. Rather, the data simply don't fit the two-site model, so it should be rejected. The problems we had fitting the model provided useful information – that the data do not support the two-site model.

26. Testing whether a parameter differs from a hypothetical value

The approaches for comparing models described in the previous chapters can also be used to perform statistical testing on individual model parameters. The key is to recast the problem as a choice between two models, one where the parameter of interest is estimated, and another where the same model is refitted to the data with the parameter of interest constrained to a specified value. This concept is illustrated with an example below.

Example. Is the Hill slope factor statistically different from 1.0?

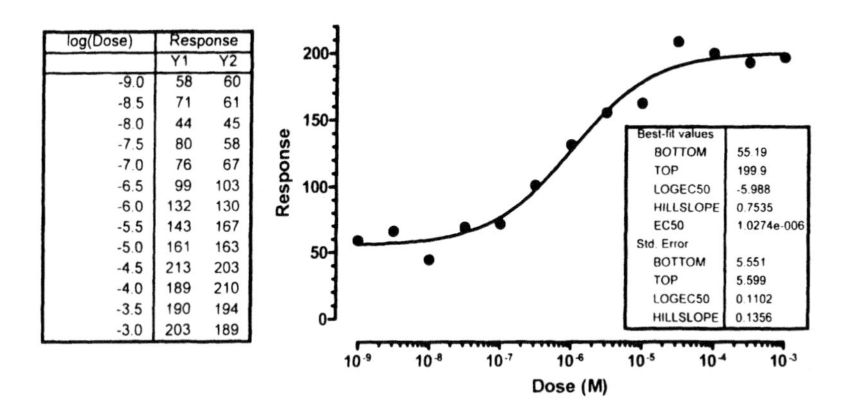

log(Dose)	Response	
	Y1	Y2
-9.0	58	60
-8.5	71	61
-8.0	44	45
-7.5	80	58
-7.0	76	67
-6.5	99	103
-6.0	132	130
-5.5	143	167
-5.0	161	163
-4.5	213	203
-4.0	189	210
-3.5	190	194
-3.0	203	189

Dose-response curves are defined by four parameters – the top plateau, the bottom plateau, the EC_{50} (the dose of drug that cause a response midway between top and bottom plateaus) and the slope factor, or Hill slope (which defines how steep the curve is). When the response is closely coupled to occupancy of a receptor, you may expect the slope factor to equal 1.0.

In this example, we ask whether the Hill slope is significantly different from 1.0.

Compare models with the F test

The null hypothesis is that the standard Hill slope of 1.0 is correct. To fit the model corresponding to the null hypothesis (simpler model), you constrain the slope to equal 1.0; you are thus fitting a three-parameter Hill equation (*Top, Bottom*, and $LogEC_{50}$ being estimated). The alternative hypothesis is that the Hill slope has some other value. To fit this model, fit the variable slope sigmoidal dose-response curve equation (also called a four-parameter Hill equation). The program will find best-fit values of *Top, Bottom,* $LogEC_{50}$ and *HillSlope*. For these data, the best-fit Hill slope is 0.7535. Is the difference between the fitted Hill slope value of 0.7535 and the hypothetical value of 1.0 more than we'd expect by chance?

Use the F test to find out.

Model	SS	df
Null hypothesis (Hill=1.0)	3504	23
Alternative hypothesis (Hill unconstrained)	3080	22
Difference	76	1
Fractional difference	13.77%	4.55%
Ratio (F)	3.03	

The data fit better (had a lower sum-of-squares) when the Hill slope was unconstrained. The program had one more way to adjust the curve to fit the data, so of course the sum-of-squares is better. In this example, the relative difference is 13.77%. The corresponding difference in df is 4.55%.

The sum-of-squares increased relatively more than the degrees of freedom. The ratio of the two relative differences (13.77%/4.55%) is the F ratio, which equals 3.03. Allowing the program to find the best-fit value of the Hill slope led to a better fit than fixing it to the standard value of 1.0. The improvement in goodness-of-fit was better than expected by chance, but only three times as good.

The P value is 0.0958. The P value tells us that we'd see this much improvement in sum-of-squares, or more, 9% of the time by chance alone if the true Hill slope really was 1.0. Since this is larger than the conventional cut-off of 5%, we conclude that fitting the Hill slope does not lead to a significantly better fit than fixing the Hill slope to the conventional value of 1.0. Our data do not provide us with convincing evidence to reject the null hypothesis that the Hill slope equals 1.0.

Compare models with AIC_c

We can also compare the models with AIC_c.

Model	SS	N	Pars	AIC_c	Probability	Evidence Ratio
Slope fixed to 1.0	3504	26	3	137.4	46.7%	1.11
Slope unconstrained	3080	26	4	137.1	53.3%	

The more complicated model (Hill slope fit to the data) has the lower AIC_c score, so it is more likely to be correct. But the difference in AIC_c is small, so the evidence ratio is only 1.11. With such a low ratio, you really can't be at all sure which model will turn out to be correct (if you repeated the experiment many times). All you can say is that the model with the slope free to vary is slightly more likely to be correct.

The AIC_c method does not follow the paradigm of statistical hypothesis testing. It is not designed to make a decision for you. Rather it tells you which model is more likely to be correct, and how much more likely. For this example, the model with the variable Hill slope is more likely, but not much more likely. The best conclusion from these results is that the data simply don't define the Hill slope well enough to know whether it is indistinguishable from 1.0.

Compare with t test

There is an alternative approach that is sometimes used to test whether a parameter value differs significantly from a hypothetical value. You can test this using a one-sample t test.

The best-fit value of the Hill slope is 0.7535 with a standard error of 0.1356 (reported by the nonlinear regression program). Calculate t using this equation:

$$t=\frac{|\text{Best-fit value-Hypothetical value}|}{\text{SE}}=\frac{|0.7535-1.00|}{0.1356}=1.816$$

Determine a two-tailed P value for t=1.691 and 22 degrees of freedom (26 data points minus 4 parameters). Use a statistics table, the GraphPad QuickCalc web calculator, or this Excel equation.

```
=TDIST(1.6816,22,2)
```

The P value is 0.1068. Since this is higher than the traditional cut-off of 0.05, you conclude that the Hill slope does not differ significantly from its standard value of 1.0.

This t test is only valid if two dubious assumptions are true:

- The distribution of the Hill slope is Gaussian. In other words, if you collected (or simulated) many data sets and fit each one, you are assuming that the distribution of Hill Slopes would follow a Gaussian distribution. If it doesn't, then the t test won't be reliable and the P value won't be exactly right.

- The standard error of the Hill slope reported by the nonlinear regression program is correct. In fact, nonlinear regression programs report approximate or asymptotic standard errors. In most cases, they are quite accurate. In some cases they are not.

Because these two assumptions may not be entirely appropriate, the P value obtained from the t test is not exactly the same as the P value obtained from the F test. The F test assumes that the scatter of the data points around the curve follows a Gaussian distribution. The t test only sees the Hill slope and its standard error, so it assumes that the Hill slope would follow a Gaussian distribution (if the experiment were repeated many times). This is a more dubious assumption.

We suggest that you avoid using the t test to compare parameter values. Instead use the F test (or AIC) to compare a model where the parameter's value is fixed to a model where the parameter's value is fit.

27. Using global fitting to test a treatment effect in one experiment

The logic underlying model comparisons described in the previous two chapters can also be extended to instances where you might want to compare one or more parameters of the same model applied to different data sets. Some examples are shown below.

Does a treatment change the EC_{50}?

Below is a graph of a dose-response curve in control and treated conditions. We want to know if the treatment changes the EC_{50}. Are the two best-fit $\log EC_{50}$ values statistically different?

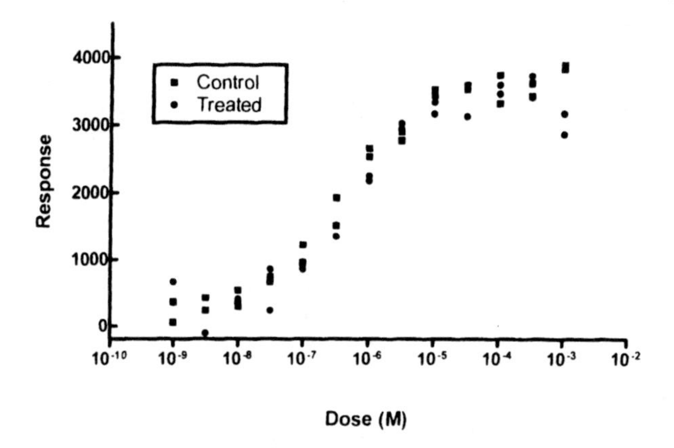

One hypothesis is that both data sets have the same EC_{50}. We fit that model by doing a global fit of both data sets. We fit two dose-response curves, while sharing the best-fit value of the $\log EC_{50}$. Fitting this model requires a program that can do global fits with shared parameters.

Here is the graph showing the curves with distinct top and bottom plateaus, distinct Hill slopes, but a single (shared) $\log EC_{50}$.

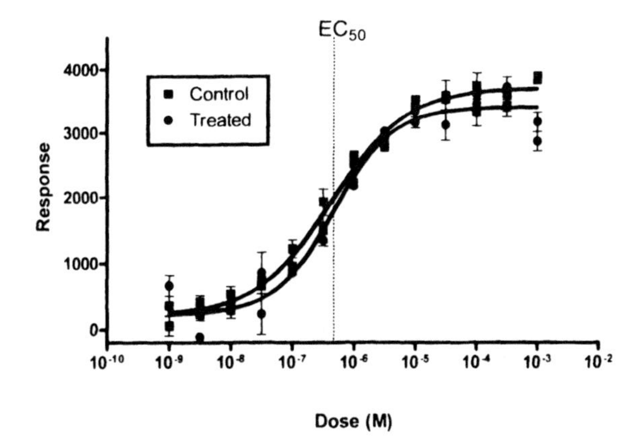

The alternative hypothesis is that the EC_{50} values are distinct. We fit that model by fitting entirely separate dose-response curves to each data set. The two dotted lines show the two EC_{50} values, which are quite close together.

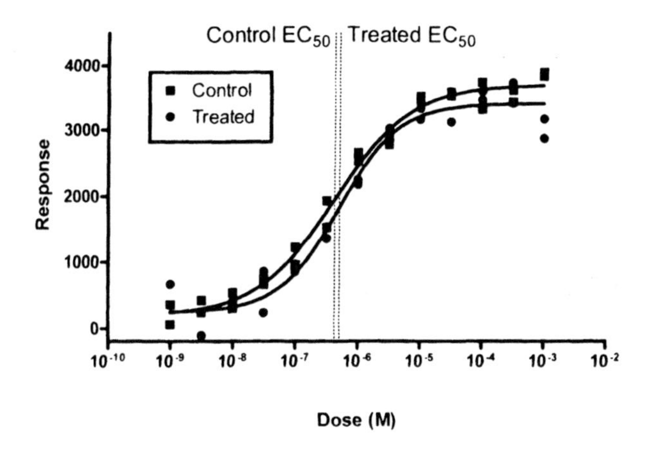

Compare with F test

Here is the comparison via the F test:

Model	SS	df
Null hypothesis (shared $\log EC_{50}$)	2,097,294	45
Alternative hypothesis (separate $\log EC_{50}$)	2,077,538	44
Difference	19,756	1
Relative difference	0.951%	2.273%
Ratio (F)	0.42	

The relative change in SS is smaller than the relative change in df, so the P value must be high. In fact, it is 0.52. If you performed many experiments where the treatment really did not change the EC_{50}, you'd find that the model with separate distinct $logEC_{50}$ values fit the data this much better (by chance) in 52% of the experiments. We conclude that there is no significant difference between $logEC_{50}$ values.

Compare with AIC_c

Here is the comparison via AIC_c.

Model	SS	N	K	AIC_c	Probability	Ratio
Shared $logEC_{50}$	2,097,294	52	7	570.80	77.25%	3.40
Separate $logEC_{50}$	2,077,538	52	8	573.25	22.75%	

The model with a shared $logEC_{50}$ has a lower AIC_c, so is more likely to be correct. But the shared model is only 3.40 times more likely to be correct than a model with separate $logEC_{50}$ values for each curve. Both models are supported by the data. The data are simply ambiguous.

Compare with t test

If your nonlinear regression program cannot fit models with shared parameters, you won't be able to use either the F test or the AIC_c approaches discussed above. Here is an alternative approach that you can use. First let's tabulate the best-fit value and SE from each data set when the data sets were fit separately.

Data set	Best-fit $logEC_{50}$	SE	df
Control	-6.373	0.07034	22
Treated	-6.295	0.09266	22

You can compare the two best-fit values for the $logEC_{50}$ using a t test. It works much like the unpaired t test for comparing means. The t ratio is calculated as:

$$t=\frac{|logEC_{50control}-logEC_{50treated}|}{\sqrt{SE^2_{control}+SE^2_{treated}}}=\frac{0.078}{0.1163}=0.67$$

You can find the associated P value using Excel. There are 44 degrees of freedom, and we want a two-tailed P value.

```
=TDIST(0.67,44,2)
```

The P value is 0.51. The null hypothesis is that the two $logEC_{50}$ values are really the same, and any discrepancy is due to random variation. If this null hypothesis were true, there is a 51% chance (the P value) of obtaining as large (or larger) a difference as you observed. Since this P value is so high, you conclude that there is no evidence to persuade you that the two $logEC_{50}$ values are different. The preferred model is one with a shared $logEC_{50}$.

This t test is only accurate if you assume that the value of the parameter follows a Gaussian distribution (if you were to collect or simulate many data sets, and look at the distribution of best-fit values). You must also assume that the standard errors of the best-fit values are accurate. Since these two assumptions can be dubious in some situations, we suggest you avoid using the t test approach to compare data sets.

Does a treatment change the dose-response curve?

Below is a graph of the same dose-response curves used in the previous example. We want to know if the treatment does anything. Are the two curves statistically different?

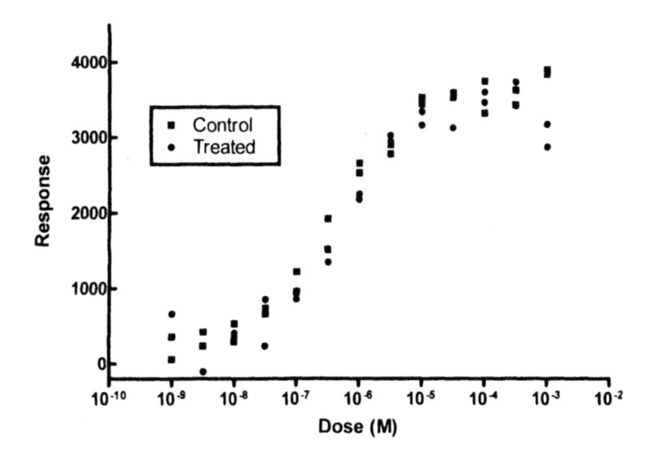

In the previous example, we addressed this problem by focusing on a particular parameter of interest that we expected to be changed by the treatment ($logEC_{50}$).

If you aren't looking for a change in a particular parameter, you can recast the question to globally compare two different data sets to the same model. We'll compare the curves by comparing two versions of the same model, the four-parameter Hill equation (often called four-parameter logistic equation).

The null hypothesis (first, simpler, model) is that one curve fits all the data points (*both* data sets) and the difference we see is purely due to chance. In essence, we ignore which points are controls and which are treated, and fit a single curve to all the data and obtain *one* estimate for *each* of the four parameters in the model (*Top*, *Bottom*, $logEC_{50}$ and *HillSlope*).

The easiest way to fit the null hypothesis is to constrain the four parameters so the values are shared between the two data sets.

> GraphPad note: This is easy with Prism. Go to the constraints tab of the nonlinear regression dialog, and choose to make all four parameters shared.

If your computer program does not allow you to share parameters, you can still fit the model. Enter all the values into one large data set, and then fit a single curve to that combined data set. Each experiment was performed with 13 doses and duplicate determination, so had 26 values. The combined data set, therefore, had twice that or 52 values.

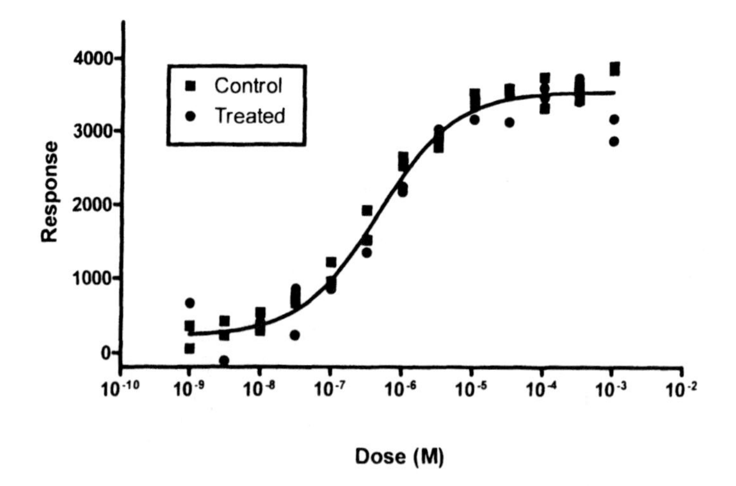

The alternative model is that the curves are distinct. We fit each data set separately to obtain *two* distinct curves with four parameters for *each* data set.

To fit the alternative hypothesis, we separately fit the control data and treated data. The sums-of-squares are 608,538 and 1,469,000, respectively. Add those two together to get the sum-of-squares for the alternative hypothesis (which is 2,077,538). Each experiment was performed with 13 doses and duplicate determination, so had 26 values. Since nonlinear regression fit four parameters to each data set, each curve had 26-4, or 22 degrees of freedom. Since the two fits are separate, the overall alternative model has 22+22=44 degrees of freedom.

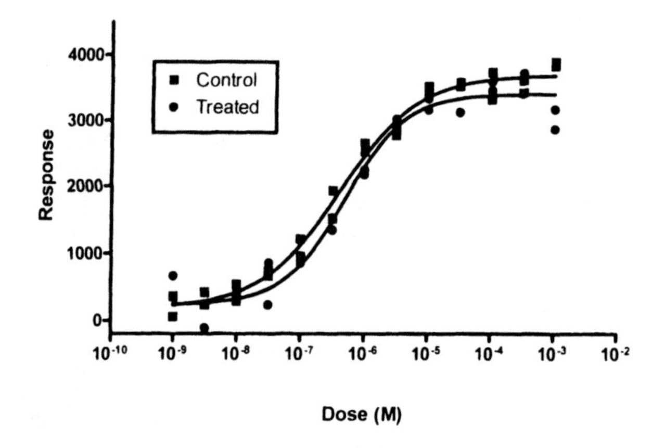

Comparing with an F test

The models are nested, so it is appropriate to use the F test. Here are the results:

Model	SS	df
Null hypothesis (one curve for all data sets)	2,602,000	48
Alternative hypothesis (separate curves)	2,077,538	44
Difference	524,4462	4
Relative difference	25.24%	9.09%
Ratio (F)	2.77	

The relative difference in SS was larger than the relative difference in df, so the F ratio is greater than 1.0. For this example, F=2.77, with 4 and 44 degrees of freedom. The corresponding P value is 0.0386.

The P value is easy to interpret. If the treatment were entirely ineffective so the two dose-response curves are really identical, such a large a difference in the sum-of-squares (or larger) would occur in only 3.86% of the experiments. Since this is less than the traditional threshold of 5%, we reject the null hypothesis and conclude that the separate (not shared) fits are significantly better. The effect of the treatment on the dose-response relationship was statistically significant.

Reality check: Look at the curves, which are very similar. In most experimental contexts, you wouldn't care about such a small difference, even if you are sure that it is real and not due to chance. Yes, the P value is small, so the difference is unlikely to be due to chance. But there is more to science than comparing P values with arbitrary thresholds to decide whether something is "significant".

Comparing with AIC_c

Here are the same data, compared with AIC_c.

Model	SS	N	K	AIC_c	Probability	Ratio
One shared curve	2602000	52	4	573.97	41.05%	1.44
Separate curves	2077538	52	8	573.25	58.95%	

The model with separate curves for each data set has the smaller AIC_c value, so it is preferred. It is 1.44 times more likely to be correct than the simpler model with one shared curve for both data sets. Because the evidence ratio is so low, we aren't very sure about the conclusion. If we repeated the experiment, it wouldn't be surprising to find that the model with the shared curve was more likely.

Given the number of data points and the scatter, our data are simply ambiguous. The model with separate curves is more likely to be correct, so we'll choose that model. But the evidence is not strong, so the conclusion is very weak.

28. Using two-way ANOVA to compare curves

Situations where curve fitting isn't helpful

The graph below shows a typical situation. You've measured a response at four time points, and compared control and treated conditions.

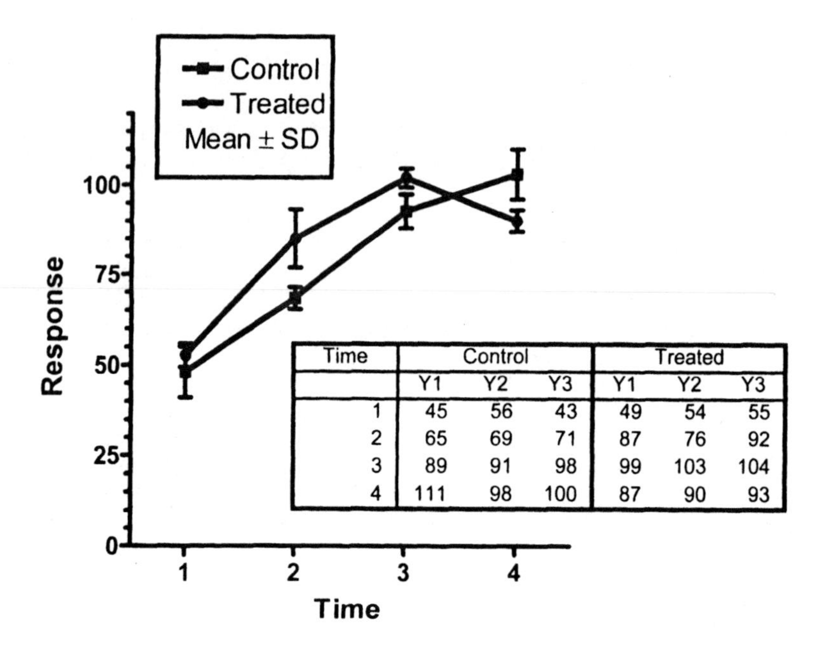

Time	Control			Treated		
	Y1	Y2	Y3	Y1	Y2	Y3
1	45	56	43	49	54	55
2	65	69	71	87	76	92
3	89	91	98	99	103	104
4	111	98	100	87	90	93

You could fit to a simple model, say fit a straight line, but the response may not be linear with time. You'd need some theory to help pick a model, but may not have any theoretical reasons to pick a model. Even if you can pick a sensible model, you may have too few data points for nonlinear regression to work.

In this situation, many biologists analyze the data with two-way analysis of variance (ANOVA) as explained below.

Introduction to two-way ANOVA

Two-way ANOVA, also called two-factor ANOVA, determines how a response is affected by two distinct factors. For example, you might measure a response to three different drugs in both men and women. In this situation, two-way ANOVA partitions the overall variability of the data into four parts:

- Due to differences between men and women.
- Due to differences among the three drugs.
- Due to interaction. This is variation due to inconsistent differences between men and women with different drugs.

- Due to random variability.

ANOVA compares the first three sources of variation with the amount of variation due to random variability, and uses that to determine a P value. Thus it reports three different P values testing three different null hypotheses.

The first null hypothesis is that there is no systematic difference between the different levels of the first factor. In this example, the null hypothesis is that there is no difference between the average result with men and woman, and any difference you see is due to chance. If this P value is low, you conclude that there is a significant difference.

The second null hypothesis is that there is no systematic difference between the levels of the second factor. In our example, the null hypothesis is that the average result is the same with all three drugs, and that any difference you see is due to chance. If this P value is small, you conclude that there is a significant variation among drugs.

The third null hypothesis is that the two factors don't interact. In our example, this hypothesis is that any difference between men and women are the same for all three drugs. Or equivalently, any difference among the drugs is the same for men and for women. If this P value is small, you reject this null hypothesis of no interaction.

How ANOVA can compare "curves"

ANOVA analyzes data where the outcome (dependent variable) is a continuous variable such as enzyme activity, weight, or concentration. Each of the two factors is usually categorical, for example: male or female; drug A, B or C; or wild type or mutant. ANOVA is not usually used to assess the effects of a continuous variable, such as time or dose. Such data are usually analyzed by regression.

But in some situations, such as the one shown at the beginning of the chapter, you only have collected data at a few time points or a few concentrations of a drug, and you really can't use any curve-fitting procedure. An alternative is to use ANOVA, and treat the time points (or dosages) as if they were a categorical variable.

Two-way ANOVA partitions the overall variability of the data shown at the beginning of the chapter into four parts:

- Due to differences between control and treated.
- Due to differences among time points. Note that ANOVA treats different time points just like it would treat different drugs or different genotypes – there is no concept of trend.
- Due to interaction. This is variation due to inconsistent differences between controls and treated at different time points.
- Due to random variability.

Here are the ANOVA results, showing how the variability is partitioned, along with the three P values.

Source of Variation	% of total variation	P value
Treatment	1.16	0.0572
Time	87.28	P<0.0001
Interaction	7.12	0.0013
Random	4.43	

The first P value is just above the traditional significance threshold of 5%. So the effect of the treatment does not appear to be statistically significant overall. The second P value is tiny, so we reject the null hypothesis that time doesn't matter. No surprise there. The third P value is also small. This tells us that there is a statistically significant interaction between treatment and time – the effect of the treatment is not consistent at all time points. Since this interaction is statistically significant, it makes sense to look further. This is done by performing Bonferroni post tests at each time point, as explained in the next section.

For completeness, here is the entire ANOVA table:

Source of Variation	df	SS	MS	F
Interaction	3	716.5	238.8	8.568
Treatment	1	117.0	117.0	4.199
Time	3	8782	2927	105.0
Residual	16	446.0	27.88	

Post-tests following two-way ANOVA

The Bonferroni post tests at each time point determine whether the difference between control and treated is statistically significant at each time point, and generate a 95% confidence interval for the difference between means at each time point. Here are the results:

Time	Difference	95% CI of diff.	P value
1	4.667	-7.460 to 16.79	P>0.05
2	16.67	4.540 to 28.79	P<0.01
3	9.33	-2.793 to 21.46	P>0.05
4	-13.00	-25.13 to -0.8733	P<0.05

The significance levels in the last column correct for multiple comparisons. The response is significantly higher at time point 2, significantly lower at time point 4, and not significantly different at the other two time points.

If your program doesn't perform the post-test automatically (Prism does), it is quite easy to calculate them by hand. The method we summarize below is detailed in pages 741-744 and 771 in J Neter, W Wasserman, and MH Kutner, *Applied Linear Statistical Models*, 3rd edition, Irwin, 1990.

For each time point or dose, calculate:

$$t=\frac{|\text{mean}_1\text{-mean}_2|}{\sqrt{MS_{\text{residual}}\left(\frac{1}{N_1}+\frac{1}{N_2}\right)}}$$

The numerator is the difference between the mean response in the two data sets (usually control and treated) at a particular dose or time point. The denominator combines the number of replicates in the two groups at that dose with the mean square of the residuals (sometimes called the mean square of the error), which is a pooled measure of variability at all doses. Your ANOVA program will report this value, which it might call MS_{error}.

Statistical significance is determined by comparing the t ratio with a critical value from t distribution, which we call t*. Its value depends on the number of df shown in the ANOVA table for $MS_{residual}$, and on the number of comparisons you are making. You can get the critical t value from Excel using the formula below, where DF is the number of degrees of freedom for the residual or error (as reported by the ANOVA program), Ncomparisons is the number of comparisons you are making (usually equal to the number of time points or doses in the analysis), and 0.05 is the significance level (traditionally set to 0.05, but you could pick another value).

```
=TINV(0.05/Ncomparisons,DF)
```

For example data, there are four comparisons and 16 degrees of freedom (from the ANOVA table for residual or error). Using Excel, t* equals 2.813

If a t ratio is higher than the threshold value computed by the Excel equation above, then the difference is statistically significant. If a t ratio is less than the threshold, then that difference is not statistically significant.

The equation for finding the critical value of t includes the Bonferroni correction by dividing the significance level (the P value that you consider to be significant, usually 0.05) by the number of comparisons. This correction ensures that the 5% probability applies to the entire family of comparisons, and not separately to each individual comparison.

Confidence intervals are computed using this equation:

$$\text{Span}=t^*\cdot\sqrt{MS_{\text{residual}}\left(\frac{1}{N_1}+\frac{1}{N_2}\right)}$$

$$95\%\text{ CI: }\left[(\text{mean}_2\text{ - mean}_1)\text{ - Span}\right]\text{ to }\left[(\text{mean}_2\text{ - mean}_1)+\text{Span}\right]$$

The critical value of t is abbreviated t* in that equation (not a standard abbreviation). Its value does not depend on your data, only on your experimental design. It depends on the number of degrees of freedom and the number of rows (number of comparisons) and can be calculated using the Excel equation presented above.

Post-tests following repeated-measures two-way ANOVA use exactly the same equation if the repeated measures are by row. If the repeated measures are by column, use $MS_{subject}$

rather than $MS_{residual}$ in both equations above, and use the degrees of freedom for subjects (matching) rather than the residual degrees of freedom.

The problem with using two-way ANOVA to compare curves

It isn't easy to make sense of the post-tests from our sample data. What does it mean that you get significant differences at time points 2 and 4, but not at time points 1 and 3? This is an intrinsic problem of this kind of analysis. Focusing on differences at particular time points (or doses) does not really address the scientific question you care about. It would be better to compare rate constants, plateaus, area-under-the-curve, or some other integrated measure of the effect of the treatment. Asking about differences at individual time points is not always helpful, and can lead to confusing results.

The fundamental problem is that the ANOVA calculations treat different time points (or different doses) just like they treat different species, different genotypes, or different countries. ANOVA treats different time points, to use statistical lingo, as different levels of a factor. The fact that the time points have an order is totally ignored. In fact, we'd get exactly the same two-way ANOVA results if we randomly scrambled the time points. The figure below rearranges the time points, yet the ANOVA results are identical.

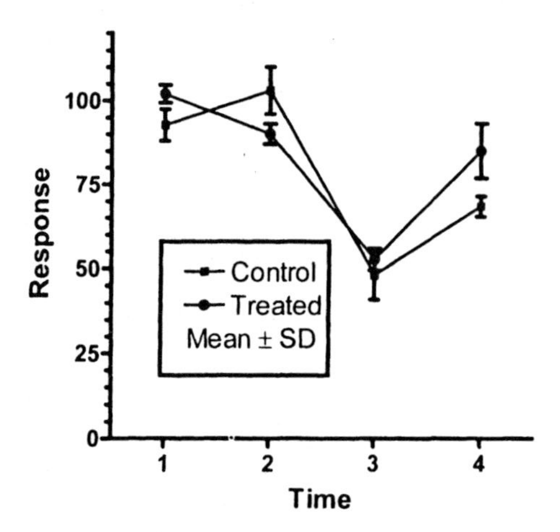

Two-way ANOVA can be a useful way to compare curves in some situations. But it rarely answers the questions you asked when you designed the experiment. Use this technique with caution.

> Tip: Beware of using two-way ANOVA to analyze data where one of the factors is time or dose. The ANOVA calculations ignore the concept of a trend. In other words, the ANOVA calculations ignore the entire point of the experiment. Whenever possible, use curve fitting rather than ANOVA.

29. Using a paired t test to test for a treatment effect in a series of matched experiments

The advantage of pooling data from several experiments

Chapter 25 explained how to compare two curves obtained in one experiment. But if you really want to know how a treatment changes the parameters of a curve, you'll want to repeat the experiment several times. This chapter explains how to pool the results from several experiments.

The first step is to focus on what you really want to know. In most cases, you'll care most about a best-fit parameter from curve fitting. For example, with dose-response curves, you will want to test whether the $\log EC_{50}$ is altered significantly by a treatment. With kinetic curves, you'll ask about changes in the rate constant.

In some cases, you may wish to quantify the result in some other way rather than the best-fit value of a parameter determined by curve fitting. You may wish to tabulate the maximum response, the minimum response, or the time to maximum. Or you may tabulate the area under the curve as an overall measure of cumulative response. Since these values are not determined by nonlinear regression, you won't be able to use the methods based on global fitting described in the next chapter. However, you may still use the methods based on the t test.

An example. Does a treatment change $\log EC_{50}$? Pooling data from three experiments.

Here is an example. We performed a dose-response experiment three times, each time before and after a treatment. To keep things simple (but not too unrealistic), all data were normalized from 0 to 100, and we assumed that the slope would be standard (Hill slope =1). So we ask the program to fit only one parameter for each curve, the $\log EC_{50}$.

Here are the data:

	X Values	A		B		C		D		E		F	
	Log(Conc)	Experiment 1 Control		Experiment 1 Treated		Experiment 2 Control		Experiment 2 Treated		Experiment 3 Control		Experiment 3 Treated	
	X	A:Y1	A:Y2	B:Y1	B:Y2	C:Y1	C:Y2	D:Y1	D:Y2	E:Y1	E:Y2	F:Y1	F:Y2
1	-9.0	-1.1	-1.6	2.6	0.4	2.4	-9.8	8.5	16.2	7.8	12.8	5.6	3.9
2	-8.5	2.7	-12.3	-0.7	-6.2	-5.6	-4.0	-6.8	21.4	-6.7	2.2	5.5	9.6
3	-8.0	-6.6	7.1	13.3	2.3	16.9	2.7	-4.3	-8.6	20.1	0.7	-5.1	10.3
4	-7.5	11.1	4.6	-1.4	8.2	2.9	5.2	-8.4	8.0	9.0	12.4	7.2	5.7
5	-7.0	7.3	8.8	17.0	12.9	16.5	5.3	-0.1	0.3	-7.5	18.3	9.3	0.4
6	-6.5	34.8	36.8	16.4	14.8	29.0	24.6	9.7	2.0	12.9	23.9	-0.7	8.1
7	-6.0	60.7	67.3	36.9	34.3	58.4	56.7	19.7	30.0	47.4	45.7	27.3	45.1
8	-5.5	84.2	83.8	77.2	62.4	91.6	89.9	59.1	55.7	63.3	69.1	58.2	61.9
9	-5.0	89.5	81.2	94.0	98.1	88.6	93.7	89.1	83.6	90.5	83.9	78.6	82.4
10	-4.5	95.7	91.1	104.1	90.5	116.2	89.4	83.4	93.4	100.4	102.9	93.1	95.1
11	-4.0	96.6	113.3	104.8	92.5	100.8	110.2	91.9	109.9	93.5	106.4	100.7	98.7
12	-3.5	97.9	102.7	96.0	104.0	96.4	93.2	105.9	98.6	107.0	90.2	92.4	97.6
13	-3.0	101.5	108.9	106.4	105.3	95.9	96.3	96.9	113.4	107.3	107.8	108.6	95.6

And here is a graph of the data with a curves fit individually (six separate curve fits).

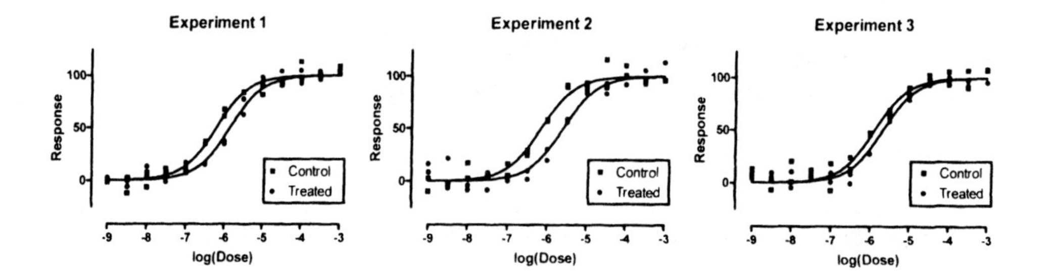

Using the technique of Chapter 27, we can compare controls vs. treated in each experiment individually. For the first two experiments, this comparison yielded a P value of less than 0.0001. For the third experiment, the P value is 0.0135. So in each experiment individually, it is clear that the treatment significantly altered the $logEC_{50}$.

Comparing via paired t test

The simplest approach to pooling the data is to perform a paired t test on the parameter of interest.

> Note: As you'll see in the next chapter, this approach is not the most powerful way to analyze the data.

Before choosing any statistical test, you should always review its assumptions. The paired t test assumes that your experiment was designed with the pairing in mind, that the pairs are independent, and that the distribution of differences between pairs (if you repeated the experiment many times) is Gaussian. These assumptions all seem reasonable for our data.

The figure below shows how the $logEC_{50}$ values of our example are compared by paired t test.

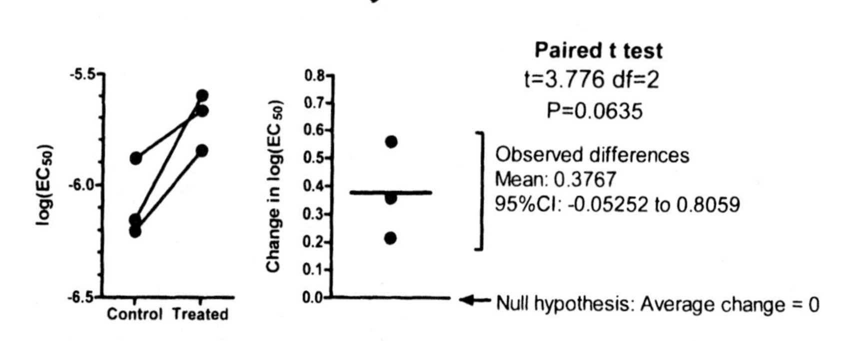

For this approach, only a single number was recorded for each experiment– the $logEC_{50}$. The paired results are shown in the left panel above, as a before-after graph. The paired t test works to simplify the data even more by subtracting the control $logEC_{50}$ from the treated $logEC_{50}$. These differences are plotted in the right panel above. The paired t test is computed entirely from these three values. Divide the mean of these differences by the standard error of the mean to compute the t ratio. To compute a P value, you have to know the number of degrees of freedom. In a paired t test, the number of degrees of freedom equals the number of pairs minus 1. So in this experiment, there are two degrees of freedom. From the t ratio and the number of df, you can find the P value using a statistical table or this Excel formula (the third parameter is 2 because you want a two-tailed P value):

```
=TDIST(3.776,2,2)
```

The result, the P value, is 0.0635. To interpret any P value, you need to state the null hypothesis. Here, the null hypothesis is that the treatment has no effect so the mean difference is really zero. In other words, the null hypothesis is that individual animals are equally likely to have increases or decreases in the $\log EC_{50}$. If the null hypothesis were true, the chance of observing a mean difference as large or larger than we observed is just over 6%.

The P value of 0.06 is slightly higher than the traditional threshold of 0.05. Using the traditional significance level, we would say that the difference observed in this set of experiments is not statistically significant. We do not have sufficient evidence to reject the null hypothesis.

The mean difference between control and treated $\log EC_{50}$ is 0.3767.The 95% confidence interval for that difference extends from -0.05252 to 0.8059. These numbers are for the difference between the $\log EC_{50}$ of control and the $\log EC_{50}$ in the treated group. The difference between two logarithms is the same as the logarithm of the ratio. So that value (0.3767) is also the log of the ratio of the treated EC_{50} divided by the control EC_{50}. Take the antilog, to convert to the ratio. The antilog ($10^{0.3767}$) is 2.381. So the treatment increases the EC_{50} by a factor of 2.4. Take the antilog of both ends of the confidence interval of the difference in $\log EC_{50}$ to express the confidence interval as a ratio. The 95% confidence interval for the ratio ranges from 0.8861 to 6.396. The 95% confidence interval includes 1 (no change), which is consistent with a P value greater than 0.05.

Why the paired t test results don't agree with the individual comparisons

When we analyzed each experiment independently, we found that the effect of the treatment in each of the three experiments was statistically significant. In two of the experiments, the P value was less than 0.0001. In the remaining experiment it was 0.017.

Now we pooled the three experiments using a paired t test, and the P value is 0.06. This doesn't make sense. The data in each experiment were sufficient to convince us that the treatment worked. Put all three experiments together and now the data are not so convincing. What's wrong?

The problem is that the paired t test only takes into account best-fit values in each experiment. The paired t test doesn't extend the method used to compare one pair of curves. Rather, it throws away most of the information, and only looks at the best-fit value from each curve. The next chapter explains how to account for all the information in the series of experiments.

30. Using global fitting to test for a treatment effect in a series of matched experiments

Why global fitting?

The previous chapter used a paired t test to pool the results of three experiments. But the results were not satisfying. When we compared treated with control for each experiment individually, each P value was less than 0.05. With each individual experiment, the evidence is convincing. The effect of the treatment was statistically significant in each of the three experiments. But when we pooled all three experiments and analyzed with a paired t test, the result was not statistically significant.

The problem is that the paired t test analyzes only the best-fit values, in our example the $\log EC_{50}$. It doesn't take into account the other information in the experiment.

What we want is a method that extends the comparison we use to compare control and treated groups in a single experiment to compare a family of experiments. Global curve fitting lets us do this.

Setting up the global model

Here is one way to write a global model that fits all the data and provides a best-fit value for the average treatment effect.

```
<A>logEC50=logEC50A
<B>Effect1 = MeanEffect + Rand1
<B>logEC50=logEC50A + Effect1

<C>logEC50=logEC50C
<D>Effect2 = MeanEffect + Rand2
<D>logEC50=logEC50C + Effect2

<E>logEC50=logEC50E
<F>Effect3 = MeanEffect - Rand1 - Rand2
<F>logEC50=logEC50E + Effect3

Y=100/(1+10^(LogEC50-X))
```

The last line in the model defines a sigmoidal dose-response curve that begins at 0, ends at 100, and has a standard Hill slope of 1.0. Therefore the equation defines Y as a function of X and only one parameter, the $\log(EC_{50})$. However, we need to find a different $\log EC_{50}$ for each data set. The rest of the model (the first nine lines) defines the $\log EC_{50}$ separately for each of the six data sets.

The first line is preceded by "<A>" which (in GraphPad Prism) means it applies only to data set A. It defines the $\log(EC_{50})$ for that data set to be a parameter named logEC50A. (Because "logEC50" is used first on the left side of an equals sign, it is an intermediate variable, and is not a parameter that Prism fits. Because "logEC50A" is used on the right side of the equals sign, it is a parameter that Prism will fit.)

The second line defines the Effect (difference between $\log EC_{50}$ values) for the first experiment. It equals the mean treatment effect plus a random effect for that experiment, so we define the effect to equal MeanEffect + Rand1.

The third line defines the $\log(EC_{50})$ of data set B to equal the $\log(EC_{50})$ of data set A plus the effect of the first experiment.

The next three lines define the second experiment (data sets C and D). This has a new control value (logEC50C) and a different random effect (Rand2).

The final three lines define the third experiment (data sets E and F). Again there is a new control value (logEC50E). The effect is defined a bit differently for this experiment, since it is the last. The effect is defined as the mean effect minus the random effects for the first two experiments. Now if you average the effects for all three experiments, the random effects cancel out and the average effect really is the same as the parameter we call MeanEffect.

$$\text{Average}=\frac{(\text{MeanEffect+Rand1})+(\text{MeanEffect+Rand2})+(\text{MeanEffect-Rand1-Rand2})}{3}$$

$$=\text{MeanEffect}$$

If we wrote the model with three different random effects, the model would be ambiguous. It only requires three parameters (the mean effect, and two random effects) to completely define the three treatment effects.

The full model still has six parameters, as it must. Each experiment has only one parameter, the $\log EC_{50}$, and there are six experiments. This model is written so there are three $\log EC_{50}$ values for the controls, one parameter for the mean effect of the treatment, and two parameters that quantify how each experiment deviates from the average.

Fitting the model to our sample data

To fit this model, we use global curve fitting, sharing the values of all six parameters. This means we'll fit one value for logEC50A that applies to data sets A (before) and B (after), one value of logEC50C value that applies to data sets C (before) and D (after), one value of logEC50E value that applies to data sets E (before) and F (after), one MeanEffect value that applies to all three pairs of experiments, one value of Rand1 that applies to experiments 1 and 3, and one value of Rand2 that applies to experiments 2 and 3.

Here are graphs of the resulting fits:

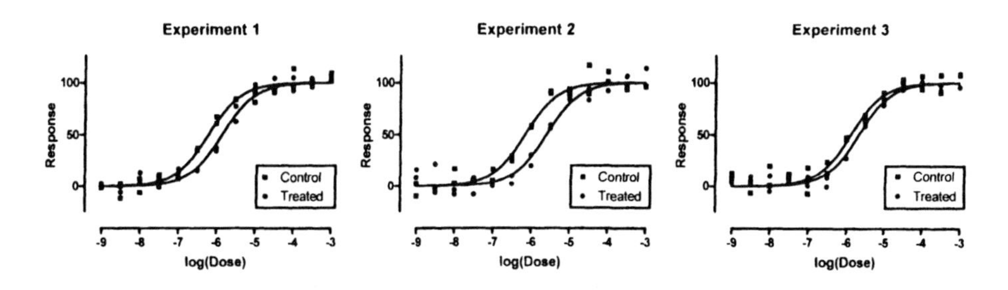

The sum-of-squares equals 7886 with 150 degrees of freedom. Why 150? There are 13 concentrations, with Y values determined in duplicate, so there are 26 data points per curve. There are 6 curves, so there are 156 data points altogether. From that subtract the number of parameters we are fitting (6), resulting in 150 degrees of freedom.

The graphs are identical to fitting the data sets individually as shown in the previous Chapter. This is what we expect since each of the three control curves has its own $\log EC_{50}$, and each of the three experiments has its own effect (difference between control and treated). Rearranging the model to report the parameters arranged differently does not change the appearance of the curves, the sum-of-squares, or the number of degrees of freedom.

Here are the results in tabular form:

Parameter	Best-fit value	SE	95% CI
LogEC50A	-6.205	0.05852	-6.320 to -6.090
LogEC50C	-6.154	0.05852	-6.269 to -6.039
LogEC50E	-5.881	0.05852	-5.996 to -5.767
Rand1	-0.01904	0.06758	-0.1515 to 0.1134
Rand2	0.1812	0.06758	0.04879 to 0.3137
MeanEffect	0.3771	0.04778	0.2835 to 0.4708

Interpreting the results

The best-fit value of MeanEffect is 0.3771. This is the same value calculated in the last chapter when each curve was fit individually. But now we get a 95% confidence interval based on all the data in the experiment. We are 95% sure the treatment effect is between 0.2835 to 0.4708. Because we are now fitting all the data, this confidence interval is tighter (narrower) than the one we computed in the last chapter, which was based solely on the three separate treatment effects.

The MeanEffect is the difference between the $\log(EC_{50})$ values. While that can be hard to interpret, it is easy to convert to a form that is easier to understand. The difference between two logs is exactly the same as the logarithm of the ratio. So the MeanEffect is the logarithm of the ratio of the two EC_{50} values. The antilog of the MeanEffect, which is 2.38, is the mean ratio of the EC_{50} for the treated divided by the EC_{50} of control. In other words, these data show that the treatment increased the EC_{50} by about a factor of two and a half. Now, take the antilog of both ends of the confidence interval of MeanEffect to obtain the confidence interval of the fold change in EC_{50}, which ranges from 1.921 to 2.957.

To summarize: We observed that the EC_{50} increased by a factor of about two and a half, and are 95% sure that change is somewhere between approximately a doubling and tripling.

Why are three standard errors of the logEC50 values all the same?

> Note: This topic is here just to satisfy your curiosity. If you aren't curious, skip it!

Turn your attention to the standard errors of the best-fit parameters. Note that the standard errors of all three $\log EC_{50}$ values are identical. If you display the results to more decimal places, you'll find they are not exactly identical. But they are very close.

To understand why these standard errors have values so close together, let's review how the standard errors of best-fit parameters are computed. The computation is complicated and involves inverting a matrix, but depends on three terms:

- The sum-of-squares of the vertical distances of the data points from the curves. In a global fit, it is the global (total) sum-of-squares that enters the computation. With six curves computed at once, this term is the sum of the sum-of-squares for each curve. There is only one sum-of-square value for all the confidence intervals.

- The number of degrees of freedom, computed as the number of data points minus the number of parameters. In a global fit, it uses the total number of data points and the total number of parameters. Therefore, there is only one df value for all the confidence intervals.

- A term that comes from the variance-covariance matrix. Its value is determined by computing how the Y value predicted by the model changes when you change the value of each parameter a little bit. More specifically, it sums the partial derivative of Y as a function of each parameter, holding the parameters equal to their best-fit value, at all values of X. The result depends on the model, the parameter, the number of data points, and the values of X. In our example, the three control curves all have the same number of data points, with the same X values. These X values are spaced so each data set covers the bottom plateau, the middle, and the top of the curve. So this term is almost the same for all three curves.

The first two terms must be identical for all confidence intervals computed from a global fit. The third term will vary between parameters, but will be almost identical between experiments for a particular parameter (so long as the range of X values covers all important parts of the curve for all treatments). So it makes sense mathematically that the three standard errors for $\log EC_{50}$ values would be nearly identical.

Does it make sense intuitively? Yes. Each experiment has the same number of data points. In each experiment, we have enough X values to define the top, middle, and bottom of the curve. And it is reasonable to assume that the average amount of random scatter is identical in all three experiments. Therefore, it makes sense that the $\log EC_{50}$ is determined with equal precision in each experiment. The global curve-fitting program effectively averaged together the scatter for all three curves, to give one uncertainty value that applies to all three $\log EC_{50}$ values.

Was the treatment effective? Fitting the null hypothesis model.

Was the treatment effective? Are we convinced that the treatment is real? One way to address this question is to look at the 95% confidence interval for the mean effect. This did not include zero (no effect) so we are at least 95% sure the treatment did something. Looking at confidence intervals is a very good way to look at treatment effects, and one could argue the best way. But it is more traditional to compute P values, so we'll do so here.

To compute a P value (or use information theory and find the AICc), we need to compare two models. One model is that the control and treated $logEC_{50}$ values are different – the treatment worked. This model was explained in the previous section. The other model, the null hypothesis, is that the treatment didn't work, so the treated $logEC_{50}$ values, on average, are identical to the control values.

We can fit the null hypothesis model using the same model as before, but constraining the MeanEffect parameter to be constant value of zero. Here are the graphs of the experiments fit this way.

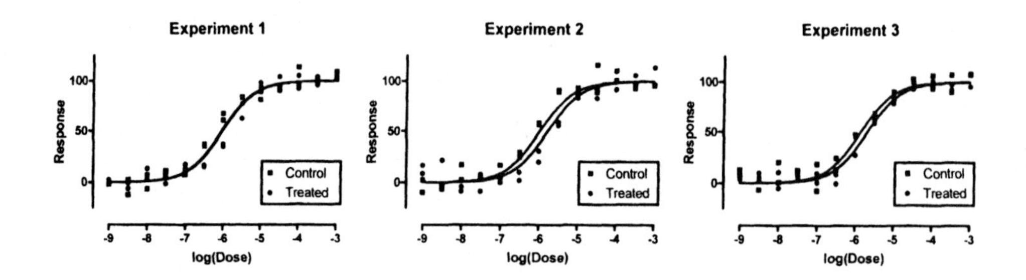

In the first experiment, the control and treated curves are almost identical, but they are not identical in experiments 2 and 3. Our null hypothesis is not that control and treated are identical in each experiment. Rather, the null hypothesis is that, on average, the control and treated EC_{50} values are identical.

The curves don't follow the data very well. That is because we have constrained the mean effect to equal zero.

Here are the results reported by nonlinear regression.

Parameter	**Best-fit value**	**SE**	**95% CI**
LogEC50A	-6.007	0.06361	-6.131 to -5.882
LogEC50C	-5.975	0.06361	-6.100 to -5.851
LogEC50E	-5.682	0.06362	-5.806 to -5.557
Rand1	-0.02808	0.08047	-0.1858 to 0.1296
Rand2	0.2034	0.08047	0.04572 to 0.3612
MeanEffect	0.000	(constant)	

To check that these results are consistent with the null hypothesis, let's compute the treatment effect in each experiment. In experiment 1, the model defines the treatment effect to equal the mean effect (which we now fix to zero) plus Rand1. So the treatment effect (difference between control and treated EC50 values) is -0.02808. For experiment 2, the model defines the treatment effect to equal the mean effect plus Rand 2, which is 0.2034. For experiment 3, the model defines the treatment effect to equal the mean effect minus the sum of Rand1 and Rand2, which is -0.17532. So the average effect of the treatment in all three experiments is (-0.02808 +0.2034 -0.17532)/3, which equals zero as it must under the null hypothesis.

Again, the standard errors are all the same. We fit all three experiments in one global fit. Since we are fitting the same model to experiments with the same number of data points and same X values, all the SE values are the same.

The total sum-of-squares equals 11257 with 151 degrees of freedom. Why 151? There are 13 concentrations, with Y values determined in duplicate, so there are 26 data points per curve. There are 6 curves, so there are 156 data points altogether. From that subtract the number of parameters we are fitting (5), resulting in 151 degrees of freedom.

Comparing models with an F test

We used global fitting to fit the data to two models. One assumes the treatment worked. The other (the null hypothesis) constrained the mean treatment to equal zero.

Here are the calculations of an F test to compare the two models.

Model	SS	df
Null hypothesis (treatment effects are zero)	11257	151
Alternative hypothesis (treatment not zero)	7886	150
Difference	3371	1
Relative difference	0.4275	0.67
Ratio (F)	64.12	

With such a high F ratio, the P value is tiny, way less than 0.0001. If the null hypothesis were true (that all treatment effects were really zero), there is only a slim chance (way less than 0.01%) of seeing such large treatment effects in a series of three experiments of this size. In other words, it is exceedingly unlikely that the difference we observed between control and treated dose-response curves is due to chance.

We conclude that the treatment effects are statistically significant, extremely so.

Comparing models with AIC_c

Alternatively, here are the results analyzed by AIC_c.

Model	SS	N	K	AIC_c	Probability	Ratio
Treatment has no effect	11257	156	5	680.07	<0.01%	3.8×10^{11}
Treatment is effective	7886	156	6	626.74	>99.99%	

The AIC_c for the second model (treatment does have an effect) is much lower, so this model is much more likely to be true. In fact, given our data and these two alternatives, the second model is many billions times more likely to be correct. We have substantial evidence that the treatment altered the $\log EC_{50}$ in our series of three experiments.

Reality check

The result of this set of experiments convinces us that the treatment effect is not zero. Rather, we are fairly sure the treatment effect is somewhere between a doubling and tripling of the EC_{50}. Statistically, we are convinced this is distinct from zero. We are quite sure the treatment did something -- the difference we observed is extremely unlikely to be due to random chance. In other words, we can say the results are statistically significant.

You should always interpret results in the context of the experiment. Do you care about a doubling or tripling of an EC_{50}? In some contexts, this would be considered a small, trivial effect not worth pursuing. Those judgments cannot be made by statistical calculations. It depends on why you did the experiment. Statistical calculations can tell us that the effect is very unlikely to be due to chance. This does not mean it is large enough to be scientifically interesting or worth reporting.

31. Using an unpaired t test to test for a treatment effect in a series of unmatched experiments

An example

We measured enzyme velocity as a function of substrate concentration in the presence and absence of a treatment. We had three control animals and three treated. We obtained tissue from each, and measured enzyme activity at various concentrations of substrate. All six curves were obtained at one time, with no pairing or matching between particular control and treated animals.

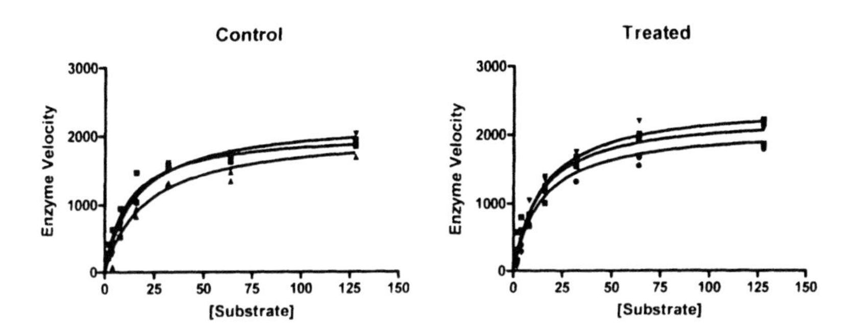

Fitting the curves individually yields these results.

	A:Control	B:Control	C:Control	D:Treated	E:Treated	F:Treated
V_{MAX}	2190	2132	2428	2897	2449	2689
K_m	14.10	22.42	17.04	20.73	16.64	21.96

Using the unpaired t test to compare best-fit values of V_{max}

Conceptually the easiest way to analyze the data is to use an unpaired t test. The graph below shows the six V_{max} values, along with the results of the unpaired t test.

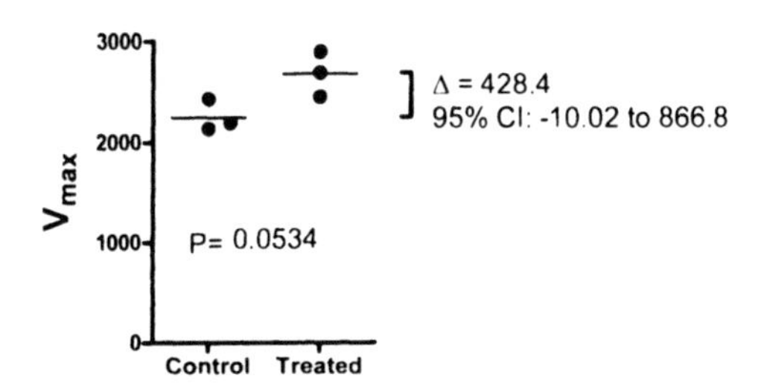

The P value is just above the traditional threshold of 0.05, so the difference is said to not be statistically significant. If the treatment did nothing to the V_{max} of the enzyme, and you repeated this kind of experiment many times (each time with the same doses, the same average amount of scatter, and with three animals in each group), you'd find this large difference (or larger) in a bit more than 5% of the experiments.

The confidence interval for the mean difference between control and treated V_{max} ranges from a decrease of 10 to an increase of 867. This confidence interval spans zero, as it must to be consistent with a P value greater than 0.05.

This t test tells you that the evidence is not strong enough to persuade you that the treatment changes V_{max}. But the results do look intriguing.

The problem with the t test is that it analyzes only the best-fit value of the parameter of interest. It doesn't look at all the data in the experiment. To take into account all the data, we need to use global fitting as explained in the next chapter.

32. Using global fitting to test for a treatment effect in a series of unmatched experiments

Setting up a global fitting to analyze unpaired experiments

Let's continue the example from the previous chapter.

Here is the model to fit the data with global fitting. It uses the syntax of GraphPad Prism.

```
<A>Vmax = MeanCon + Rand1
<B>Vmax = MeanCon + Rand2
<C>Vmax = MeanCon - Rand1 - Rand2

<D>Vmax = MeanCon + Effect + RandT1
<E>Vmax = MeanCon + Effect + RandT2
<F>Vmax = MeanCon + Effect - RandT1 - RandT2

Y=Vmax*X/(Kd+X)
```

The bottom line in the model defines the enzyme kinetics curve.

The first six lines in the model define the V_{max} term for each data set. Data sets A, B, and C were controls. We want to fit the mean V_{max} for the control curves. Curve A has a V_{max} defined to be equal to the mean value for the control (MeanCon) plus a random factor (Rand1). Curve B is defined similarly with a different random factor. Curve C is defined as the mean control value minus the two random factors. If you take the average of the three V_{max} values for curves A, B and C, you'll see that the average really is MeanCon.

$$\text{Average}=\frac{(\text{MeanCon+Rand1})+(\text{MeanCon+Rand2})+(\text{MeanCon-Rand1-Rand2})}{3}$$

$$=\text{MeanCon}$$

Curves D, E, and F are for the treated condition. The mean of the treated is defined to be the mean of the controls plus a variable called Effect. This is the difference between the mean of the controls and the mean of the treated. The variation around this mean is modeled by RandT1 and RandT2.

The model looks complicated, but it needs to be that way to fit a single parameter, Effect, for the mean difference between the control and treated animals. The model has seven parameters to fit: K_m, MeanCon, Rand1, Rand2, Effect, RandT1, and RandT2.

When fitting this model, all the parameters that define the V_{max} and the treatment effect must be shared. But what about K_m? You could argue that it should be shared, since there is no evidence that the treatment changes the K_m, and no theory suggesting it might. Or you could argue that the K_m might really vary between animals based on subtle differences in the pH or ionic composition of the preparations. We chose the latter approach, although the results below would be very similar if we had designated the K_m to be a shared parameter.

Fitting our sample data to the global model

We fit the data using global curve fitting, sharing all parameters except the K_m. Here are the results for the other parameters. Since they are shared, we get one best-fit value for each for the entire model.

Parameter	Best-fit value	SE	95% CI
MeanCon	2250.00	65.50	2119 to 2381
RandC1	-60.27	87.15	-234.2 to 113.7
RandC2	-118.10	99.17	-316.1 to 79.81
Effect	428.30	95.26	238.2 to 618.5
RandT1	218.40	99.07	20.62 to 416.1
RandT2	-228.90	93.29	-415.1 to -42.74

The first thing to do is make sure that the results match the results of the individual fits. This is a way to make sure that the model is correct. The V_{max} for the first data set (A) is defined as MeanCon+Rand1, or 2250 – 60.27, which equals (to four digits of precision) 2190. This is identical to the value we obtained when we fit each curve independently. Similarly, the other V_{max} values all match the values obtained when we fit the curves independently. Using the global model did not change the best-fit curves, just how we organize the results. The sum-of-squares (1145088) of the global fit accordingly matches the sum of the sum-of-squares of the six individual curve fits. Similarly, the degrees of freedom of the global fit (72) match the sum of the degrees of freedom of the six individual curves.

The whole point of this fit is to determine the best-fit value of the treatment effect. The best-fit value is 428. This is the same as the mean difference we calculated for the unpaired t test. No surprise here. The curves are the same, so the difference between the average control V_{max} and the average treated V_{max} has to match. If it didn't, we know there was something wrong with how we set up or fit the global model.

Now let's turn our attention to the confidence interval of the treatment effect. The global fit tells us that we can be 95% sure that the range 228 to 619 contains the true treatment effect. This interval was calculated by considering all of our data. In contrast, the t test only took into account the six V_{max} values so its confidence interval was much wider (-10 to 867).

Finally, let's compute a P value testing the null hypothesis that the treatment really doesn't affect the V_{max} at all. To do this, we compare the fit we just did with one constrained so Effect=0. Here are the results.

Parameter	Best-fit value	SE	95% CI
MeanCon	2505	55.15	2395 to 2615
RandC1	-164.5	110.4	-384.9 to 55.79
RandC2	28.20	137.8	-246.9 to 303.3

Effect	0.0 (fixed)		
RandT1	206.4	100.3	6.248 to 406.5
RandT2	-200.7	94.83	-390.0 to -11.49

Let's check that this fit makes sense.

Parameter	**Calculation**	**Value**	**Mean**
Control A	2505 - 165	2340	
Control B	2505 + 28	2533	2505
Control C	2505 + 165 - 28	2642	
Treated D	2505 + 206	2711	
Treated E	2505 - 201	2304	2505
Treated F	2505 – 206 + 201	2500	

Each curve is fit its own V_{max} value, but the global fitting was constructed to ensure that the average of the control V_{max} values matched the average of the treated V_{max} values (since Effect was constrained to equal 0). The table shows that the fit worked as intended. The two means are the same.

Comparing models with an F test

Here are the calculations of an F test to compare the two models.

Model	**SS**	**df**
Null hypothesis (treatment effect is zero)	1,442,275	73
Alternative hypothesis (treatment not zero)	1,145,087	72
Difference	297,188	1
Relative difference	0.25953	0.01389
Ratio (F)	18.68	

With such a high F ratio, the P value is tiny, much less than 0.0001. If the null hypothesis (that the treatment doesn't change V_{max}) were true, there is only a slim chance (way less than 0.01%) of seeing such large treatment effects in our series of six experiments. In other words, it is exceedingly unlikely that the difference we observed between control and treated V_{max} is due to chance.

We conclude that the treatment effect is statistically significant.

Comparing models with AIC_c

Here are the results analyzed by AIC_c.

Model	SS	N	K	AICc	Probability	Ratio
Treatment has no effect	1,442,275	84	11	847.5	0.03%	3977
Treatment is effective	1,145,087	84	12	830.9	99.97%	

The AIC_c for the second model (treatment does have an effect) is much lower, so this model is much more likely to be true. In fact, given our data and these two alternatives, the second model is almost four thousand times more likely to be correct. We have substantial evidence that the treatment altered the V_{max} in our series of experiments.

Reality check

The global fit convinced us that the difference we observed in V_{max} between control and treated animals is very unlikely to be due to chance. The P value is tiny, so the results are statistically significant. The AIC_c criterion tells us that a model where control and treated have the same average V_{max} is extremely unlikely to be correct.

When interpreting data, one should not focus on statistical significance. Instead, concentrate on the size of the difference. Our treatment effect was 428, compared to control mean of 2228. This means the treatment increased V_{max}, on average, by 428/2228 or only 19.2%. Putting aside statistical calculations of P values, you cannot really interpret these results except by thinking about the reason you did the experiment. Is a 20% change worth caring about? The answer depends on why you did the experiment.

H. Fitting radioligand and enzyme kinetics data

33. The law of mass action

What is the law of mass action?

Analyses of radioligand binding experiments and enzyme kinetics are based on a simple molecular model called the *law of mass action*. This "law" states that the velocity of a chemical reaction is proportional to the product of the concentrations (or mass) of the reactants.

In the simplest case, one kind of molecule is transformed into a different kind of molecule. The rate at which the reaction occurs is proportional to the concentration of the starting material.

$$A \rightarrow Z$$

$$\frac{dA}{dt} = -\frac{dZ}{dt} = k \cdot [A]$$

If two molecules combine, the reaction rate is proportional to the product of the concentrations.

$$A + B \rightarrow Z$$

$$\frac{dZ}{dt} = -\frac{dA}{dt} = -\frac{dB}{dt} = k \cdot [A] \cdot [B]$$

If a molecule dimerizes, the reaction rate is proportional to the square of its concentration.

$$A + A \rightarrow Z$$

$$\frac{dZ}{dt} = k \cdot [A] \cdot [A] = k \cdot [A]^2$$

The law of mass action applied to receptor binding

Binding between a ligand and its receptor is described by a simple, reversible, bimolecular reaction. ("Ligand" is just a fancy word for a molecule that binds to something.)

$$\text{Receptor} + \text{Ligand} \underset{\text{Koff}}{\overset{\text{Kon}}{\rightleftarrows}} \text{Receptor} \cdot \text{Ligand}$$

According to the law of mass action, the rate of association is:

Number of binding events per unit of time = k_{on}· [Ligand]·[Receptor].

[Ligand] and [Receptor] represent the *free* (not total) concentrations of each reactant, respectively, and k_{on} denotes the association rate constant.

The law of mass action says that the reverse reaction (the dissociation of the complex) occurs at a rate proportional to the concentration of the complex.

Number of dissociation events per unit time = [Ligand·Receptor]·k_{off}.

Here k_{off} is the dissociation rate constant, in units of inverse time.

You can also make sense of these equations intuitively. Forward binding occurs when ligand and receptor collide due to diffusion, so the rate of binding is proportional to the concentration of each. The rate constant takes into account the fact that not every collision leads to binding -- the collision must have the correct orientation and enough energy. Once binding has occurred, the ligand and receptor remain bound together for a random amount of time. The probability of dissociation is the same at every instant of time. Therefore the rate of dissociation is only proportional to the concentration of the receptor ligand complex.

After dissociation, the ligand and receptor are the same as they were before binding. If either the ligand or receptor is chemically modified, then the binding does not follow the law of mass action.

Mass action model at equilibrium

Equilibrium is reached when the rate at which new ligand·receptor complexes are formed equals the rate at which the ligand·receptor complexes dissociate. At equilibrium:

$$[\text{Ligand}]\cdot[\text{Receptor}]\cdot k_{on} = [\text{Ligand}\cdot\text{Receptor}]\cdot k_{off}$$

Rearrange that equation to define the equilibrium dissociation constant K_d.

$$\frac{[\text{Ligand}]\cdot[\text{Receptor}]}{[\text{Ligand}\cdot\text{Receptor}]} = \frac{k_{off}}{k_{on}} = K_d$$

The K_d has a meaning that is easy to understand. Set [Ligand] equal to K_d in the equation above. The K_d terms cancel out, and you'll see that the ratio [Receptor]/ [Ligand·Receptor] equals 1.0, so [Receptor] equals [Ligand·Receptor]. Since all the receptors are either free or bound to ligand, this means that half the receptors are free and half are bound to ligand. In other words, when the concentration of ligand equals the K_d, at equilibrium ligand will be bound to half the receptors.

If the receptors have a high affinity for the ligand, even a low concentration of ligand will bind a substantial number of receptors, so the the K_d will be low. If the receptors have a lower affinity for the ligand, you'll need more ligand to get binding, so the K_d will be high.

Don't mix up K_d, the equilibrium dissociation constant, with k_{off}, the dissociation rate constant. Even though they both include the word "dissociation", they are not the same, and aren't even expressed in the same units.

The table below reviews the units used to express the rate and equilibrium constants.

Constant	Name	Units
k_{on}	Association rate constant, or on-rate constant	$M^{-1}min^{-1}$ (or $M^{-1}sec^{-1}$)
k_{off}	Dissociation rate constant, or off-rate constant	min^{-1} (or sec^{-1})
K_d	Equilibrium dissociation constant	M

Fractional occupancy predicted by the law of mass action at equilibrium

The law of mass action predicts the fractional receptor occupancy at equilibrium as a function of ligand concentration. Fractional occupancy is the fraction of all receptors that are bound to ligand.

$$\text{Fractional occupancy} = \frac{[\text{Ligand} \cdot \text{Receptor}]}{[\text{Total Receptor}]} = \frac{[\text{Ligand} \cdot \text{Receptor}]}{[\text{Receptor}]+[\text{Ligand} \cdot \text{Receptor}]}$$

This equation is not useful because you don't know the concentration of unoccupied receptor, [Receptor]. A bit of algebra creates a useful equation. First, rearrange the definition of the K_d.

$$K_d = \frac{[\text{Ligand}] \cdot [\text{Receptor}]}{[\text{Ligand} \cdot \text{Receptor}]}$$

$$[\text{Ligand} \cdot \text{Receptor}] = \frac{[\text{Ligand}] \cdot [\text{Receptor}]}{K_d}$$

Now substitute that expression for [Ligand·Receptor] in the equation that defines fractional occupancy.

$$\text{Fractional occupancy} = \frac{[\text{Ligand} \cdot \text{Receptor}]}{[\text{Receptor}]+[\text{Ligand} \cdot \text{Receptor}]} = \frac{[\text{Ligand}] \cdot [\text{Receptor}]/K_d}{[\text{Receptor}]+([\text{Ligand}] \cdot [\text{Receptor}]/K_d)} = \frac{[\text{Ligand}]}{[\text{Ligand}+K_d]}$$

This equation assumes equilibrium. To make sense of it, think about a few different values for [Ligand].

[Ligand]	Occupancy
0	0%
$1 \cdot K_d$	50%
$4 \cdot K_d$	80%
$9 \cdot K_d$	90%
$99 \cdot K_d$	99%

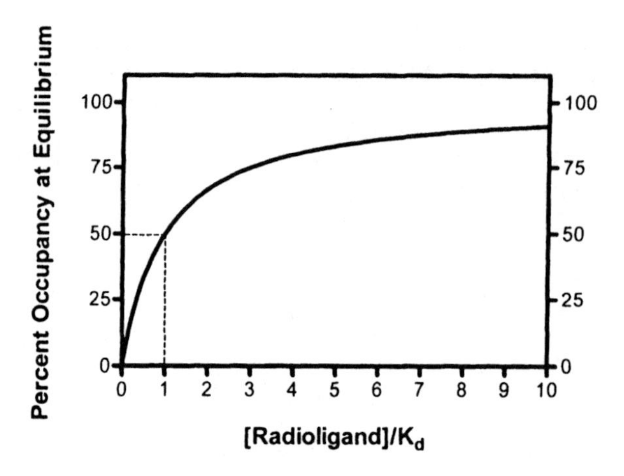

Note that when [Ligand] is equal to the K_d, half the receptors are occupied at equilibrium.

Assumptions of the law of mass action

Although termed a "law", the law of mass action is simply a model that can be used to explain some experimental data. Because it is so simple, however, the model is not useful in all situations. The model assumes:

- All receptors are equally accessible to ligands.
- Receptors are either free or bound to ligand. It doesn't allow for more than one affinity state, or states of partial binding.
- Binding does not alter the ligand or receptor.
- Binding is reversible.

Despite its simplicity, the law of mass action has proven to be very useful in describing many aspects of receptor pharmacology and physiology.

The law of mass action was developed long ago, before we knew much about the chemical nature of molecular bonds. Now we know more. And for some systems, we know precise molecular details about exactly how a ligand binds to a receptor. Clearly the law of mass action is a very simple approximation of a ligand-receptor binding that may involve conformational changes, hydrogen bonds, ionic forces, and more. But it is a useful approximation, and that is the test of any model. A more complete model, that accounted for all the chemical details, would have lots of parameters and be impossible to fit data to.

Hyperbolas, isotherms, and sigmoidal curves

The graphs below show binding at equilibrium, as predicted by the law of mass action. The two graphs show the same thing, but have different shapes.

- The graph on the left plots occupancy as a function of [Ligand]. It follows a mathematical form called a *rectangular hyperbola*, which is also called a *binding isotherm.*

- The graph on the right plots occupancy as a function of the logarithm of [Ligand] (or equivalently, plots occupancy as a function of [Ligand] plotted on a logarithmic axis). This graph follows a sigmoidal shape. The solid portion of this curve corresponds to the curve on the left. The dotted portion of the curve goes to higher concentrations of [Ligand] not shown on the other graph.

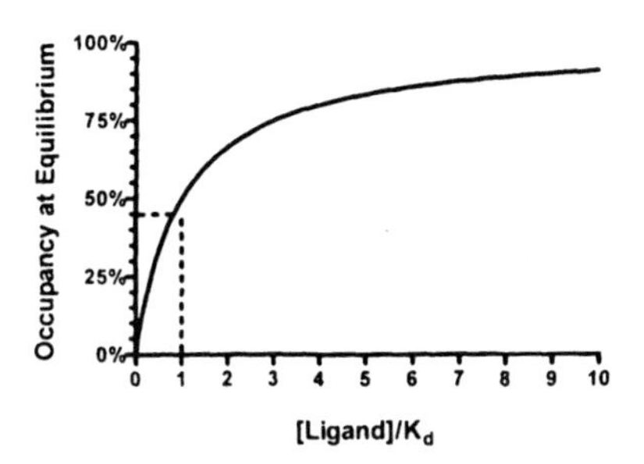

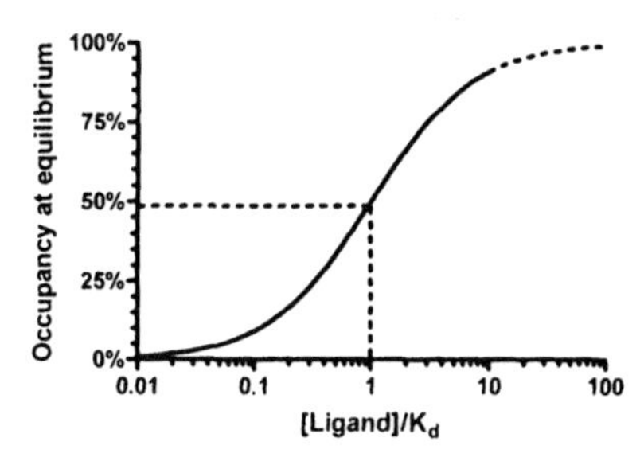

34. Analyzing radioligand binding data

Introduction to radioligand binding

A radioligand is a radioactively labeled drug that can associate with a receptor, transporter, enzyme, or any site of interest. Measuring the rate and extent of binding provides information on the number of binding sites, and their affinity and accessibility for various drugs. There are three kinds of experimental protocols for radioligand binding, discussed in the next three chapters.

Saturation binding experiments measure equilibrium binding of various concentrations of the radioligand. Analyze the relationship between binding and ligand concentration to determine the number of sites, B_{max}, and the ligand affinity, K_d. See "Analyzing saturation radioligand binding data" on page 199.

Competitive binding experiments measure equilibrium binding of a single concentration of radioligand at various concentrations of an unlabeled competitor. Analyze these data to learn the affinity of the receptor for the competitor. See "Analyzing competitive binding data" on page 211.

Kinetic experiments measure binding at various times to determine the rate constants for radioligand association and dissociation. See "Analyzing kinetic binding data" on page 233.

Nonspecific binding

In addition to binding to receptors of interest, radioligands also bind to other sites. Binding to the receptor of interest is called *specific binding*, while binding to the other sites is called *nonspecific binding*. This means that nonspecific binding can represent several phenomena:

- In most cases, the bulk of nonspecific binding represents some sort of interaction of the ligand with membranes. The molecular details are unclear, but nonspecific binding depends on the charge and hydrophobicity of a ligand – but not its exact structure.

- Nonspecific binding can also be binding to receptors, transporters, or other proteins not of interest to the investigator. For example binding of the adrenoceptor agonist, epinephrine, to serotonin receptors or metabolic enzymes can be considered "nonspecific".

- Nonspecific binding can also be binding to the filters used to separate bound from free ligand.

Nonspecific binding is usually (but not necessarily) proportional to the concentration of radioligand (within the range it is used). Add twice as much radioligand, and you'll see twice as much nonspecific binding.

Nonspecific binding is detected by measuring radioligand binding in the presence of a saturating concentration of an unlabeled drug that binds to the receptors. Under those conditions, virtually all the receptors are occupied by the unlabeled drug so the radioligand can only bind to nonspecific sites. Subtract the nonspecific binding at a particular concentration of radioligand from the total binding at that concentration to calculate the specific radioligand binding to receptors.

Which unlabeled drug should you use for determining nonspecific binding? The obvious answer is to use the same compound as the radioligand, but in its unlabeled form. In many cases this is necessary, as no other drug is known to bind to the receptors. But most investigators avoid using the same compound as the hot and cold ligand and prefer to define nonspecific binding with a drug that is chemically distinct from the radioligand but which binds to the same receptor.

What concentration of unlabeled drug should you use? You want to use enough to block virtually all the specific radioligand binding, but not so much that you cause more general physical changes to the membrane that might alter binding. If you are studying a well-characterized receptor, a useful rule-of-thumb is to use the unlabeled compound at a concentration equal to 100 times its K_d for the receptors, or 100 times the highest concentration of radioligand, whichever is higher.

Ideally, you should get the same results defining nonspecific binding with a range of concentrations of several drugs, and you should test this when possible. In many assay systems, nonspecific binding is only 10-20% of the total radioligand binding. If the nonspecific binding makes up more than half of the total binding, you'll find it hard to get quality data. If your system has a lot of nonspecific binding, try different kinds of filters, a larger volume of washing buffer, warmer washing buffer, or a different radioligand.

Ligand depletion

In many experimental situations, you can assume that a very small fraction of the ligand binds to receptors (or to nonspecific sites). In these situations, you can also assume that the free concentration of ligand is approximately equal to the concentration you added. This assumption vastly simplifies the analysis of binding experiments, and the standard analysis methods depend on this assumption.

In other situations, a large fraction of the ligand binds to the receptors (or binds nonspecifically). This means that the concentration of ligand free in the solution does not equal the concentration you added. The discrepancy is not the same in all tubes or at all times. The free ligand concentration is depleted by binding.

Many investigators use this rule of thumb: If less than 10% of the ligand binds, don't worry about ligand depletion; if more than 10% of the ligand binds, you have three choices:

- Change the experimental conditions. Increase the reaction volume without changing the amount of tissue. The problem with this approach is that it requires more radioligand, which is usually very expensive.

- Measure the free concentration of ligand in every tube. This is possible if you use centrifugation or equilibrium dialysis, but is quite difficult if you use vacuum filtration to remove free radioligand.

- Use analysis techniques that adjust for the difference between the concentration of added ligand and the concentration of free ligand. The next few chapters explain several such methods.

35. Calculations with radioactivity

Efficiency of detecting radioactivity

It is not possible to detect each radioactive disintegration. The fraction of radioactive disintegrations detected by your counter is called *efficiency*. Determine efficiency by counting a standard sample under conditions identical to those used in your experiment.

It is relatively easy to detect gamma rays emitted from isotopes such as ^{125}I, so efficiencies are usually over 90%, depending on the geometry of the counter. The efficiency is not quite 100% because the detector doesn't entirely surround the tube, and a small fraction of gamma rays (photons) miss the detector.

With ^{3}H, the efficiency of counting is much lower, often about 40%. When a tritium atom decays, a neutron converts to a proton, and the reaction shoots off both an electron and a neutrino. The energy released is always the same, but it is randomly partitioned between the neutrino (not detected) and an electron (that we try to detect). When the electron has sufficient energy, it can travel far enough to encounter a fluor molecule in the scintillation fluid. This fluid amplifies the signal and gives of a flash of light detected by the scintillation counter. If the electron has insufficient energy, it is not captured by the fluor and is not detected.

Since a large fraction of tritium atoms do not decay with enough energy to lead to a detectable number of photons, the efficiency of counting is much less than 100%. This low efficiency is a consequence of the physics of decay, and you can't increase the efficiency very much by using a better scintillation counter or a better scintillation fluid. However, poor technique can reduce the efficiency. Electrons emitted by tritium decay have very little energy, so can only be detected when they immediately encounter a fluor molecule. Radioactivity trapped in tissue chunks won't be detected. Nor will radioactivity trapped in an aqueous phase not well mixed into the scintillation fluid.

Specific radioactivity

When you buy radioligands, the packaging usually states the specific radioactivity as Curies per millimole (Ci/mmol). Since you measure counts per minute (cpm), the specific radioactivity is more useful if you change the units to be in terms of cpm rather than Curie. Often the specific radioactivity is expressed as cpm/fmol (1 fmol = 10^{-15} mole).

To convert from Ci/mmol to cpm/fmol, you need to know that 1 Ci equals 2.22×10^{12} disintegrations per minute dpm). The following equation converts Z Ci/mmol to Y cpm/fmol when the counter has an efficiency (expressed as a fraction) equal to E.

$$
\begin{aligned}
Y\frac{\text{cpm}}{\text{fmol}} &= Z\frac{\text{Ci}}{\text{mmole}} \cdot 2.22\text{x}10^{12}\frac{\text{dpm}}{\text{Ci}} \cdot 10^{-12}\frac{\text{mmole}}{\text{fmole}} \cdot E\frac{\text{cpm}}{\text{dpm}} \\
&= Z\frac{\text{Ci}}{\text{mmole}} \cdot 2.22 \times E
\end{aligned}
$$

In some countries, radioligand packaging states the specific radioactivity in Gbq/mmol. A Becquerel, abbreviated bq, equals one radioactive disintegration per second. A Gbq is 10^9 disintegrations per second. To convert from Gbq/mmol to cpm/fmol, use this equation:

$$Y\frac{\text{cpm}}{\text{fmol}} = Z\frac{\text{Gbq}}{\text{mmole}}\cdot\frac{10^9\text{bq}}{\text{Gbq}}\cdot 60\frac{\text{dpm}}{\text{bq}}\cdot 10^{-12}\frac{\text{mmole}}{\text{fmole}}\cdot E\frac{\text{cpm}}{\text{dpm}}$$

$$= Z\frac{\text{Gbq}}{\text{mmole}}\cdot 0.06\cdot E$$

Calculating the concentration of the radioligand

Rather than trust your dilutions, you can accurately calculate the concentration of radioligand in a stock solution. Measure the number of counts per minute in a small volume of solution and use the equation below. C is cpm counted, V is volume of the solution you counted in ml, and Y is the specific activity of the radioligand in cpm/fmol (calculated in the previous section).

$$\text{Concentration in pM} = \frac{\text{C cpm}/\text{Y cpm/fmol}}{\text{V ml}}\cdot\frac{0.001\ \text{pmol/fmol}}{0.001\ \text{liter/ml}}$$

$$= \frac{C/Y}{V}$$

Radioactive decay

Radioactive decay is entirely random. A particular atom doesn't "know" how old it is, and it can decay at any time. The probability of decay at any particular interval is the same as the probability of decay during any other interval. If you start with N_0 radioactive atoms, the number remaining at time t is:

$$N_t = N_0 \cdot e^{-K_{decay}\cdot t}$$

K_{decay} is the rate constant of decay expressed in units of inverse time. Each radioactive isotope has a different value of K_{decay}.

The half-life ($t_{1/2}$) is the time it takes for half the isotope to decay. Half-life and decay rate constant are related by this equation:

$$t_{1/2} = \frac{\ln(2)}{K_{decay}} = \frac{0.693}{K_{decay}}$$

The table below shows the half-lives and rate constants for commonly used radioisotopes. The table also shows the specific activity assuming that each molecule is labeled with one isotope. (This is often the case with ^{125}I and ^{32}P. Tritiated molecules often incorporate two or three tritium atoms, which increases the specific radioactivity.)

Isotope	Half life	K_{decay}	Specific Radioactivity
^{3}H	12.43 years	0.056/year	28.7 Ci/mmol
^{125}I	59.6 days	0.0116/day	2190 Ci/mmol
^{32}P	14.3 days	0.0485/day	9128 Ci/mmol
^{35}S	87.4 days	0.0079/day	1493 CI/mmol

You can calculate radioactive decay from a date where you (or the manufacturer) knew the concentration and specific radioactivity using this equation.

$$\text{FractionRemaining} = e^{-K_{decay} \cdot \text{Time}}$$

For example, after ^{125}I decays for 20 days, the fraction remaining equals 79.5%. Although data appear to be scanty, most scientists assume that the energy released during decay destroys the ligand so it no longer binds to receptors. Therefore the specific radioactivity does not change over time. What changes is the concentration of ligand. After 20 days, therefore, the concentration of the iodinated ligand is 79.5% of what it was originally, but the specific radioactivity remains 2190 Ci/mmol. This approach assumes that the unlabeled decay product is not able to bind to receptors and has no effect on the binding. Rather than trust this assumption, you should always try to use newly synthesized or repurified radioligand for key experiments.

Counting errors and the Poisson distribution

The decay of a population of radioactive atoms is random, and therefore subject to a sampling error. For example, the radioactive atoms in a tube containing 1000 cpm of radioactivity won't give off exactly 1000 counts in every minute. There will be more counts in some minutes and fewer in others. This variation follows a distribution known as the Poisson distribution. The variability is intrinsic to radioactive decay and cannot be reduced by more careful experimental controls.

After counting a certain number of disintegrations in your tube, you want to know what the "real" number of counts is. Obviously, there is no way to know that. Just by chance, you may have chosen a time interval to count when there were more (or fewer) radioactive disintegrations than usual. But you can calculate a range of counts that is 95% certain to contain the true average value. If the number of counts, C, is greater than about 50 you can calculate the confidence interval using this approximate equation:

$$95\%\ \text{Confidence Interval:}\ \left(C - 1.96\sqrt{C}\right)\ \text{to}\ \left(C + 1.96\sqrt{C}\right)$$

For example, if you measure 100 radioactive counts in an interval, you can be 95% sure that the true average number of counts ranges approximately between 80 and 120 (using the equation here) or between 81.37 and 121.61 (using an exact equation not shown here).

When calculating the confidence interval, you must set C equal to the total number of counts you measured experimentally, which is probably *not* the same as the number of counts per minute.

Example: You placed a radioactive sample into a scintillation counter and counted for 10 minutes. The counter tells you that there were 225 counts per minute. What is the 95% confidence interval? Since you counted for 10 minutes, the instrument must have detected 2250 radioactive disintegrations. The 95% confidence interval of this number extends from 2157 to 2343. This is the confidence interval for the number of counts in 10 minutes, so the 95% confidence interval for the average number of counts per minute extends from 216 to 234. If you had attempted to calculate the confidence interval using the number 225 (counts per minute) rather than 2250 (counts detected), you would have calculated a wider (and incorrect) interval.

The Poisson distribution explains the advantages of counting your samples for a longer time. For example, the table below shows the confidence interval for 100 cpm counted for various times. When you count for longer times, the confidence intervals (when you express counts as cpm) will be narrower.

	1 minute	10 minutes	100 minutes
Counts per minute (cpm)	100	100	100
Total counts	100	1000	10000
95% CI of counts	81.4 to 121.6	938 to 1062	9804 to 10196
95% CI of cpm	81.4 to 121.6	93.8 to 106.2	98.0 to 102.0

The table below shows percent error as a function of the number of counts. Percent error is defined as the width of the confidence interval divided by the number of counts.

Counts	Percent Error
50	27.72%
100	19.60%
200	13.86%
500	8.77%
1000	6.20%
2000	4.38%
5000	2.77%
10000	1.96%
25000	1.24%
50000	0.88%
100000	0.62%
C	$100 \cdot \frac{1.96 \cdot \sqrt{C}}{C} = \frac{196}{\sqrt{C}}$

Note that the percent error is inversely proportional to the square root of the counts. If you are trying to decide how long to count a tube, this means the percent error is inversely proportional to the square root of the counting time. If you count for four times as long, the results will be twice as precise. If you count for nine times as long, the results will be three times as precise.

The GraphPad radioactivity web calculator

GraphPad Software provides a free radioactivity calculator on our web site at http://www.graphpad.com. Click on QuickCalcs to see this, and other, free calculators.

Use it to perform these seven common calculations:

Calculation	Description
Isotope decay	Calculates radioactive decay during a specified number of days. Select one of the common isotopes, or enter the half-life of another isotope.
Concentration of stock	Enter mCi/ml and Ci/mmole, which should be on the label. If you are using a molecule labeled with ^{125}I, the specific activity equals 2200 Ci/mmole if each molecule is labeled with one iodine. Also enter the percent of the original isotope remaining (calculated above). The calculations assume that the decay product is not biologically active, so the concentration of stock that is biologically active decreases over time.
Dilution of stock	Enter the concentration in your stock solution, after accounting for decay. Also enter the concentration and volume you want. The result is the volume of stock you need to use.
Specific activity (cpm/fmol)	Enter the specific radioactivity as Ci/mmol which should be on the label. If you are using a molecule labeled with ^{125}I, the specific activity equals 2200 Ci/mmol if each molecule is labeled with one iodine. Also enter the counter efficiency - the fraction of radioactive disintegrations that are detected. The efficiency depends on the isotope and instrumentation. With low energy isotopes such as tritium, the efficiency also depends on the experimental details such as the choice of scintillation fluid, the amount of water in the sample, and the presence of any colored substances in the sample.
Cpm to fmol/mg	Enter the specific radioactivity as cpm/fmol, the number of cpm counted, and the protein content of the sample in mg. The result is the number of binding sites in fmol/mg protein.
Cpm to sites/cell	Enter the specific radioactivity as cpm/fmol, the number of cpm counted, and the cell count. The result is the number of binding sites per cell.
Cpm to nM	Enter the specific radioactivity as cpm/fmol, the number of cpm counted, and the volume counted. The result is the concentration of radioligand in nM.

36. Analyzing saturation radioligand binding data

Introduction to saturation binding experiments

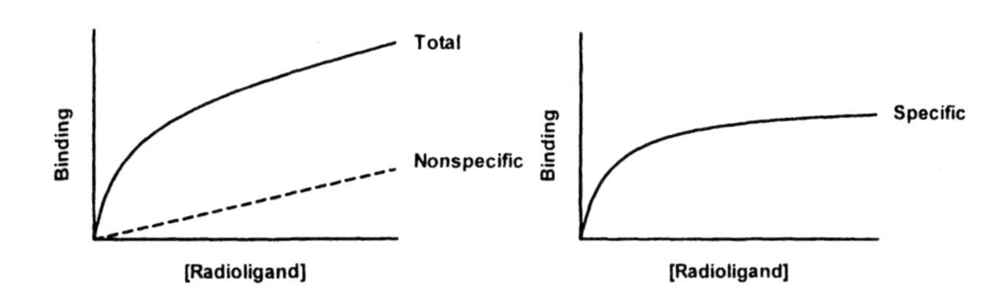

Saturation radioligand binding experiments measure specific radioligand binding at equilibrium at various concentrations of the radioligand. Nonlinear regression analysis of saturation binding data allows you to determine receptor number and affinity. Because this kind of experiment used to be analyzed with linear Scatchard plots (more accurately attributed to Rosenthal), they are sometimes called "Scatchard experiments".

Analyses of saturation binding data depend on the assumption that you have allowed the incubation to proceed to equilibrium. This can take anywhere from a few minutes to many hours, depending on the ligand, receptor, temperature, and other experimental conditions. The lowest concentration of radioligand will take the longest to equilibrate. When testing equilibration time, therefore, use a low concentration of radioligand (perhaps 10-20% of the K_d).

Fitting saturation binding data

The figure below illustrates a saturation binding assay at human M_2 muscarinic acetylcholine receptors stably expressed in Chinese hamster ovary (CHO) cell membranes. The radioligand is the antagonist, [^{3}H]N-methylscopolamine ([^{3}H]NMS) and non-specific binding of this radioligand was determined as that obtained in the presence of a saturating concentration of the unlabeled muscarinic receptor antagonist, atropine.

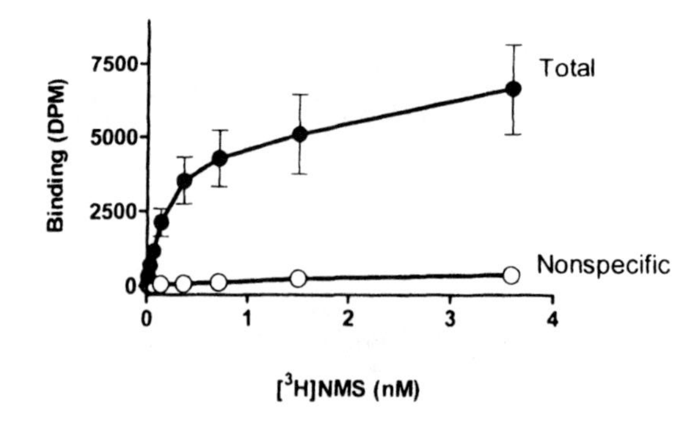

The standard equation for specific radioligand binding to a receptor is based on the law of mass action and describes a hyperbolic relationship as follows:

$$Y = \frac{B_{max} \cdot X}{X + K_d}$$

where B_{max} denotes the maximal density of receptor sites and K_d denotes the radioligand equilibrium dissociation constant. Because nonspecific binding is generally a linear function of radioligand concentration (see figure above), its equation is that of a straight line that intercepts the axes at a value of 0:

$$Y = NS \cdot X$$

Therefore, the equation for the total binding curve (see figure above) represents both specific binding together with non-specific binding:

$$Y = \frac{B_{max} \cdot X}{X + K_d} + NS \cdot X$$

Fitting a curve to determine B_{max} and K_d from specific binding

The standard approach for analyzing saturation binding experiments is to subtract the non-specific curve from the total binding curve to determine the specific (receptor) binding for each concentration of the radioligand, and then fitting the resulting points to the first equation (above) describing specific binding. This is shown in the figure below for our muscarinic receptor binding example:

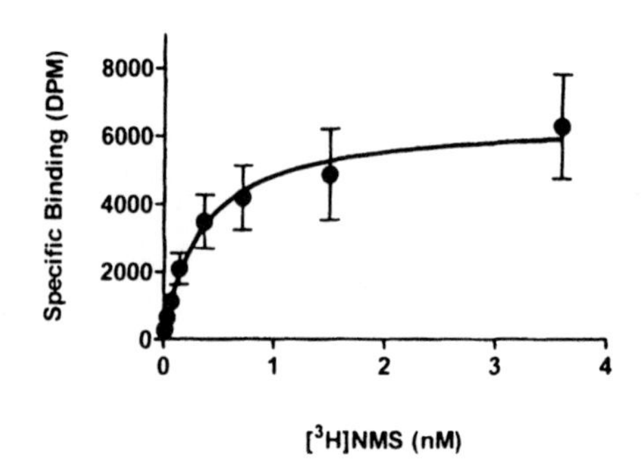

	B_{max}	K_d
Value	6464	0.343
95% CI	4751 to 8177	0.0389 to 0.647

B_{max} is expressed in the same units as the Y values (commonly dpm, cpm, sites/cell or fmol/mg protein). K_d is expressed in the same units as the X values, usually nM or pM. If the K_d is low, the affinity is high.

This method is only valid when a small fraction of the ligand binds to the receptor. If this assumption is true, then the free concentration of ligand equals the added concentration in both the tubes used to measure total binding and the tubes used to measure nonspecific binding. If the assumption is not valid, then the free concentration of ligand will differ in the two sets of tubes. In this case subtracting the two values makes no sense, and determining specific binding is difficult.

Even when the above assumption is met, another point of contention among researchers relates to how one actually goes about subtracting nonspecific binding from total binding to derive the specific binding. Usually, you will have determined total binding in one set of assay tubes and non-specific binding in another set. Furthermore, you will probably have determined each point in duplicate, triplicate or even more replicates. How do you subtract the non-specific binding from the total binding under these conditions?

If the replicates are matched, you can subtract each nonspecific data point from its paired total binding data point. This is rare, as most binding assays do not match total and nonspecific replicates.

You can take the average value of each set of nonspecific binding replicates, and subtract this from the average value of the corresponding total binding replicates. However, this approach reduces the total number of data points that constitute the specific binding curve.

You can take the average value of each set of nonspecific binding replicates, and subtract this from each *individual* corresponding total binding replicate. This is probably the most common approach, and results in a larger number of data points in the resulting specific binding curve.

If you assume that non-specific binding is a linear function of radioligand concentration, you can perform a linear regression on the nonspecific binding data to derive the corresponding line of best-fit. Then, you subtract the ideal value of nonspecific binding described by the line from each corresponding total binding replicate.

None of these approaches are ideal, as they all involve data transformation of some kind. A less ambiguous approach is to use global fitting, as described in the next section.

The problem with the standard approach to fitting specific saturation binding data

In the standard approach to analyzing saturation binding data (above), you basically combine two real, experimentally determined, data sets (total binding and nonspecific binding) to yield a "composite" curve (specific binding). The data you analyze, the resulting specific binding values, were not actually experimentally determined but rather calculated by combining other data. Although this approach is generally valid for high-quality data, it is nevertheless suboptimal for two reasons:

The curve-fitting procedure will always have fewer degrees of freedom (fewer data points) after the transformation. For example, the 16 data points describing the total and nonspecific binding of [^{3}H]NMS in the original data set above are reduced to 8 points describing the specific binding of the radioligand.

Whenever one set of data is used to perform a transformation of another set of data, the potential always exists for the propagation of error from either data set to the resulting transform. For example, you may have made a mistake in the determination of nonspecific binding in one set of tubes (see figure below). After subtracting the nonspecific binding from the total binding, the resulting specific binding curve may provide a poor fit to the saturation binding model, and thus provide poor estimates of B_{max} and/or K_d.

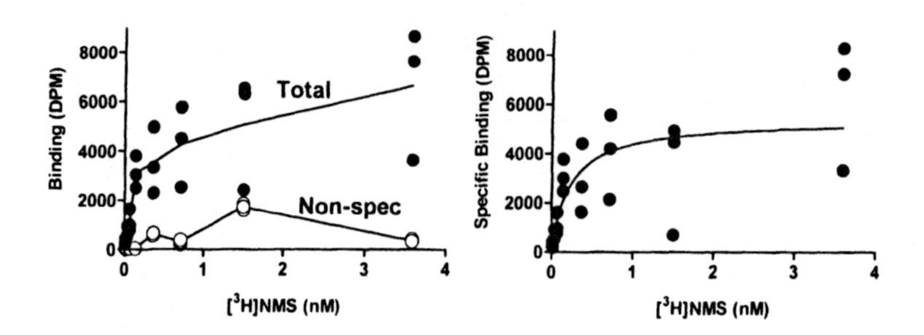

All these problems associated with nonlinear regression analysis of transformed data sets can be overcome if the regression procedure fits the equations describing total and nonspecific binding to the raw data at the same time.

Using global fitting to determine total and nonspecific binding at one time

Using a global fitting approach, you can to fit total and nonspecific binding at one time to the raw, experimentally determined, saturation binding data. There is thus no need to subtract one data set from another to get specific binding. Just fit total and nonspecific at once.

Depending on your nonlinear regression program, you would most likely need to place your total binding data into one column (e.g., column A for in our case), and nonspecific binding data into another column (e.g., column B). The following is in Prism 4 syntax equation, and defines one equation for column A and another for column B. When you fit this model, set the parameter NS to be shared between both data sets

```
Nonspecific=NS*X
Specific=Bmax*X/(KD + X)
<A> Y = Specific + Nonspecific
<B> Y = Nonspecific
```

Variable	Units	Comments
X	Usually nanomolar	Concentration of unlabeled drug.
Y	Cpm; dpm; fmol/mg protein	Total binding of radioligand.
Bmax	Same as Y axis; i.e., cpm; dpm; fmol/mg protein	Maximal density of binding sites for the radioligand.
KD	Same as X axis; i.e. concentration units	Equilibrium dissociation constant of the radioligand.
NS	Same as Y axis; i.e., cpm; dpm; fmol/mg protein	Fraction of the total binding that is nonspecific binding; constrained to be greater than zero.

Note that the above syntax is specific to GraphPad Prism, particularly the <A>/<B> format, which refers to specific data set columns. Depending on the program you use, you will have to adapt the syntax to assign one equation for the total binding data set and another for the nonspecific binding data set. Unfortunately, not all nonlinear regression programs offer global curve fitting.

The figure below shows the best-fit curve for the binding of [^{3}H]NMS to the M_2 muscarinic acetylcholine receptor, as well as its nonspecific binding. The best-fit parameters are also shown:

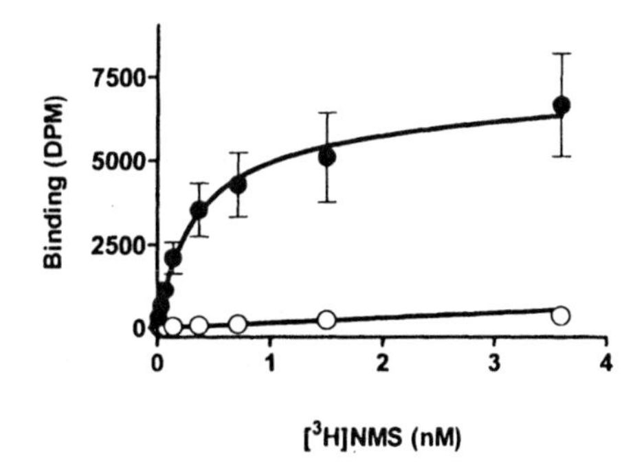

Parameter	Best-fit value	95% CI
B_{max}	6291	4843 to 7739
K_d	0.317	0.109 to 0.525
NS	149.1	0.0 to 402.1

If you compare the results from this analysis to the standard method shown in the previous section, you will note that the current global fit yields narrower 95% confidence intervals for the B_{max} and K_d parameters. This highlights the greater reliability in parameter estimation provided by sharing parameters across data sets.

Determining K_d and B_{max} for two classes of binding sites

If the radioligand binds to two classes of binding sites, fit the specific binding data to the following two-site binding equation. This equation is simply an extension of the one-site equation shown previously, and sums the binding to two sites, each with its own B_{max} and K_d.

$$Y=\frac{B_{max1}\cdot X}{K_{d1}+X}+\frac{B_{max2}\cdot X}{K_{d2}+X}$$

This equation assumes that the radioligand binds to two independent noninteracting binding sites, and that the binding to each site follows the law of mass action.

You will only get meaningful results from a two-site fit if you have ten or more (preferably a lot more) data points spaced over a wide range of radioligand concentrations. Be sure that you measure binding at radioligand concentrations below the high-affinity K_d and above the low affinity K_d.

> Note: The two-site saturation binding model is simply an extension of the one-site model. Thus, the same procedures described earlier for using global-fitting to analyze total and nonspecific binding experiments are readily applied to both models.

Checklist for saturation binding

When evaluating results of saturation binding analyses, ask yourself these questions:

Question	Comment
Did only a small fraction of the radioligand bind?	The analysis assumes that the free concentration is almost identical to the concentration you added. You can test this by comparing the total counts that bound to the total counts added to the tube. If more than 10% of the ligand bound (at any ligand concentration), then the standard analysis won't work. Either change the experimental protocol (increase the volume) or use a method that accounts for depletion of radioligand. See "Analyzing saturation binding with ligand depletion" on page 208.
Did the binding equilibrate?	The tubes with the lowest concentration of radioligand take the longest to equilibrate. So test equilibration time with a low concentration of radioligand.
Did you use high enough concentrations of radioligand?	Calculate the ratio of the highest radioligand concentration you used divided by the K_d reported by the program (both in nM or pM). The highest concentration should be at least 10 times the K_d, so that occupancy exceeds 90%.
Is the B_{max} reasonable?	Typical values for B_{max} are 10-1,000 fmol binding sites per milligram of membrane protein, 100-10,000 sites per cell or 1 receptor per square micron of membrane. If you use cells transfected with receptor genes, then the B_{max} may be many times larger than these values.

Question	Comment
Is the K_d reasonable?	Typical values for K_d of useful radioligands range between 10 pM and 10 nM. If the K_d is much lower than 10 pM, the dissociation rate is probably very slow and it will be difficult to achieve equilibrium. If the K_d is much higher than 10 nM, the dissociation rate will probably be fast, and you may be losing binding sites during separation of bound ligand from free radioligand.
Are the standard errors too large? Are the confidence intervals too wide?	Divide the SE of the B_{max} by the B_{max}, and divide the SE of the K_d by the K_d. If either ratio is much larger than about 20%, look further to try to find out why.
Is the nonspecific binding too high?	Divide the nonspecific binding at the highest concentration of radioligand by the total binding at the highest concentration. Nonspecific binding should usually be less than 50% of the total binding.

Scatchard plots

What is a Scatchard plot?

In the days before nonlinear regression programs were widely available, scientists transformed data into a linear form, and then analyzed the data by linear regression. There are several ways to linearize binding data, including the methods of Lineweaver-Burk and Eadie-Hofstee. However, the most popular method to linearize binding data is to create a Scatchard plot (more accurately attributed to Rosenthal), shown in the right panel below.

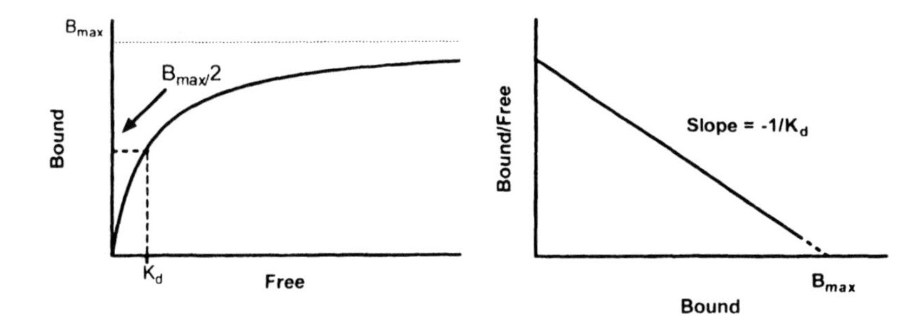

In the Scatchard plot, the X axis is specific binding and the Y axis is specific binding divided by free radioligand concentration. It is possible to estimate the B_{max} and K_d from a Scatchard plot (B_{max} is the X intercept; K_d is the negative reciprocal of the slope). However, the Scatchard transformation distorts the experimental error, and thus violates several assumptions of linear regression. The B_{max} and K_d values you determine by linear regression of Scatchard transformed data may be far from their true values.

After analyzing your data with nonlinear regression, however, it is often useful to *display* data as a Scatchard plot. The human retina and visual cortex are wired to detect edges (straight lines), not rectangular hyperbolas. Scatchard plots are often shown as insets to the saturation binding curves. They are especially useful when you want to show how a treatment changes B_{max} or K_d.

> Tip: You should analyze saturation binding data with nonlinear regression, not with Scatchard plots. Use Scatchard plots to display data, not to analyze data.

When making a Scatchard plot, you have to choose what units you want to use for the Y-axis. Some investigators express both free ligand and specific binding in cpm or dpm so the ratio bound/free is a unitless fraction. While this is easy to interpret (it is the fraction of radioligand bound to receptors), a more rigorous alternative is to express specific binding in sites/cell or fmol/mg protein, and to express the free radioligand concentration in nM. While this makes the Y-axis hard to interpret visually, it provides correct units for the slope (which equals $-1/K_d$).

> GraphPad note: It is easy to transform saturation data to form a Scatchard plot with Prism. From your data table, click Analyze, choose Transforms, switch to the list of Pharmacology and biochemistry transforms, and then choose the Scatchard transform.

Plotting the line that corresponds to nonlinear regression analyses

If there is one class of receptors, the Scatchard plot will be linear. Some people use linear regression to draw a line through the points. However, the linear regression line should NOT be used to analyze the data. The X-intercept of the regression line will be near the B_{max}, and the negative inverse of the slope will be near the K_d. However, the B_{max} and K_d values determined directly with nonlinear regression will be more accurate.

It isn't hard to draw the Scatchard line that corresponds to the nonlinear regression determination of B_{max} and K_d. The discussion below assumes that the "Bound" units for the Y axis are the same units used for the X axis and in which you want to express B_{max} (sites/cell or fmol/mg,), and the "free" units are the same as the units you want to use for the K_d (nM or pM). Since the X intercept of the Scatchard line is the B_{max}, the Scatchard line ends at $X=B_{max}$, $Y=0$. Since the slope of the Scatchard line equals $-1/K_d$, the Y-intercept equals the B_{max} divided by the K_d. The Scatchard line begins at $X=0$, $Y= B_{max}/K_d$.

To create a Scatchard line corresponding to the nonlinear regression fit:

1. Create a new data table, with numerical X values and single Y values.
2. Into row 1 enter $X=0$, $Y= B_{max}/K_d$ (both previously determined by nonlinear regression).
3. Into row 2 enter $X= B_{max}$ (previously determined by nonlinear regression) and $Y=0$.
4. Superimpose the points contained in this new data table on the Scatchard plot of your binding data.
5. Connect the symbols of the new data table with a straight line. If your program allows you, hide the data point symbols but leave the interconnecting straight line.

The result is a straight line, based on the nonlinear regression of your binding data, superimposed on the Scatchard transformation of the same binding data.

Scatchard plots of binding to two sites

The appearance of a two-site Scatchard plot

The left panel below shows binding of a radioligand to two independent binding sites present in equal concentrations, but with a tenfold difference in K_d. The two individual curves are shown as dotted and dashed curves. When you do the experiment, you can't observe the individual components, but observe the sum, which is shown as a solid curve. Note that this curve is not obviously biphasic.

The right panel shows the same data plotted on a Scatchard plot. The binding to each receptor is showm as a straight line (dotted or dashed). The total binding, plotted on a Scatchard plot, is curved. Note that the two lines that represent binding to each type of receptor are NOT the asymptotes of the curve.

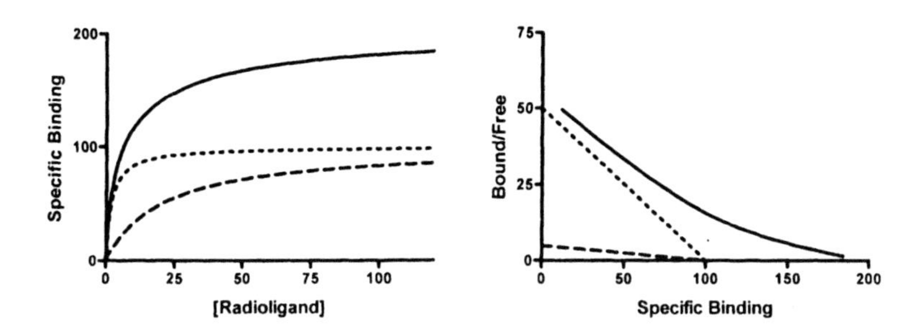

Graphing the two lines of a two-site Scatchard plot

To plot the two straight lines that correspond to the nonlinear regression fit, adapt the instructions (previous section) for plotting a Scatchard plot for one-site binding. Create a new data table that defines the two lines as shown below, using B_{max} and K_d values determined by nonlinear regression.

X	A	B
0	B_{max1}/K_{d1}	
B_{max1}	0	
0		B_{max2}/K_{d2}
B_{max2}		0

Go to the graph of the Scatchard transformed data and add the new table to that graph. Plot the two data sets from the table using connecting lines but no symbols.

Analyzing saturation binding with ligand depletion

Total binding with ligand depletion

The standard methods for analyzing saturation binding assume that a tiny fraction of the radioligand binds to receptors. This means that the concentration of radioligand added is very similar to the concentration of radioligand free in solution.

In some experimental situations, where the receptors are present in high concentration and have a high affinity for the ligand, that assumption is not true. A large fraction of the radioligand binds to receptors, so the free concentration in solution is quite a bit lower than the concentration you added. The free ligand concentration is depleted by binding to receptors.

If possible you should avoid experimental situations where the free ligand concentration is far from the total concentration. You can do this by increasing the volume of the assay without changing the amount of tissue. The drawback is that you'll need more radioligand, which is usually expensive or difficult to synthesize.

If you can't avoid radioligand depletion, you need to account for the depletion in your analyses. The obvious way to do this is to subtract the number of cpm (counts per minute) of total binding from the cpm of added ligand to calculate the number of cpm free in solution. This can then be converted to the free concentration in molar. There are four problems with this approach:

- If you used this method, experimental error in determining specific binding would affect the free ligand concentration you calculate. Error in Y would affect X, which violates an assumption of nonlinear regression.

- Since the free concentration in the nonspecific tubes is not the same as the free concentration in the total tubes, it is difficult to deal with nonspecific binding using this approach. You cannot calculate specific binding as the difference between the total binding and nonspecific binding.

- This method works only for saturation binding experiments, and cannot be extended to analysis of competition curves.

- You cannot implement this method with many commercially available nonlinear regression programs, because they do not let you subtract Y from X as part of the nonlinear regression process.

S. Swillens (*Molecular Pharmacology*, 47: 1197-1203, 1995) developed an equation that defines total binding as a function of added ligand, accounting for nonspecific binding and ligand depletion. By analyzing simulated experiments, that paper shows that fitting total binding gives more reliable results than you would get by calculating free ligand by subtraction. The equations shown below are not exactly the same as in Swillens' paper, but the ideas are the same.

From the law of mass action, total binding follows this equation.

$$\text{Total Binding} = \text{Specific} + \text{Nonspecific}$$

$$\text{Total Binding} = \frac{B_{max} \cdot [\text{Free Ligand}]}{K_d + [\text{Free Ligand}]} + [\text{Free Ligand}] \cdot \text{NS}$$

The first term is the specific binding, which equals fractional occupancy times B_{max}, the total number of binding sites. The second term is nonspecific binding, which is assumed to be proportional to free ligand concentration. The variable NS is the fraction of the free ligand that binds to nonspecific sites.

This equation is not useful, because you don't know the concentration of free ligand. What you know is that the free concentration of ligand equals the concentration you added minus the concentration that bound (specific and nonspecific). Defining X to be the amount of ligand added and Y to be total binding, the system is defined by two equations:

$$Y = \frac{B_{max} \cdot [\text{Free Ligand}]}{K_d + [\text{Free Ligand}]} + [\text{Free Ligand}] \cdot NS$$

$$[\text{Free Ligand}] = X\text{-}Y$$

Combining the two equations:

$$Y = \frac{B_{max} \cdot (X\text{-}Y)}{K_d + (X\text{-}Y)} + (X\text{-}Y) \cdot NS$$

X, Y and B_{max} are expressed in units of cpm per tube. To keep the equation consistent, therefore, K_d must also be converted to cpm units (the number of cpm added to each tube when the total concentration equals the K_d).

Unfortunately, you cannot enter the equation for total binding with ligand depletion into most commercially-available nonlinear regression programs because Y appears on both sides of the equal sign. This is called an *implicit* equation, because it doesn't really define Y but the definition is implicit in the equation. But simple algebra rearranges it into a quadratic equation, whose solution is shown below:

$$Y = \frac{-b + \sqrt{b^2 - 4 \cdot a \cdot c}}{2 \cdot a}$$

where

$$a = -1\text{-}NS$$

$$b = X(2 \cdot NS + 1) + K_d(NS + 1) + B_{max}$$

$$c = -X(NS(K_d + X) + B_{max})$$

Nonspecific binding with ligand depletion

The method described above fits total binding data to an equation that includes both specific and nonspecific components. It does not require that you experimentally determine nonspecific binding. While this is convenient, many investigators would feel uneasy trusting those results without determining nonspecific binding experimentally.

You can experimentally determine nonspecific binding by including a large concentration of an unlabeled ligand in your incubations. This will bind to virtually all the receptors, leaving only nonspecific sites free to bind radioligand. The conventional approach is to measure total and nonspecific binding at each ligand concentration, and to define specific binding as the difference. This approach cannot be used when a high fraction of ligand binds, because the free concentration of ligand in the total tubes is not the same as the free concentration of ligand in the nonspecific tubes.

We assume that nonspecific binding is a constant fraction of the concentration of free ligand.

$$\text{Nonspecific binding} = Y = NS \bullet [\text{Ligand}]$$

We also assume that the free concentration of ligand equals the added concentration (X) minus the amount that bound (Y).

$$[\text{Ligand}] = X - Y$$

Combining the two equations:

$$Y = (X - Y) \cdot NS$$

$$Y = X \cdot \frac{NS}{NS + 1}$$

Fitting total and nonspecific binding with ligand depletion

The most ideal approach to analyzing saturation binding data with depletion is to use the combination of total and nonspecific binding equations described above to simultaneously fit both total and nonspecific binding data sets, if your nonlinear regression program allows you to do global fitting with parameter-sharing. Shown below is the programming (using the Prism 4 syntax) for entering these equations into a computer program.

```
KdCPM=KdnM * Vol * 1000 * SpecAct
a = -1-NS
b = X*(2*NS+1) + KdCPM*(NS+1) + Bmax
c= -1*X*(NS*(KdCPM + X)+Bmax)
Total= (-b + SQRT(b*b - 4*a*c))/(2*a)
Nonspec=X*NS/(NS+1)
<A>Y=Total
<B>Y=Nonspec
```

The <A> & <B> syntax defines one equation for column A (total binding) and a different equation for column B (nonspecific binding). If you use a program other than Prism, you will have to adapt the equation.

In the above equation, X is total ligand added in cpm, Y is total binding in cpm, SpecAct is specific radioactivity in cpm/fmol, and Vol is reaction volume in ml. You must fix both of these last two parameters to constant values based on your experimental design. You must share the parameter NS between the two data sets. Nonlinear regression will find the best-fit value for that as well as for KdnM (the Kd in nM) and the Bmax (in cpm). The remaining variables (KdCPM, a, b, c, total, nonspec) are intermediate variables used to allow the equation to be written over many lines. They are not fit.

In situations of ligand depletion, the free concentration of radioligand will differ between the total binding assay tubes and the nonspecific binding assay tubes. Subtracting the two values makes no sense, and determining specific binding is difficult. This is clearly a situation where global curve-fitting is the preferred analytical approach.

37. Analyzing competitive binding data

What is a competitive binding curve?

Competitive binding experiments measure the binding of a single concentration of labeled ligand in the presence of various concentrations of unlabeled ligand. Ideally, the concentration of unlabeled ligand varies over at least six orders of magnitude.

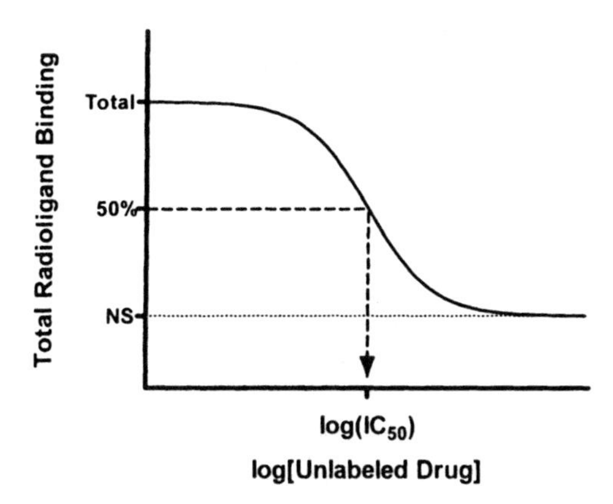

The top of the curve is a plateau at a value equal to radioligand binding in the absence of the competing unlabeled drug. The bottom of the curve is a plateau equal to nonspecific binding. The concentration of unlabeled drug that produces radioligand binding halfway between the upper and lower plateaus is called the IC_{50} (inhibitory concentration 50%) or EC_{50} (effective concentration 50%).

If the radioligand and competitor both bind reversibly to the same binding site, binding at equilibrium follows this equation (where *Top* and *Bottom* are the Y values at the top and bottom plateau of the curve).

$$Y = \text{Bottom} + \frac{(\text{Top} - \text{Bottom})}{1 + 10^{(X - \text{LogIC}_{50})}}$$

Entering data for competitive binding curves

Many investigators enter Y values as cpm or dpm. Other investigators transform their data to percent specific binding. The problem with this latter approach is that you need to define how many cpm (or dpm) equal 100% binding and how many equal 0% specific binding.

It is best to use the logarithm of the concentration of competitor as your X value, rather than the concentration itself. For example, if the competitor concentration varied from 1 nM to 1 mM, enter X values from -9 to -3. A log axis cannot accommodate a concentration of zero (log 0 is undefined). Instead, enter the logarithm of a very low competitor concentration (e.g., -10).

If you prefer to enter concentrations in the data table, rather than the logarithm of concentration, be sure to then transform the data to logarithms before performing nonlinear regression.

Decisions to make before fitting the data

Weighting

When analyzing the data, you need to decide whether to minimize the sum of the squares of the absolute distances of the points from the curve or to minimize the sum of the squares of the relative distances. See "Weighting method" on page 307. The choice depends on the source of the experimental error. Follow these guidelines:

If the bulk of the error comes from pipetting, the standard deviation of replicate measurements will be, on average, a constant fraction of the amount of binding. In a typical experiment, for example, the highest amount of binding might be 2000 cpm with an SD of 100 cpm. The lowest binding might be 400 cpm with an SD of 20 cpm. With data like this, you should evaluate goodness-of-fit with relative distances. The details on how to do this are in the next section.

In other experiments, there are many contributions to the scatter, and the standard deviation is not related to the amount of binding. With this kind of data, you should evaluate goodness-of-fit using absolute distances.

You should only consider weighting by relative distances when you are analyzing total binding data. When analyzing specific binding (or data normalized to percent inhibition), you should evaluate goodness-of-fit using absolute distances, as there is no clear relationship between the amount of scatter and the amount of specific binding.

Constants

To find the IC_{50}, the concentration that inhibits 50% of the binding, the nonlinear regression algorithm needs to first define 100% and 0%.

Ideally your data span a wide range of concentrations of unlabeled drug, and clearly define the bottom or top plateaus of the curve. If this is the case, the computer program can fit the 0% and 100% values from the plateaus of the curve, and you don't need to do anything special.

In some cases, your competition data may not define a clear bottom plateau, but you can define the plateau from other data. All drugs that bind to the same receptor should compete all specific radioligand binding and reach the same bottom plateau value. This means that you can define the 0% value (the bottom plateau of the curve) by measuring radioligand binding in the presence of a standard drug known to block all specific binding. If you do this, make sure that you use plenty of replicates to determine this value accurately. If your definition of the lower plateau is wrong, the values for the IC_{50} will be wrong as well. You can also define the top plateau as binding in the absence of any competitor.

If you have collected enough data to clearly define the entire curve, let your program fit all the variables and find the top and bottom plateaus based on the overall shape of your data. If your data don't define a clear top or bottom plateau, you should define one or both of these values to be constants fixed to values determined from other data.

Competitive binding data with one class of receptors

Fitting data to a one-site competitive binding curve

If labeled and unlabeled ligand compete for binding to a single class of receptors and the system is at equilibrium, then the one-site competitive binding equation applies:

```
Y=Bottom+(Top-Bottom)/(1+10^(X-LogIC50))
```

This is equivalent to a sigmoidal dose-response curve with a slope factor (Hill slope) of -1.0.

Checklist for competitive binding results

When evaluating results of competitive binding, ask yourself these questions:

Question	Comment
Is the $\log IC_{50}$ reasonable?	The $\log IC_{50}$ should be near the middle of the curve, with at least several concentrations of unlabeled competitor on either side of it.
Are the standard errors too large? Are the confidence intervals too wide.	The SE of the $\log IC_{50}$ should be less than 0.5 log unit (ideally a lot less).
Are the values of *Top* and *Bottom* reasonable?	*Top* should be near the binding you observed in the absence of competitor. *Bottom* should be near the binding you observed in the presence of a maximal concentration of competitor. If the best-fit value of *Bottom* is negative, consider fixing it to a constant value equal to nonspecific binding determined in a control tube.
Has binding reached equilibrium?	Competitive binding incubations take longer to equilibrate than saturation binding incubations. You should incubate for 4-5 times the half-life for radioligand dissociation.
Does only a small fraction of the radioligand bind?	The equations are based on the assumption that the free concentration of labeled ligand is essentially identical to the concentration you added. Compare the total binding in the absence of competitor in cpm to the amount of ligand added in cpm. If the ratio is greater than 10% at any concentration, then you've violated this assumption. Try to revise your experimental protocol, perhaps using a large incubation volume.
Does the curve have the expected steepness?	The competitive binding curve has a Hill slope (or slope factor) of –1. If your data form a curve shallower than this, see "Shallow competitive binding curves" on page 215.

K_i from IC_{50}

The three-parameter, one-site competitive binding equation, given above, fits your binding curve to find the IC_{50}, the concentration of competitor that competes for half the specific binding. This is a measure of the competitor's *potency* for interacting with the receptor against the radioligand, but it should *not* be confused as a measure of the competitor's *affinity*.

The value of the IC_{50} is determined by three factors:

- The affinity of the receptor for the competing drug. If the affinity is high, the IC_{50} will be low. The affinity is usually quantified as the equilibrium dissociation constant, K_i. The subscript "i" is used to indicate that the competitor inhibited radioligand binding. You can interpret it the same as you interpret a K_d. The K_i is the concentration of the competing ligand that will bind to half the binding sites at equilibrium, in the absence of radioligand or other competitors. If the K_i is low, the affinity of the receptor for the inhibitor is high.

- The concentration of the radioligand. If you choose to use a higher concentration of radioligand, it will take a larger concentration of unlabeled drug to compete for half the radioligand binding sites.

- The affinity of the radioligand for the receptor (K_d). It takes more unlabeled drug to compete for a tightly bound radioligand (low K_d) than for a loosely bound radioligand (high K_d).

You calculate the K_i, using the equation of Y. Cheng and W. H. Prusoff (*Biochem. Pharmacol.* 22: 3099-3108, 1973).

$$K_i = \frac{IC_{50}}{1+\frac{[\text{Radioligand}]}{K_d}}$$

Fitting the K_i directly

The previous discussion assumes that you fit a model to find the best-fit value of the IC_{50} and then compute the K_i from that. It is not difficult to rewrite the model to directly determine the K_i. Use this equation:

$$Y = \text{Bottom} + \frac{(\text{Top} - \text{Bottom})}{1+10^{(X-\text{LogIC}_{50})}}$$

where

$$\text{LogIC}_{50} = \text{Log}\left(\frac{10^{\text{LogK}_i}}{1+10^{(\text{Log[Radioligand]}-\text{LogK}_d)}}\right)$$

As a user-defined equation for programming into a nonlinear regression program:

```
IC50=(10^LogKi)/(1+(10^(LogRadioligand-LogKd)))
LogIC50=Log(IC50)
Y=Bottom+(Top-Bottom)/(1+10^(X-LogIC50))
```

For this analysis, be sure to constrain the values of *LogRadioligand* (logarithm of radioligand concentration) and *LogKd* (logarithm of radioligand K_d value) as constants.

IC_{50} or K_i versus log IC_{50} or log K_i

The standard competitive binding equations above are defined in terms of the logarithm of the IC_{50} or the logarithm of the K_i, so the nonlinear regression algorithm will find the best-fit value of the log IC_{50} or log K_i along with its SE and 95% CI. Depending on your program, the algorithm may also report the IC_{50} (or K_i) and its 95% CI. It does this by taking the antilog of the log IC_{50} (or log K_i) and of both ends of the 95% CI. Since the confidence interval is symmetrical on the log scale, it is not symmetrical when converted to IC_{50} or K_i.

If the concentrations of unlabeled compound are equally spaced on a log scale, the uncertainty of the log IC_{50} or log K_i will be symmetrical, but the uncertainty of the IC_{50} or K_i will not be. That is why the equations are written in terms of log IC_{50} or log K_i. This is a general feature of models that deal with measures of drug affinity and/or potency. If you average together results from several experiments, it is better to average the log IC_{50} or log K_i values, rather than the IC_{50} or K_i values. If you average IC_{50} or K_i values, one value that is far from the rest will have too much influence on the mean. See "Why you should fit the logEC_{50} rather than EC_{50}" on page 263.

Shallow competitive binding curves

The slope factor (Hill slope)

If the labeled and unlabeled ligands compete for a single class of binding site, the competitive binding curve will have a shape determined by the law of mass action. In this case, the curve will descend from 90% specific binding to 10% specific binding over an 81-fold increase in the concentration of the unlabeled drug. More simply, virtually the entire curve will cover two log units (100-fold change in concentration).

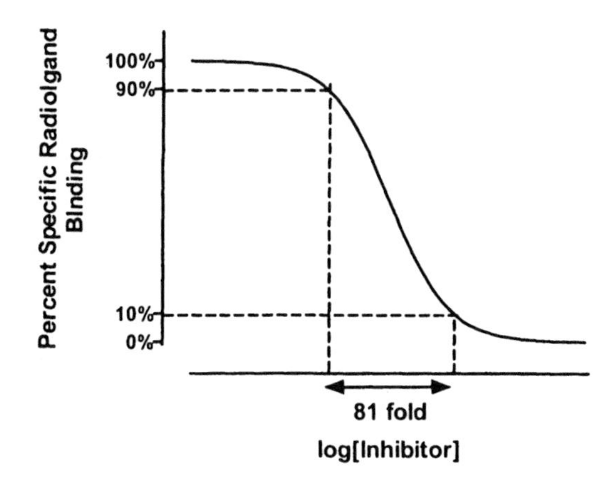

To quantify the steepness of a competitive binding curve, you must fit your data to a sigmoid equation that contains a parameter that specifically describes the effect of changing the midpoint slope. One such example is the four-parameter *Hill equation*, which is described in Chapter 41. The Hill equation is identical to the three-parameter competitive binding equation described above, except for the addition of a fourth parameter (usually called the *HillSlope* or the *slope factor*), which allows the curve to be either shallower or steeper than a hyperbola with a slope of 1.

$$Y=\text{Bottom}+\frac{(\text{Top-Bottom})}{1+10^{(X-\text{LogIC}_{50})\cdot\text{HillSlope}}}$$

The nonlinear regression program will then fit the top and bottom plateaus, the (log)IC_{50}, and the Hill slope factor). A standard competitive binding curve that follows the law of mass action has a slope of -1.0. If the slope is shallower, the Hill slope factor will be a negative fraction, perhaps -0.85 or -0.60. Shown below is the syntax for programming this equation into a nonlinear regression program.

```
Y=Bottom+(Top-Bottom)/(1+10^(X-LogIC50)*HillSlope)
```

The slope factor describes the steepness of a curve. In most situations, there is no way to interpret the value in terms of chemistry or biology. If the slope factor is far from 1.0, then the binding does not follow the law of mass action at a single site.

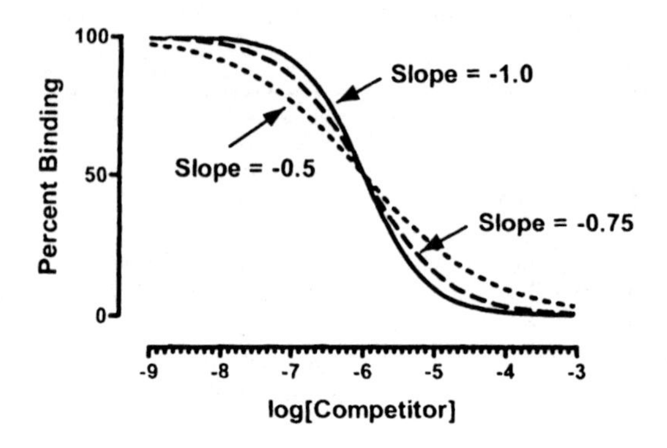

Explanations for shallow binding curves include:

Explanation	Comment
Experimental problems	If the serial dilution of the unlabeled drug concentrations was done incorrectly, the slope factor is not meaningful.
Curve-fitting problems	If the top and bottom plateaus are not correct, then the slope factor is not meaningful. Don't try to interpret the slope factor unless the curve has clear top and bottom plateaus. You may need to set the variables *Top* and *Bottom* to constant values.
Negative cooperativity	You will observe a shallow binding curve if the binding sites are clustered (perhaps several binding sites per molecule) and binding of the unlabeled ligand to one site causes the remaining site(s) to bind the unlabeled ligand with lower affinity.
Heterogeneous receptors	The receptors do not all bind the unlabeled drug with the same affinity.
Ternary complex mechanism	If agonist-bound receptors interact with an accessory protein to form a ternary complex, agonist curves can become shallow if the availability of the accessory protein becomes limiting relative to the concentration of receptor and agonist.

Competitive binding curves can also (but rarely) be steep, with Hill slopes significantly *more negative* (smaller) than -1. Explanations include:

Explanation	Comment
Wrong concentrations	If the serial dilution of the unlabeled drug concentrations was done incorrectly, the slope factor is not meaningful.
Not at equilibrium	Before equilibrium is reached, you don't expect a standard Hill slope of 1.0.
Curve-fitting problems	You may have collected too many points at the top and bottom of your curve, but very few in the middle, even though this is the portion of the curve that is most important in defining the inhibition.
Positive cooperativity	You will observe a steep binding curve if the binding sites are clustered (perhaps several binding sites per receptor) and binding of the unlabeled ligand to one site causes the remaining site(s) to bind the unlabeled ligand with higher affinity.

Note: If the slope factor of a competitive binding curve is significantly different from 1, then it is difficult to interpret the meaning of the IC_{50} in terms of any standard mechanistic mass-action models. Therefore, you should *not* attempt to derive a K_i value from the IC_{50}, as it would be meaningless. Rather, you can simply quote the IC_{50} as an empirical measure of competitor potency.

Competitive binding with two sites (same K_d for hot ligand)

It is quite common to see competitive binding curves that deviate significantly from the expectations of simple mass-action binding to a single site. The most common reason for this is because the drugs are interacting with two distinct classes of receptors. For example, a tissue could contain a mixture of β_1 and β_2 adrenoceptors. In this instance, the one-site competitive binding model can be readily extended to accommodate two binding sites, as follows:

$$Y=\text{Bottom}+(\text{Top-Bottom})\left[\frac{\text{Fraction}}{1+10^{X-\text{LogIC50}_1}}+\frac{1\text{-Fraction}}{1+10^{X-\text{LogIC50}_2}}\right]$$

In order to use this equation, the following assumptions must also be met:

- The unlabeled ligand has distinct affinities for the two sites.
- The labeled ligand has equal affinity for both sites. (If you are not willing to make this assumption, see "Competitive binding with two receptor types (" on page 219.)
- Binding has reached equilibrium.
- A small fraction of both labeled and unlabeled ligand bind. This means that the concentration of labeled ligand that you added is very close to the free concentration in all tubes.

This equation has five variables: the top and bottom binding plateau, the fraction of the receptors of the first class, and the IC_{50} for competition of the unlabeled ligand at both classes of receptors. If you know the K_d of the labeled ligand and its concentration, you can convert the IC_{50} values to K_i values.

When you look at the competitive binding curve, you will only see a biphasic curve in unusual cases where the affinities are extremely different. More often you will see a shallow curve with the two components blurred together. For example, the graph below shows competition for two equally abundant sites with a tenfold (one log unit) difference in IC_{50}. If you look carefully, you can see that the curve is shallow (it takes more than two log units to go from 90% to 10% competition), but you cannot see two distinct components.

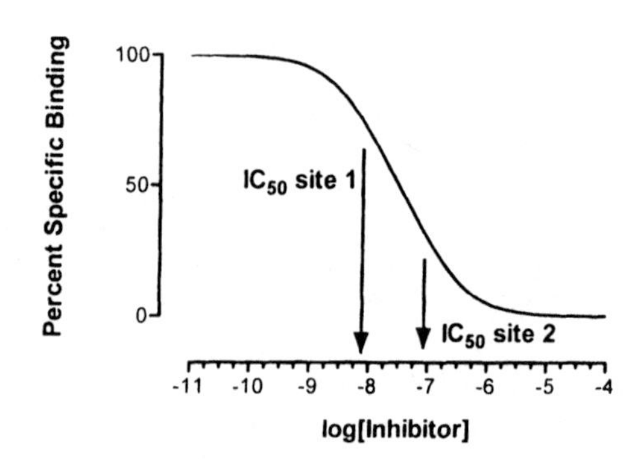

Comparing one- and two-site models

Often, it is a good idea to fit your competitive binding data to both the one-site competitive binding model and a two-site competitive binding model, and then compare the fits. Since the two-site model has two extra parameters, and thus the curve has extra inflection points, the model almost always fits the data better than the one site model. And a three-site model fits even better. Before accepting the more complicated model(s), you need to ask whether the improvement in goodness of fit is more than you'd expect by chance. The most common way to answer this question is to use an F test or Akaike's Information Criterion. Model comparison using these methods was discussed in Section G of this book.

Before comparing the two models, you should look at the results of the actual curve fits yourself. Sometimes the two-site fit gives results that are clearly nonsense. For example, disregard a two-site fit when:

- The two IC_{50} values are almost identical.
- One of the IC_{50} values is outside the range of your data.
- The best-fit value for *Fraction* is close to 1.0 or to 0.0. In this case, virtually all the receptors have the same affinity, and the IC_{50} value for the other site will not be reliable.
- The best-fit value for *Fraction* is negative or greater than 1.0.
- The best-fit values for *Bottom* or *Top* are far from the range of Y values observed in your experiment.

If the results don't make sense, don't believe them. Only pay attention to the comparison of two fits when each of the fits makes sense.

Competitive binding with two receptor types (different K_d for hot ligand)

The standard equation for competitive binding to two sites assumes that the labeled ligand has equal affinity for both sites. It is easy to derive an equation for situations where the labeled ligand binds differently to the two sites.

This is one form of the standard equation for competitive binding to one site:

$$Y=\frac{[\text{Hot Ligand}]\cdot B_{max}}{[\text{Hot Ligand}]+K_d+\frac{K_d}{K_i}\cdot[\text{Cold Ligand}]}+\text{Nonspecific}$$

Binding is the sum of specific and nonspecific binding. To create an equation for two sites, you simply need to create an equation with two specific binding components with different values for B_{max}, K_d, and K_i: This is shown below as a user-defined equation.

```
ColdnM=10^(X+9)
KI1nM = 10^(LogKI1+9)
KI2nM = 10^(LogKI2+9)
SITE1= HotnM*Bmax1/(HotnM + KD1*(1+coldnM/Ki1nM))
SITE2= HotnM*Bmax2/(HotnM + KD2*(1+coldnM/Ki2nM))
Y = SITE1 + SITE2 + NS
```

Make sure you enter data with X=log[unlabeled] and Y=cpm or dpm.

Variable	Units	Comments
X	log(Molar)	Concentration of unlabeled drug.
Y	cpm	Total binding of radioligand.
HotnM	nM	Concentration of labeled ligand added to each tube. Set to a constant value.
KD1	nM	K_d of the labeled ligand for the first site. Set to a constant value based on other experiments.
KD2	nM	K_d of the labeled ligand for the second site. Set to a constant value .
logKI1	log(Molar)	Affinity of the unlabeled drug for the first site. Try an initial value of 1.2 x your mid-range X value.
logKI2	log(Molar)	Affinity of the unlabeled drug for the second site. Try an initial value of 0.8 x your mid-range X value.
Bmax1	Units of Y axis, usually cpm	Try an initial value of 2 x your maximum Y value. (This assumes that you've used a concentration of radioligand that binds to half of the receptors. You may wish to adjust this.)
Bmax2	Units of Y axis, usually cpm	Try an initial value of 2 x your maximum Y value. (This assumes that you've used a concentration of radioligand that binds to one tenth of the receptors.)
NSCPM	Units of Y-axis, usually cpm.	Nonspecific binding. Try an initial value of 1 x your minimum Y value.

Notes:

- This equation does not account for ligand depletion. It assumes that the free concentration equals the added concentration.

- When using this equation to fit data, you will need to assign constant values to *KD1* and *KD2*, the *KD* of the hot ligand for the two sites. You will need to obtain these values from other experiments. Perhaps you can isolate tissue with only one of the receptor types and measure K_d in that preparation.

Heterologous competitive binding with ligand depletion

The standard sigmoidal equations used to fit competitive binding data assume that a small fraction of the radioligand binds relative to what is added. This means that the free concentration of radioligand is almost equal to the concentration you added, and that the free concentration is the same in all tubes in the assay.

If a large (say greater than 10%) fraction of the radioligand binds to receptors, then the free concentration will be less than the added concentration of radioligand. The discrepancy between free and added radioligand concentration depends on the concentration of the unlabeled drug. The standard equation for competitive binding, shown below, needs two corrections to account for ligand depletion.

$$Y = \frac{[\text{Free Ligand}] \cdot B_{max}}{[\text{Free Ligand}] + K_d \left(1 + \frac{[\text{Cold Ligand}]}{K_i}\right)} + \text{Nonspecific}$$

The free concentration of labeled ligand equals the amount you added minus the amount that bound.

$$[\text{Free ligand}] = [\text{Added ligand}] - Y$$

The nonspecific binding is not the same for all tubes. As you increase the concentration of cold ligand, less radioligand binds to receptors so the free concentration of radioligand increases. Since nonspecific binding is assumed to be proportional to the free concentration of radioligand, there will be more nonspecific binding in the tubes with higher concentrations of unlabeled drug.

$$\text{Nonspecific binding} = \text{NS} \cdot [\text{Free ligand}]$$

Y, [Free ligand], and [Added ligand] are expressed in units of cpm. To be consistent, therefore, the K_d also needs to be expressed in cpm units. [Cold ligand] and K_i are expressed in the same units (molar), so the ratio is unitless.

Combine these equations, and you end up with a complicated quadratic equation whose solution is shown here:

$$Y = \frac{-b + \sqrt{b^2 - 4 \cdot a \cdot c}}{2 \cdot a}$$

where

$$a=-(NS+1)$$

$$b=(NS+1)\cdot\left([1+10^{(LogCold-LogK_i)}]\cdot K_d+[\text{Added Ligand}]\right)+NS\cdot[\text{Added Ligand}]+B_{max}$$

$$c=-[\text{Added Ligand}]\cdot\left[\left(([1+10^{(LogCold-LogK_i)}]\cdot K_d+[\text{Added Ligand}]\right)\cdot NS+B_{max})\right]$$

```
KdCPM=KdnM*SpAct*vol*1000
R=NS+1
S=[1+10^(X-LogKi)]*KdCPM+Hot
a=-1*R
b=R*S+NS*Hot + Bmax
c= -1*Hot*(S*NS + Bmax)
Y= (-1*b + sqrt(b*b-4*a*c))/(2*a)
```

Variable	Units	Comments
X	log(Molar)	Logarithm of the concentration of unlabelled competitor.
Y	CPM	Bound labeled ligand.
Hot	CPM	Amount of labeled ligand added to each tube. Set to a constant value.
SpAct	cpm/fmol	Specific radioactivity of labeled ligand. Set to constant value.
Vol	ml	Incubation volume. Set to a constant value.
KdnM	nM	K_d of labeled ligand. Set to a constant value.
LogKi	log(Molar)	Try an initial value of 1 x your mid-range X value
Bmax	Units of Y axis, usually cpm	Try an initial value of 10 x your maximal Y value (This assumes that you've used a concentration of radioligand that binds to one tenth of the receptors. You may wish to adjust this.)
NS	Unitless fraction	Initial value =0.01

You need to set four of the parameters to constant values. *Hot* is the number of cpm of labeled ligand added to each tube. *SpAct* is the specific activity of the radioligand in cpm/fmol, *Vol* is the incubation volume in ml, and *KdnM* is the Kd of the radioligand in nM (determined from other experiments). The nonlinear regression algorithm then fits this equation to your data to determine the *logKI*. It also fits two other variables, which are of less interest: *Bmax*, which is the maximum binding of radioligand (if present at a saturating concentration) in cpm, and *NS*, which is the fraction of the free ligand that binds nonspecifically.

Note that this equation accounts for depletion of the hot ligand only, and does *not* adjust for depletion of the unlabeled compound. Also note that this equation is not easily extended to a situation with two binding sites.

38. Homologous competitive binding curves

Introducing homologous competition

The most common way to determine receptor number and affinity is to perform a saturation binding experiment, where you vary the concentration of radioligand. An alternative is to keep the radioligand concentration constant, and compete for binding with the same ligand, but not radioactively labeled. Since the hot (radiolabeled) and cold (unlabeled) ligands are chemically identical, this is called a *homologous competitive binding* experiment. It is also sometimes called a *cold saturation experiment*. An advantage of performing homologous competition binding experiments for determining radioligand K_d and B_{max} values, rather than the more standard saturation binding method, is that you use less radioligand in the homologous competition experiment and generally have lower nonspecific binding than in the saturation assay.

Most analyses of homologous competition data are based on these assumptions:

Assumption	Comments
The receptors have identical affinity for the labeled and unlabeled ligand.	This is a nontrivial assumption. With tritiated ligands, there is no reason to doubt it, since tritium doesn't greatly alter the conformation of a molecule. However, iodination can change conformation and alter the binding affinity. Don't assume that iodinated and noniodinated compounds bind with the same affinity. If your radioligand is labeled with radioactive iodine, then you should use a competitor that is the same compound iodinated with nonradioactive iodine.
There is no cooperativity.	This means that binding of ligand to one binding site does not change its affinity at other site(s).
Ligand is not depleted.	The basic analysis methods assume that only a small fraction of ligand binds. In other words, the method assumes that free concentration of hot (and cold) equals the concentration you added. Since homologous competition curves are best performed with low concentrations of radioligand, it may be difficult to comply with this assumption. If a large fraction of radioligand binds, you can lower the fractional binding by increasing the incubation volume (without increasing the amount of tissue). A later section in this chapter explains how to analyze data when this assumption is not valid.
Nonspecific binding is proportional to the concentration of labeled ligand.	We assume that a certain fraction of hot ligand binds nonspecifically, regardless of the concentration of unlabeled ligand. This assumption has proven to be true in many systems.

Theory of homologous competition binding

Start with the equation for equilibrium binding to a single class of receptors.

$$\text{Specific Binding}=\frac{[\text{Ligand}]\cdot B_{max}}{[\text{Ligand}]+K_d}$$

Set [Ligand] equal to the sum of the labeled (hot) and unlabeled (cold) ligand. Specific binding you measure (specific binding of the labeled ligand) equals specific binding of all ligand times the fraction of the ligand that is labeled. This fraction, hot/(hot+cold), varies from tube to tube. Therefore specific binding of labeled ligand (Y) follows this equation:

$$\begin{aligned}Y&=\text{Sp. binding of all ligand}\cdot\text{Fraction of ligand that is hot}\\&=\frac{([\text{Hot}]+[\text{Cold}])\cdot B_{max}}{[\text{Hot}]+[\text{Cold}]+K_d}\cdot\frac{[\text{Hot}]}{[\text{Hot}]+[\text{Cold}]}\\&=\frac{B_{max}\cdot[\text{Hot}]}{[\text{Hot}]+[\text{Cold}]+K_d}\end{aligned}$$

Specific binding and B_{max} are in the same units, usually cpm, dpm, sites/cell, or fmol/mg. [Hot], [Cold], and K_d are in concentration units. Those units cancel so it doesn't matter if you use molar, nM, or some other unit, so long as you are consistent.

Maximum binding of labeled ligand occurs when the concentration of cold ligand equals zero. This is not the same as B_{max}, because the concentration of hot ligand will not saturate all the receptors. In the absence of cold ligand (set [cold]=0), the binding equals

$$\text{Specific Binding}_{[\text{Cold}]=0}=\frac{B_{max}\cdot[\text{Hot}]}{[\text{Hot}]+K_d}$$

The IC_{50} in a homologous binding experiment is the concentration of [Cold] that reduces specific binding of labeled ligand by 50%. So the IC_{50} is the concentration of cold that solves the equation below. The left side of the equation is half the maximum binding with no cold ligand. The right side is binding in the presence of a particular concentration of cold ligand. We want to solve for [Cold].

$$0.5\cdot\frac{B_{max}\cdot[\text{Hot}]}{[\text{Hot}]+K_d}=\frac{B_{max}\cdot[\text{Hot}]}{[\text{Hot}]+K_d+[\text{Cold}]}$$

Solve this equation for [Cold], and you'll find that you achieve half-maximal binding when [Cold] = [Hot] + K_d. In other words,

$$IC_{50}=[\text{Hot}]+K_d$$

Why homologous binding data can be ambiguous

Since the IC_{50} equals [Hot] + K_d, the value of the K_d doesn't affect the IC_{50} very much when you use a high concentration of radioligand. This means that you'll see the same IC_{50} with a large range of K_d values. For example, if you use a hot ligand concentration of 10 nM, the IC_{50} will equal 10.1 nM if the K_d is 0.1 nM (dashed curve below), and the IC_{50} will equal 11 nM if the K_d is 1 nM (solid curve below).

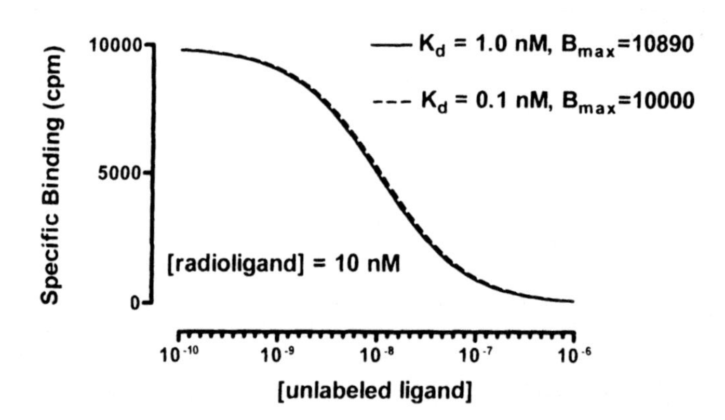

If you used 10 nM of radioligand and your data followed the course of the curves shown above, the analysis would simply be ambiguous. You wouldn't know if you have a low K_d with lots of receptors or a high K_d with few receptors.

If the concentration of hot (radiolabeled) ligand greatly exceeds the K_d, the curve is ambiguous. An infinite number of curves, defined by different K_d and B_{max} values, are almost identical. The data simply don't define the K_d and B_{max}. No curve-fitting program can determine the K_d and B_{max} from this type of experiment – the data are consistent with many K_d and B_{max} values.

Using global curve fitting to analyze homologous (one site) competition data

The difficulties associated with fitting homologous binding data discussed above can be minimized by performing the experiment with two, or more, concentrations of radioligand and using global curve-fitting (parameter sharing) to simultaneously fit the entire family of curves to the homologous binding model.

> Tip: We recommend that you perform homologous competition experiments with at least two radioligand concentrations, and analyze the data using global curve fitting.

To fit homologous binding data, we also need to account for the effects of nonspecific binding. The total measured binding equals specific binding plus nonspecific binding. Nonspecific binding is the same for all tubes since it only depends on the concentration of hot ligand, which is constant. The equation for specific binding is derived in "Theory of homologous competition binding" above. We simply need to add a nonspecific binding term to define total binding:

$$\text{Total Binding} = \frac{B_{max} \cdot [\text{Hot}]}{[\text{Hot}]+[\text{Cold}]+K_d} + \text{NS}$$

Here is an equation for homologous competition binding.

```
ColdNM=10^(x+9)
KdNM=10^(logKD+9)
Y=(Bmax*HotnM)/(HotnM + ColdNM + KdNM) + NS
```

This equation assumes that you have entered X values as the logarithm of the concentrations of the unlabeled ligand in molar, so 1 nM (10^{-9} molar) is entered as -9. The first line in the equation adds 9 to make it the logarithm of the concentration in nM, and then takes the antilog to get concentration in nM.

Y is the total binding of the radioligand, expressed in cpm, dpm, fmol/mg, or sites/cell. The Bmax will be expressed in the same units you use for Y.

Since the experiment is performed with the concentrations of unlabeled ligand equally spaced on a log scale, the confidence intervals will be most accurate when the K_d is fit as the logK_d. The second line converts the log of the K_d in moles/liter to nM.

The equation defines Y (total binding) as a function of X (the logarithm of the concentration of the unlabeled compound) and four parameters. We are going to use global curve fitting to fit both data sets at once, so need to tell the program which parameters to share.

Parameter	**Units**	**Comments**
Bmax	Same as Y values	You want to fit one value of B_{max} from both data sets, so set your program to share this parameter between data sets.
HotnM	nM	Concentration of labeled ligand in every tube. Set this to a constant value that you know from experimental design. This will have a different constant value for each data set. In Prism, make this a data set constant, so Prism will read its value from the column titles of the data table.
LogKd	Log(M)	You want to fit one value of K_d from both data sets, so set your program to share this parameter between data sets.
NS	Same as Y values	Nonspecific binding depends on how much radioligand we used. The equation, as written above, adds a constant term NS (in units of Y). Since we expect different amounts of nonspecific binding at each concentration of hot ligand, we do not share the value of NS. Instead, we ask the curve-fitting program to find a different best-fit value of NS for each curve.

To fit the model, you have to enter or choose initial values for each parameter.

Parameter	**Units**	**Comments**
Bmax	Same as Y values	Try an initial value of 10 × your maximal Y value (this assumes that you used a concentration of ligand that binds to about one tenth of these receptors).
LogKd	Log(moles/liter)	Try an initial value of 1.2 × your mid-range X value.
NS	Same as Y values	Try an initial value of 1 × your minimum Y value.

Shown in the figure below are results of homologous competition binding data using two concentrations of radioligand (0.3 and 1 nM). The nonlinear regression was set up so both the B_{max} and the K_dnM (K_d in nM concentration units) were shared between the data sets. This means that the program finds a single B_{max} and single K_d for both data sets. A separate estimate is derived for NS from each data set, however, as each radioligand concentration will result in a different degree of nonspecific binding. The parameter HotnM was set to a constant value for each data set (0.3 or 1).

> GraphPad note: To make a parameter have a different constant value for each data set, use Prism's constraint tab in the nonlinear regression dialog. Set the parameter to be a data set constant, and put the values into the column title of the data table. See page 300

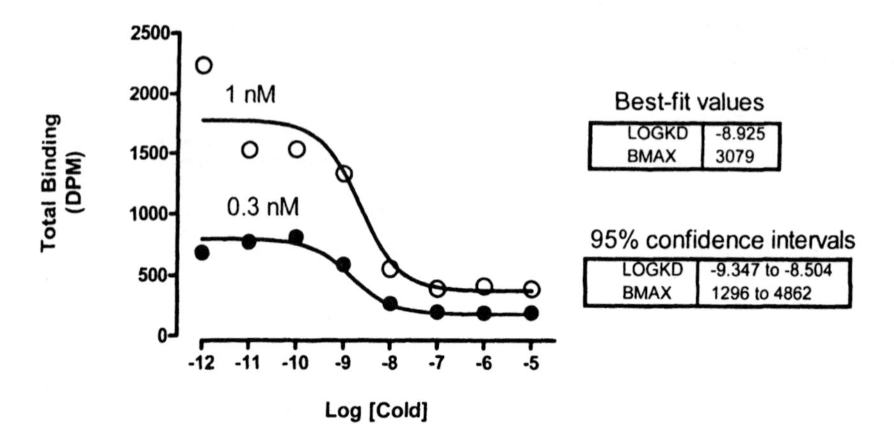

The confidence intervals of the global estimates of LogK_d and B_{max} are reasonable. By simultaneously analyzing both data sets, the results are far tighter than they would be if we fit each curve separately.

Analyzing homologous (one site) competition data without global curve fitting

There are two practical situations where you may not be able to use the global curve-fitting approach outlined above to analyze homologous competition binding data. The first situation is where your nonlinear regression program does not allow you to perform global curve-fitting with parameter-sharing across data sets. The second situation is where you have performed the experiment with only a single concentration of radioligand.

The ambiguity associated with fitting single curves to the homologous binding model has already been discussed above. If we take the same example from the previous section, but this time fit each curve individually instead of globally fitting the model to both data sets, we obtain the following results.

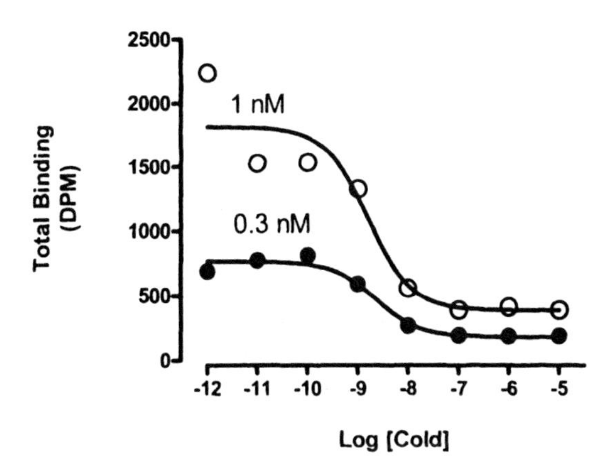

	[Hot] = 0.3 nM	[Hot] = 1 nM
Best-fit values		
$LogK_d$	-8.702	-9.238
B_{max}	4442	2259
Std. Error		
$LogK_d$	0.2967	0.6205
B_{max}	2661	1187
95% CI		
$LogK_d$	-9.319 to -8.085	-10.53 to -7.947
B_{max}	-1093 to 9978	-209.5 to 4727

Although the curves fit the data very well, both B_{max} and $logK_d$ parameters have large standard errors and accordingly wide 95% confidence intervals. This reflects the difficulty inherent in obtaining reliable parameter estimates from any one homologous competitive binding curve.

> Note: If you have a computer program that can perform global curve fitting, then we recommend you perform any homologous competition experiment using two radioligand concentrations and globally fit both data sets to the homologous binding model.

If you cannot use global curve fitting to analyze your data, or you are going to do the experiment with a single radioligand concentration, then you should follow these steps to ensure that you increase your chances of obtaining quality data.

Step 1: Determine the IC_{50}

This first step is to check that you used a reasonable concentration of radioligand.

Fit your data to the standard ***one-site competition*** equation. If your competition curve doesn't have clearly defined top and bottom plateaus, you should set one or both of these to constant values based on control experiments.

Compare the best-fit value of the IC_{50} to the concentration of hot ligand you used. Homologous competition experiments only lead to useful results when the concentration of hot ligand is less than half the IC_{50}.

If	Then do this
IC50 is more than ten times [Hot]	The concentration of [Hot] is lower than it needs to be. If you have quality data, you can continue with step 2. If you have too few cpm bound or if more than 10% of the added radioactivity binds to tissue, rerun the experiment with more radioligand
IC50 is between two and ten times [Hot]	You've designed the experiment appropriately. Continue to step 2 to determine the K_d and B_{max} with confidence limits.
IC50 is greater than [Hot] but not twice as high	You'll get better results if you repeat the experiment using less hot ligand.
IC50 is less than [Hot]	If binding follows the assumptions of the analysis, it is impossible for the IC_{50} to be less than the concentration of hot ligand. Since $K_d = IC_{50} - [Hot]$, these data would lead to a negative value for K_d, which is impossible. One possibility is that your system does not follow the assumptions of the analysis. Perhaps the hot and cold ligands do not bind with identical affinities. Another possibility is that you simply used too high a concentration of hot ligand, so the true IC_{50} is very close to (but greater than) [Hot]. Experimental scatter may lead to a best-fit IC_{50} that is too low, leading to the impossible results. Repeat the experiment using a lot less hot ligand.

Step 2: Determine K_d and B_{max}

Once you've determined that the IC_{50} is quite a bit larger than the concentration of hot ligand, fit the homologous competition binding model (see previous section) to determine the B_{max} and K_d. Set HotnM equal to the concentration of labeled ligand in nM, and set it to be a constant value, otherwise the nonlinear regression procedure will not be able to converge on reasonable estimates for your B_{max} and K_d parameters.

If you have performed a series of homologous competition binding experiments using different radioligand concentrations, but cannot fit your data using global-fitting, you can still derive useful information from your experiments that can optimize your fit.

From the one-site homologous competition binding model, we have shown previously that

$$IC_{50} = [Hot] + K_d$$

By performing a series of homologous competition experiments using different concentrations of radioligand, you can plot each IC_{50} value against the corresponding value of [Hot]; a linear regression of these data should yield an estimate of the K_d value. Unfortunately, you cannot determine B_{max} from this approach.

Homologous competitive binding with ligand depletion

If a large fraction of the radioligand is bound to the receptors, homologous binding will be affected. Although you add the same amount of labeled ligand to each tube, the free concentration will not be the same. High concentrations of unlabeled drug compete for the binding of the labeled ligand, and thus increase the free concentration of the labeled ligand.

Nonspecific binding is also affected by ligand depletion. Since nonspecific binding is proportional to the free concentration of labeled ligand, the amount of nonspecific binding will not be the same in all tubes. The tubes with the highest concentration of unlabeled drug have the highest concentration of free radioligand, so they will have the most nonspecific binding.

Because the free concentration varies among tubes, as does the nonspecific binding, there is no simple relationship between IC_{50} and K_d. The IC_{50} is nearly meaningless in a homologous binding curve with ligand depletion.

The equations for homologous binding with ligand depletion are quite a bit more complicated than for homologous binding without depletion. The math that follows is adapted from S. Swillens (*Molecular Pharmacology*, 47: 1197-1203, 1995).

Start with the equation for total binding in homologous competition as a function of the free concentration of radioligand.

$$\text{Specific} = \frac{B_{max} \cdot [\text{Free Radioligand, nM}]}{K_d + [\text{Free Radioligand, nM}] + [\text{Free Cold ligand, nM}]}$$

$$\text{Nonspecific} = [\text{Free Radioligand, cpm}] \cdot \text{NS}$$

$$Y = \text{Specific} + \text{Nonspecific}$$

This equation defines total binding as specific binding plus nonspecific binding. Nonspecific binding equals a constant fraction of free radioligand, and we define this fraction to be NS. To keep units consistent, the radioligand concentration is expressed in nM in the left half of the equation (to be consistent with K_d and the concentration of cold ligand) and is expressed in cpm on the right half of the equation (to be consistent with Y).

The problem with this equation is that you don't know the concentrations of free radioligand or free cold ligand. What you know are the concentrations of labeled and unlabeled ligand you added. Since a high fraction of ligand binds to the receptors, you cannot assume that the concentration of free ligand equals the concentration of added ligand.

Defining the free concentration of hot ligand is easy. You added the same number of cpm of hot ligand to each tube, which we'll call *HotCPM*. The concentration of free radioligand equals the concentration added minus the total concentration bound, or *HotCPM*-Y (both *HotCPM* and Y are expressed in cpm).

Defining the free concentration of cold ligand is harder, so it is done indirectly. The fraction of hot radioligand that is free equals (*HotCPM* - Y)/*HotCPM*. This fraction will be different in different tubes. Since the hot and cold ligands are chemically identical, the fraction of cold ligand that is free in each tube is identical to the fraction of hot ligand that is free. Define X to be the logarithm of the total concentration of cold ligand, the variable you vary in a homologous competitive binding experiment. Therefore, the total concentration of cold ligand is 10^X, and the free concentration of cold ligand is 10^X(*HotCPM* - *Y*)/*HotCPM*.

Substitute these definitions of the free concentrations of hot and cold ligand into the equation above, and the equation is still unusable. The problem is that the variable Y appears on both sides of the equal sign. Some simple, but messy, algebra puts Y on the left side of a quadratic equation, shown below as a user-defined equation:

$$Y = \frac{-b + \sqrt{b^2 - 4ac}}{2a}$$

where

$$a = -(NS + 1) \cdot [\text{Total(nM)}]$$

$$b = [\text{Hot(cpm)}] \cdot ([\text{Total(nM)}] + K_D) \cdot (NS + 1) + [\text{Total(nM)}] \cdot [\text{Hot(cpm)}] \cdot NS + B_{max} \cdot [\text{Hot(nM)}]$$

$$c = -[\text{Hot(cpm)}] \cdot ([\text{Total(nM)}] + K_D) \cdot [\text{Hot(cpm)}] \cdot NS - [\text{Hot(cpm)}] \cdot B_{max} \cdot [\text{Hot(nM)}]$$

```
ColdnM=10^(X+9)
KDnM=10^(LogKD+9)
HotnM=HotCPM/(SpAct*vol*1000)
TotalnM=HotnM+ColdnM
Q=HotCPM*(TotalnM + KDnM)
a=(NS+1)*TotalnM*-1
b=Q*(NS+1)+TotalnM*HotCPM*NS + Bmax*HotnM
c=-1*Q*HotCPM*NS - HotCPM*Bmax*HotnM
Y= (-1*b + sqrt(b*b-4*a*c))/(2*a)
```

When fitting data to this equation, you need to set three parameters to constant values. *HotCPM* is the number of cpm of hot ligand added to each tube. *Vol* is the incubation volume in ml. *SpAct* is the specific radioactivity in cpm/fmol. The nonlinear regression algorithm then fits *Bmax* in the units of the Y axis (usually cpm, which you can convert to more useful units) and *logKd* as log molar.

As with homologous binding data without ligand depletion, we recommend that you perform the experiment using more than one radioligand concentration and fit the data using global-fitting to share a single B_{max} value and single K_d value across all data sets.

Variable	Units	Comments
X	log(Molar)	
Y	cpm	
HotCPM	cpm	Amount of labeled ligand added to each tube. Set to a constant value.
SpAct	cpm/fmol	Specific radioactivity of labeled ligand. Set to a constant value.
Vol	ml	Incubation volume. Set to a constant value.
logKd	Log(Molar)	Initial value = 1 x the mid-range X value
Bmax	Units of Y axis, usually cpm	Initial value = 10 x your maximal Y value (This assumes that you've used a concentration of radioligand that binds to one tenth of the receptors. You may wish to adjust this.)
NS	Unitless fraction	This is the fraction of free ligand that binds nonspecifically. Initial value =0.01

Fitting homologous competition data (two sites)

With some systems it is possible to determine K_d and B_{max} values for two independent sites using homologous competition data. With most systems, however, you won't get reliable results.

You can only determine B_{max} and K_d from homologous binding data if you use a concentration of hot ligand that is much lower than the K_d value. If your system has two binding sites, you must choose a concentration much lower than the K_d of the high affinity site. Using such a low concentration of radioligand, you'll bind only a small fraction of low-affinity sites. You will only be able to detect the presence of the second, low-affinity, site if they are far more abundant than the high-affinity sites.

For example, imagine that the low affinity site (K_d=10 nM) is ten times as abundant as the high-affinity site (K_d=0.1 nM). You need to use a concentration of hot ligand less than 0.1 nM, say 0.05 nM. At this concentration you bind to 33.33% of the high-affinity sites, but only to 0.0049% of the low affinity sites. Even though the low-affinity sites are ten times as abundant, you won't find them in your assay (low-affinity binding will be only 0.15% of the binding).

To attempt to determine the two K_d and B_{max} values from a homologous competition curve, fit the data to the equation below. Assuming no cooperativity and no ligand depletion, the binding to each site is independent and depends on the Bmax and K_d values of each site. The binding that you measure, Y, is the sum of the binding to the two receptor sites plus nonspecific binding.

```
Site1=(Bmax1*HotnM)/(HotnM + 10^(X+9)+ 10^(logKd1+9))
Site2=(Bmax2*HotnM)/(HotnM + 10^(X+9)+ 10^(logKd2+9))
Y= site1 + site2 + NS
```

This is a difficult equation to fit, and you will almost certainly have to try many sets of initial values to converge on a reasonable solution. It is especially important to adjust the initial values of the two *Bmax* values.

Variable	Units	Comments
X	log(molar)	Concentration of the unlabeled compound.
Y	cpm or fmol/mg or sites/cell.	Total binding of the labeled compound.
Bmax1	Same as Y values.	Initial value =1 x your maximum Y value (this assumes that you used a concentration of ligand that binds to almost all of the high affinity class of receptors).
Bmax2	Same as Y values.	Initial value = 20 x your maximum Y value (this assumes that you used a concentration of ligand that binds to five percent of the second class of receptors).
LogKd1	log(molar)	Initial value = 1.2 x your mid-range X value.
LogKd2	log(molar)	Initial value = 0.8 x your mid-range X value.
NS	Same as Y values.	Initial value = = 1. x your minimum Y value.
HotnM	nM	Concentration of labeled ligand in every tube. Set this to a constant value that you know from your experimental design.

Consider this approach for analyzing homologous competitive binding data to determine the characteristics of two sites. First use a very low concentration of radioligand and fit to a single site. This will determine the B_{max} and K_d of the high affinity site. Then repeat the experiment with a higher concentration of radioligand. Fit these data to the two-site equation, but set *Kd* and *Bmax* for the high affinity site to constant values determined from the first experiment. Alternatively, and if your nonlinear regression program allows it, you can perform a global fit of both equations simultaneously to both data sets by using parameter-sharing, similar to the detailed method outlined above for homologous competition binding to one binding site.

Advantages and disadvantages of homologous binding experiments

Determining receptor number with homologous binding has one clear advantage: You need far less radioligand than you would need if you performed a saturation binding experiment. This reason can be compelling for ligands that are particularly expensive or difficult to synthesize.

The disadvantage of determining receptor number and affinity from a homologous competitive binding experiment is that it can be hard to pick an appropriate concentration of radioligand. If you use too little radioligand, you'll observe little binding and will obtain poor quality data. If you use too much radioligand, the curve will be ambiguous and you won't be able to determine B_{max} and K_d. The use of a minimum of two different radioligand concentrations in a homologous binding assay, in conjunction with the global curve-fitting approach outlined above, provides the best compromise between saturation binding assays and standard (one-curve) homologous competition binding assays.

Using homologous binding to determine the K_d and B_{max} of two binding sites with is difficult, even with multiple radioligand concentrations. You are probably better off using a saturation binding experiment.

39. Analyzing kinetic binding data

Dissociation ("off rate") experiments

A dissociation binding experiment measures the "off rate" for radioligand dissociating from the receptor. Initially ligand and receptor are allowed to bind, perhaps to equilibrium. At that point, you need to block further binding of radioligand to receptor so you can measure the rate of dissociation. There are several ways to do this:

- If the tissue is attached to a surface, you can remove the buffer containing radioligand and replace with fresh buffer without radioligand.
- Spin the suspension and resuspend in fresh buffer.
- Add a very high concentration of an unlabeled ligand. If this concentration is high enough, it will instantly bind to nearly all the unoccupied receptors and thus block binding of the radioligand.
- Dilute the incubation by a large factor, at least 100 fold dilution. This will reduce the concentration of radioligand by that factor. At such a low concentration, new binding of radioligand will be negligible. This method is only practical when you use a fairly low concentration of radioligand so its concentration after dilution is far below its K_d for binding.

You then measure binding at various times after that to determine how rapidly the radioligand falls off the receptors.

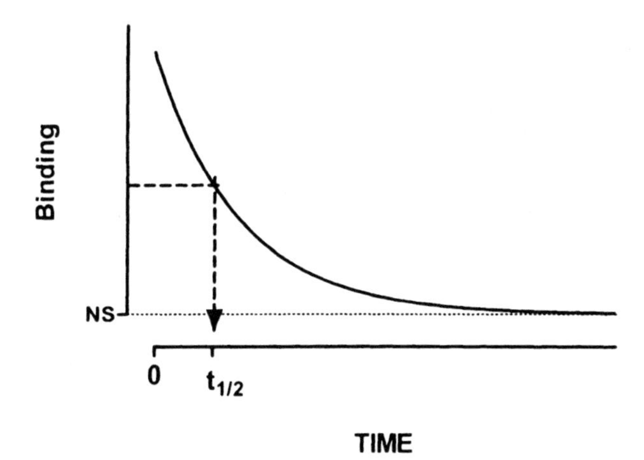

Each ligand-receptor complex dissociates at a random time, so the amount of specific binding follows an exponential dissociation.

$$Y = \text{Span} \cdot e^{-K \cdot X} + \text{Plateau}$$

Variable	Meaning	Comment
X	Time	Usually expressed in units of seconds or minutes.
Y	Total binding	Usually expressed in units of cpm, dpm, fmol/mg, or sites/cell.
Span	Difference between binding at time zero and plateau.	Specific binding (same units as Y).
Plateau	Binding that doesn't dissociate.	Nonspecific binding (same units as Y).
K	Dissociation rate constant often called k_{off}.	Expressed in units of inverse time (inverse of units of X-axis).

Analyzing dissociation data

To analyze dissociation binding data:

1. Enter the data with X equal to time after you initiated dissociation and Y equal to binding (usually total binding).

2. Perform nonlinear regression using an exponential decay equation.

3. If you entered specific (rather than total) binding as your dependent variable, make the parameter, *Plateau,* a constant equal to zero. If you have entered total binding, leave the parameter *Plateau* as to be fitted.

4. Look at the nonlinear regression results. The variable *K* is the dissociation constant (often referred to in texts as k_{off} or k_{-1}) expressed in units of inverse time. If you entered the X values as minutes, k_{off} is in units of min^{-1}. The results also show the half-life in units of time (minutes in this example).

5. You can use the k_{off} value to determine the half-life ($t_{1/2}$) of dissociation. Use the following relationship: $t_{1/2} = 0.693/k_{off}$.

Association binding experiments

Association binding experiments are used to determine the association rate constant. You add radioligand and measure specific binding at various times thereafter.

Binding follows the law of mass action:

$$\text{Receptor} + \text{Ligand} \underset{K_{off}}{\overset{K_{on}}{\rightleftarrows}} \text{Receptor}\bullet\text{Ligand}$$

At any given time, the rate at which receptor-ligand complexes form is proportional to the radioligand concentration and the number of receptors still unoccupied. The rate of dissociation is proportional to the concentration of receptor-ligand complexes.

Binding increases over time until it plateaus when specific binding equals a value we call Y_{max}. This is not the same as B_{max}. Y_{max} is the amount of specific binding at equilibrium for a certain concentration of ligand used in an association experiment. B_{max} is the maximum amount of binding extrapolated to a very high concentration of ligand. These principles let us define the model mathematically.

$$\begin{aligned}
\text{Rate of association} &= [\text{Receptor}]\bullet[\text{Ligand}]\times k_{on} \\
&= \left(Y_{max} - Y\right)\cdot[\text{Ligand}]\times k_{on} \\
\text{Rate of dissociation} &= [\text{Receptor}\cdot\text{Ligand}]\times k_{off} \\
&= Y\cdot k_{off} \\
\text{Net rate of association} &= \frac{dY}{dX} \\
&= \text{Rate of association - rate of dissociation} \\
&= \left(Y_{max} - Y\right)\cdot[\text{Ligand}]\cdot k_{on} - Y\cdot k_{off} \\
&= Y_{max}\cdot[\text{Ligand}]\cdot k_{on} - Y\left([\text{Ligand}]\cdot k_{on} + k_{off}\right)
\end{aligned}$$

Integrate that differential equation to obtain the equation defining the kinetics of association:

$$Y = Y_{max}\cdot\left(1 - e^{-([\text{Ligand}]\cdot k_{on} + k_{off})\cdot X}\right)$$

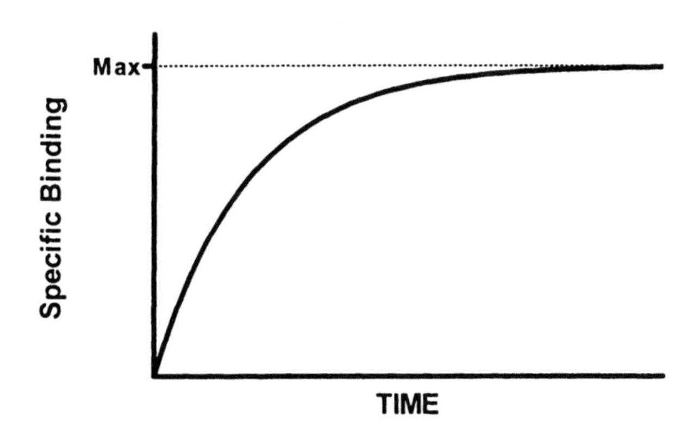

The rate at which binding increases is determined by three factors (for a constant set of experimental conditions such as pH and temperature):

- The association rate constant, k_{on} or k_{+1}. This is what you are trying to determine.
- The concentration of radioligand. If you use more radioligand, the system equilibrates faster.
- The dissociation rate constant, k_{off} or k_{-1}. Some people are surprised to see that the observed rate of association depends in part on the dissociation rate constant.

During the incubation, radioligand both binds to and dissociates from receptors. The system reaches equilibrium when the two rates are equal. The observed rate of association measures how long it takes to reach equilibrium. If the radioligand dissociates quickly from the receptor, equilibrium will be reached faster (but with less binding).

Analyzing "on rate" experiments

To analyze association (on-rate) data:

1. Enter the data with X equal to time and Y equal to specific binding. (If you enter total binding, you'll need to use a more complicated equation that accounts for the kinetics of nonspecific binding.)

2. Fit the specific binding data to an exponential association equation. The simplest case, one-phase exponential association, is shown here:

$$Y=Y_{max}\cdot\left(1-e^{-k\cdot X}\right)$$

3. The variable k in the exponential association equation is the *observed* rate constant, k_{ob}, expressed in units of inverse time. If you entered X values in minutes, then k_{ob} is expressed in min^{-1}. This is not the same as the association rate constant, k_{on}.

4. This equation assumes that a small fraction of the radioligand binds to receptors, so the concentration of free radioligand equals the amount you added and does not change over time.

5. To calculate the association rate constant (k_{on} or k_1), usually expressed in units of $Molar^{-1}$ min^{-1}, use this equation:

$$k_{on}=\frac{k_{ob}-k_{off}}{[radioligand]}$$

Variable	Units	Comment
k_{on}	$Molar^{-1}$ min^{-1}	Association rate constant; what you want to know.
k_{ob}	min^{-1}	The value of K determined by fitting an exponential association equation to your data.
k_{off}	min^{-1}	The dissociation rate constant. See the previous section.
[radioligand]	Molar	Set by the experimenter. Assumed to be constant during the experiment (only a small fraction binds).

Fitting a family of association kinetic curves

If you perform your association experiment at a single concentration of radioligand, it cannot be used to determine k_{on} unless you are able to fix the value of k_{off} from another experiment.

If you perform multiple association kinetic experiments, each with a different radioligand concentration, you can globally fit the data to the association kinetic model to derive a single best-fit estimate for k_{on} and one for k_{off}. You don't need to also perform a dissociation (off rate) experiment.

Shown below is an example of an association kinetic experiment conducted using two concentrations of radioligand. All other conditions (temperature, pH, etc.) were the same for both runs, of course. Times were entered into the X column, specific binding for one concentration of radioligand were entered into the first (A) Y column, and binding for the other concentration were entered into column B.

Here is the syntax for the global model as you would enter it into GraphPad Prism.

```
<A>Radioligand=1e-9
<B>Radioligand=3e-9
Kob=Radioligand*kon+koff
Y=Ymax*(1-exp(-1*Kob*X))
```

The first line defines the intermediate variable Radioligand for data set A. The concentration is 1nM, and it is expressed in scientific notation as 1×10^{-9} molar. The second line defines the variable Radioligand for data set B. The third line defines the observed rate constant as a function of radioligand concentration and the two rate constants. The final line computes Y (specific binding) for any value of X (time).

You could also enter the model in a simpler form into Prism.

```
Kob=Radioligand*kon+koff
Y=Ymax*(1-exp(-1*Kob*X))
```

This model does not define the parameter radioligand. Instead, enter the concentrations as the column titles of columns A and B, and define the parameter Radioligand to be a data set constant within Prism. See page 300.

Here are the curves as determined by global fitting.

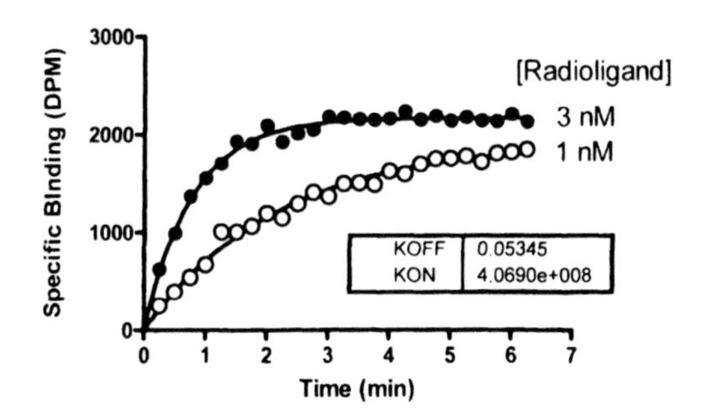

Globally fitting an association curve together with a dissociation curve

Another way to determine both association and dissociation rate constants from one experiment is to run an experiment that first measures ligand association, and then measures ligand dissociation.

For example in the experiment below, radioligand was added at time zero, and total binding was measured every five minutes for one hour. At time = 60, a very high concentration of unlabeled drug was added to the tubes. This entirely blocked new binding of radioligand to the receptors. Every five minutes, total binding was measured as a way to assess the rate of ligand dissociation.

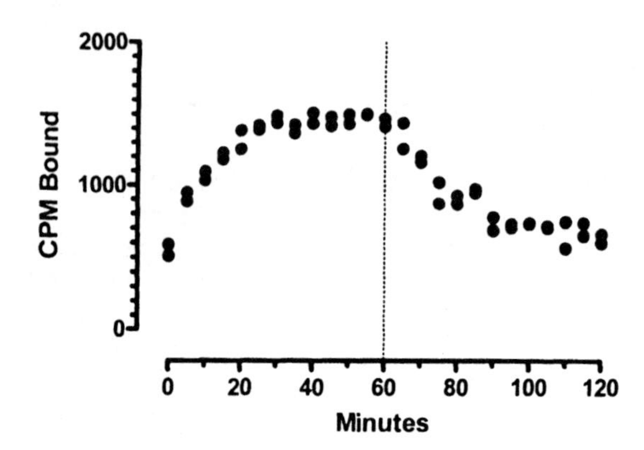

Here is the model, in the form used by GraphPad Prism.

```
radioligand=1e-10
Kob=[Radioligand]*Kon+Koff
Time0=60
Kd=Koff/Kon
Eq=Bmax*radioligand/(radioligand + Kd)

Association=Eq*(1-exp(-1*Kob*X))

YatTime0 = Eq*(1-exp(-1*Kob*Time0))
Dissociation= YatTime0*exp(-1*Koff*(X-Time0))

Y=IF(X<Time0, Association, Dissociation) + NS
```

Variable	Explanation	Units	Kind of variable
Radioligand	The concentration of radioactively labeled ligand.	Molar	Constant. We set its value within the equation. An alternative would have been to set it to a constant value in the nonlinear regression dialog.

Variable	Explanation	Units	Kind of variable
Kob	The apparent "on rate" calculated from the concentration of radioligand with the association and dissociation rate constants.	min^{-1}	Intermediate variable, defined within the equation for convenience.
Kon	Association rate constant.	$M^{-1}\ min^{-1}$	Parameter to fit.
Koff	Dissociation rate constant.	min^{-1}	Parameter to fit.
Time0	The time when dissociation was initiated.	min	Constant. We set its value within the equation. An alternative would have been to set it to a constant value in the nonlinear regression dialog.
Kd	The dissociation constant for the receptor-ligand binding, in Molar.	min^{-1}	Intermediate variable, defined within the equation for convenience.
Eq	The amount of binding (in cpm, to match the Y values) at equilibrium. Calculated from the concentration of radioligand we used and its Kd. This is an intermediate variable, defined within the equation.	cpm	Intermediate variable, defined within the equation for convenience.
Bmax	The maximum amount of binding sites. This is not the maximum bound at the concentration of ligand we chose, but the maximum number of sites that can binding ligand.	cpm	Parameter to fit.
YatTime0	The predicted binding at X=Time0. The dissociation part of the experiment begins with this amount of binding.	cpm	Intermediate variable, defined within the equation for convenience.
Association	The predicted Y values for the association part of the experiment (up to X=60). In units	cpm	Intermediate variable, defined within the equation for convenience.
Dissociation	The predicted Y values for the dissociation part of the experiment (after time X=60).	cpm	Intermediate variable, defined within the equation for convenience.
NS	Nonspecific binding. Assumed to be constant over time. In units of cpm (to match Y).	cpm	We set this to a constant value based on good control measurements.

Variable	Explanation	Units	Kind of variable
			An alternative would have been to make this a parameter to fit.
X	Time	minutes	Independent variable
Y	Total binding	cpm	Dependent variable

Association computes the association of a ligand over time. *Dissociation* computes the dissociation of a ligand after *timeo*. The final line uses an IF statement (Prism syntax) to define Y equal to *Association* before *timeo* and equal to dissociation thereafter, and then adds the nonspecific binding (*NS*).

The model has three parameters to fit: the association rate constant, the dissociation rate constant, and the maximum number of binding sites.

> Note: This is not strictly a global model, as it fits one data set and not several. But it fits one model (exponential association) to part of the data set and another model (exponential decay) to another portion of the data set. This is effectively a form of global fitting.

The graph below shows the best-fit curve and results.

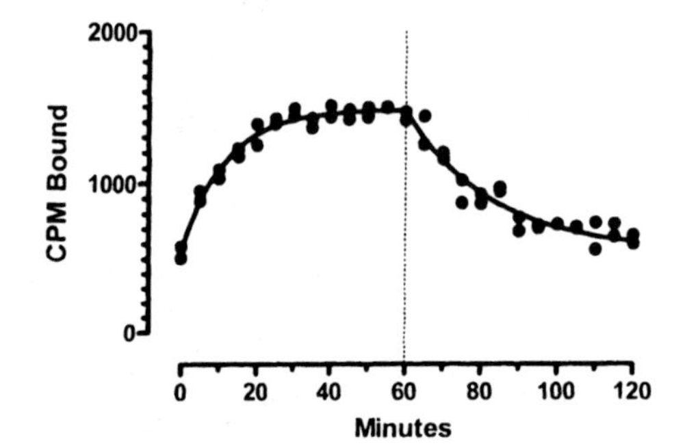

KON	4.3907e+008
KOFF	0.04358
BMAX	1858
NS	550.0
Std. Error	
KON	8.4066e+007
KOFF	0.002168
BMAX	234.1
95% Confidence Intervals	
KON	2.6979e+008 to 6.0835e+008
KOFF	0.03922 to 0.04795
BMAX	1387 to 2330

Analysis checklist for kinetic binding experiments

Question	Comment
Did you go out to a long enough time point?	Dissociation and association data should plateau, so the data obtained at the last few time points should be indistinguishable.
Is the value of k_{on} reasonable?	The association rate constant, k_{on}, depends largely on diffusion so is similar for many ligands. Expect a result of about 10^7 - 10^8 M^{-1} min^{-1}

Is the value of k_{off} reasonable?	If the k_{off} is greater than 1 min^{-1}, the ligand has a low affinity for the receptor, dissociation will occur while you are separating bound and free ligands, and you'll have a hard time obtaining quality data. If k_{off} is less than 0.001 min^{-1}, you'll have a difficult time obtaining equilibrium as the half-time of dissociation will be greater than 10 hours! Even if you wait that long, other reactions may occur that ruin the experiment.
Are the standard errors too large?	Examine the SE and the confidence intervals to see how much confidence you have in the rate constants.
Does only a tiny fraction of radioligand bind to the receptors?	The standard analyses of association experiments assume that the concentration of free radioligand is constant during the experiment. This will be approximately true only if a tiny fraction of the added radioligand binds to the receptors. Compare the maximum total binding in cpm to the amount of added radioligand in cpm. If that ratio exceeds 10% or so, you should revise your experimental protocol.

Using kinetic data to test the law of mass action

Standard binding experiments are usually fit to equations derived from the law of mass action. Kinetic experiments provide a more sensitive test than equilibrium experiments to determine whether the law of mass action actually applies for your system. To test the law of mass action, ask these questions:

Does the K_d calculated from kinetic data match the K_d calculated from saturation binding?

According to the law of mass action, the ratio of k_{off} to k_{on} is the K_d of receptor binding:

$$K_d = \frac{k_{off}}{k_{on}}$$

The units are consistent: k_{off} is in units of min^{-1}; k_{on} is in units of $M^{-1}min^{-1}$, so K_d is in units of M.

If binding follows the law of mass action, the K_d calculated this way should be the same as the K_d calculated from a saturation binding curve.

Does k_{ob} increase linearly with the concentration of radioligand?

The observed association rate constant, k_{ob}, is defined by this equation:

$$k_{ob} = k_{off} + k_{on} \cdot [\text{radioligand}]$$

Therefore, if you perform association rate experiments at various concentrations of radioligand, the results should look like the figure below. As you increase the concentration of radioligand, the observed rate constant increases linearly. If the binding is more complex than a simple mass action model (such as a binding step followed by a conformational change) the plot of k_{ob} vs. [radioligand] may plateau at higher radioligand concentrations. Also, you should extrapolate the plot back to zero radioligand to determine the intercept, which equals k_{off}. If the law of mass action applies to your system, the k_{off} determined this way should correspond to the k_{off} determined from a dissociation experiment. Finally, this kind of experiment provides a more rigorous determination of k_{on} than that obtained with a single concentration of radioligand.

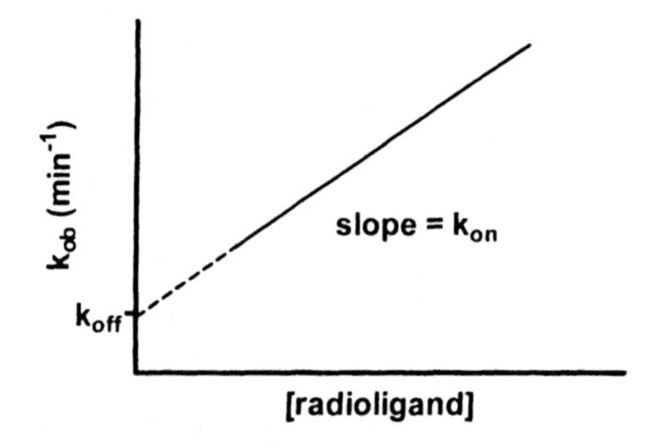

Is specific binding 100% reversible, and is the dissociated ligand chemically intact?

Nonspecific binding at "time zero" should equal total binding at the end (plateau) of the dissociation. In other words, all of the specific binding should dissociate if you wait long enough. Use chromatography to analyze the radioligand that dissociates to prove that it has not been altered.

Is the dissociation rate consistent with different experimental conditions?

Determine the dissociation constant after binding various concentrations of radioligand for various lengths of time. If your ligand binds to a single site and obeys the law of mass action, you'll obtain the same dissociation rate constant in all experiments.

Is there cooperativity?

If the law of mass action applies, binding of a ligand to one binding site does not alter the affinity of another binding site. This also means that dissociation of a ligand from one site should not change the dissociation of ligand from other sites. To test this assumption, compare the dissociation rate after initiating dissociation by infinite dilution with the dissociation rate when initiated by addition of a large concentration of unlabeled drug. If the radioligand is bound to multiple noninteracting binding sites, the dissociation will be identical in both experimental protocols as shown in the left figure. Note that the Y axis is shown using a log scale. If there were a single binding site, you'd expect the dissociation data to appear linear on this graph. With two binding sites, the graph is curved even on a log axis.

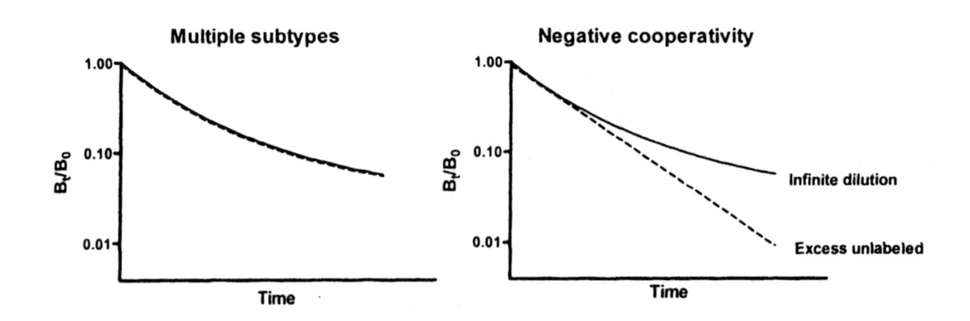

The right figure shows ideal dissociation data when radioligand is bound to interacting binding sites with negative cooperativity. The data are different depending on how dissociation was initiated. If dissociation is initiated by infinite dilution, the dissociation rate will change over time. The dissociation of some radioligand will leave the remaining ligand bound more tightly. When dissociation is initiated by addition of cold drug, all the receptors are always occupied by ligand (some hot, some cold) and dissociation occurs at its maximal unchanging rate.

Kinetics of competitive binding

The standard methods of analyzing competitive binding experiments assume that the incubation has reached equilibrium. These experiments are usually used to learn the dissociation constant of the receptors for the unlabeled compound, the K_i. The law of mass action tells us that the K_i is the ratio of the off-rate to the on-rate of the unlabeled compound. You can determine these values in a kinetics experiment as follows. Add labeled and unlabeled ligand together and measure the binding of the labeled ligand over time. This method was described by Motulsky and Mahan in *Mol. Pharmacol.* 25:1-9, 1984.

```
KA = K1*L*1E-9 + k2
KB = K3*I*1e-9 + K4
S=SQRT((KA-KB)^2+4*K1*K3*L*I*1e-18)
KF = 0.5 * (Ka + KB + S)
KS = 0.5 * (KA + KB - S)
DIFF=KF - KS
Q=Bmax*K1*L*1e-9/DIFF
Y=Q*(k4*DIFF/(KF*KS)+((K4-Kf)/KF)*exp(-KF*X)-((K4-KS)/KS)*exp(-KS*X))
```

Variable	Units	Comments
X	Minutes	Time.
Y	cpm	Specific binding.
k1	M^{-1} min^{-1}	Association rate constant of radioligand. Set to a constant value known from other experiments.
k2	min^{-1}	Dissociation rate constant of radioligand. Set to a constant value known from other experiments.
k3	M^{-1} min^{-1}	Association rate constant of unlabeled ligand. Variable to be fit. Try 10^8 as an initial value.
k4	min^{-1}	Dissociation rate constant of unlabeled ligand. Variable to be fit. Try 0.01 as an initial value.
L	nM	Concentration of radioligand. Set to a constant value you know from experimental design.
Bmax	Units of Y axis. Usually cpm.	Total number of receptors. Either leave as a variable or set to a constant you know from other experiments. If a variable, set the initial value to 100 x your maximal Y value (assumes that it binds to 1% of receptors.
I	nM	Constant set experimentally. Concentration of unlabeled ligand.

Notes:

- This equation does not account for ligand depletion. It assumes that only a small fraction of radioligand binds to receptors, so that the free concentration of radioligand is very close to the added concentration.

- This method will only give reliable results if you have plenty of data points at early time points.

40. Analyzing enzyme kinetic data

Introduction to enzyme kinetics

Living systems depend on chemical reactions which, on their own, would occur at extremely slow rates. Enzymes are catalysts that reduce the needed activation energy so these reactions proceed at rates that are useful to the cell.

Product accumulation is often linear with time

In most cases, an enzyme converts one chemical (the *substrate*) into another (the *product*). A graph of product concentration vs. time follows three phases.

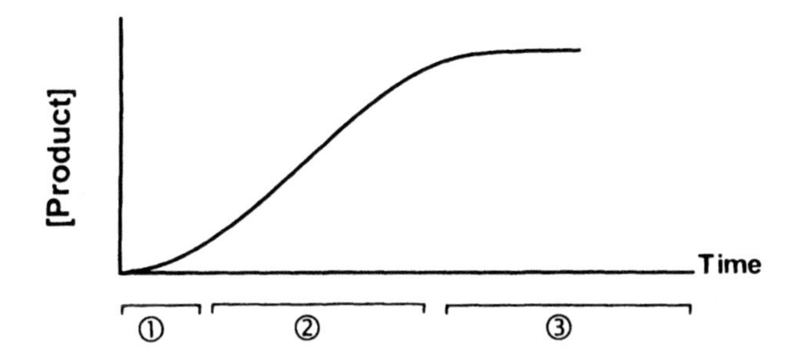

1. At very early time points, the rate of product accumulation increases over time. Special techniques are needed to study the early kinetics of enzyme action, since this transient phase usually lasts less than a second (the figure greatly exaggerates the first phase).

2. For an extended period of time, the product concentration increases linearly with time.

3. At later times, the substrate is depleted, so the curve starts to level off. Eventually the concentration of product reaches a plateau.

It is difficult to fit a curve to a graph of product as a function of time, even if you use a simplified model that ignores the transient phase and assumes that the reaction is irreversible. The model simply cannot be reduced to an equation that expresses product concentration as a function of time. To fit these kind of data (called an *enzyme progress curve*) you need to use a program that can fit data to a model defined by differential equations or by an implicit equation. For more details, see RG Duggleby, "Analysis of Enzyme Reaction Progress Curves by Nonlinear Regression", *Methods in Enzymology*, 249: 61-60, 1995.

Rather than fit the enzyme progress curve, most analyses of enzyme kinetics fit the *initial velocity* of the enzyme reaction as a function of *substrate concentration*. The velocity of the enzyme reaction is the slope of the linear phase of product accumulation, expressed as amount of product formed per time. If the initial transient phase is very short, you can simply measure product formed at a single time, and define the velocity to be the concentration divided by the time interval.

The terminology describing these phases can be confusing. The second phase is often called the "initial rate", ignoring the short transient phase that precedes it. It is also called "steady state", because the concentration of enzyme-substrate complex doesn't change. However, the concentration of product accumulates, so the system is not truly at steady state until, much later, the concentration of product truly doesn't change over time. This chapter considers data collected only in the second phase.

Enzyme velocity as a function of substrate concentration

If you measure enzyme velocity at many different concentrations of substrate, the graph generally looks like this:

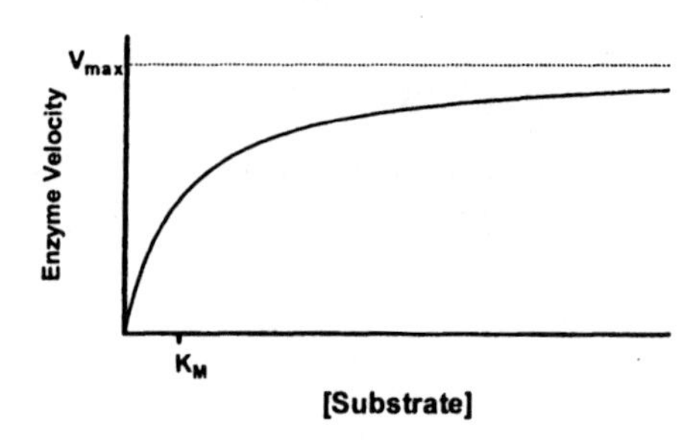

Enzyme velocity as a function of substrate concentration often follows the Michaelis-Menten equation:

$$\text{Velocity}=V=\frac{V_{max}[S]}{[S]+K_m}$$

V_{max} is the limiting velocity as substrate concentrations get very large. V_{max} (and V) are expressed in units of product formed per time. If you know the molar concentration of enzyme, you can divide the observed velocity by the concentration of enzyme sites in the assay, and express V_{max} in units of moles of product formed per second per mole of enzyme sites. This is the *turnover number,* also called k_{cat}. It is the number of molecules of substrate converted to product by one enzyme site per second. In defining enzyme concentration, distinguish the concentration of enzyme molecules from the concentration of enzyme sites (if the enzyme is a dimer with two active sites, the molar concentration of sites is twice the molar concentration of enzyme).

K_m is expressed in units of concentration, usually in molar units. K_m is the concentration of substrate that leads to half-maximal velocity. To prove this, set [S] equal to K_m in the equation above. Cancel terms and you'll see that $V=V_{max}/2$.

The meaning of K_m

To understand the meaning of K_m, you need to have a model of enzyme action. The simplest model is the classic model of Michaelis and Menten, which has proven useful with many kinds of enzymes.

$$E+S \underset{k_{-1}}{\overset{k_1}{\rightleftarrows}} ES \xrightarrow{k_2} E+P$$

The substrate (S) binds reversibly to the enzyme (E) in the first reaction. In most cases, you can't measure this step. What you measure is production of product (P), created by the second reaction.

From the model, we want to derive an equation that describes the rate of enzyme activity (amount of product formed per time interval) as a function of substrate concentration.

The rate of product formation equals the rate at which ES turns into E+P, which equals k_2 times [ES]. This equation isn't helpful, because we don't know ES. We need to solve for ES in terms of the other quantities. This calculation can be greatly simplified by making two reasonable assumptions. First, we assume that the concentration of ES is steady during the time intervals used for enzyme kinetic work. That means that the rate of ES formation equals the rate of ES dissociation (either back to E+S or forward to E+P). Second, we assume that the reverse reaction (formation of ES from E+P) is negligible, because we are working at early time points where the concentration of product is very low.

$$\text{Rate of ES formation} = \text{Rate of ES dissolution}$$

$$k_1 \cdot [S] \cdot [E] = k_{-1} \cdot [ES] + k_2 \cdot [ES]$$

We also know that the concentration of free enzyme [E] equals the total concentration of enzyme [E_{total}] minus the concentration of substrate-bound enzyme [ES]. So by substitution,

$$k_1 \cdot [S] \cdot ([E_{total}] - [ES]) = k_{-1} \cdot [ES] + k_2 \cdot [ES]$$

Solving for ES,

$$[ES] = \frac{k_1 \cdot [E_{total}] \cdot [S]}{k_1 \cdot [S] + k_2 + k_{-1}} = \frac{[E_{total}] \cdot [S]}{[S] + \frac{k_2 + k_{-1}}{k_1}}$$

The rate of product formation is

$$\text{Velocity} = k_2 \cdot [ES]$$

and, substituting the expression for [ES] above,

$$\text{Velocity} = k_2 \cdot [ES] = \frac{k_2 \cdot [E_{total}] \cdot [S]}{[S] + \frac{k_2 + k_{-1}}{k_1}}$$

Finally, define V_{max} (the velocity at maximal concentrations of substrate) as k_2 times E_{total}, and K_m, the Michaelis-Menten constant, as $(k_2+k_{-1})/k_1$. Substituting:

$$\text{Velocity} = V = \frac{V_{max} \cdot [S]}{[S] + K_m}$$

Note that K_m is not a binding constant that measures the strength of binding between the enzyme and substrate. Its value includes the affinity of substrate for enzyme, but also the rate at which the substrate bound to the enzyme is converted to product. Only if k_2 is much smaller than k_{-1} will K_m equal a binding affinity.

The Michaelis-Menten model is too simple for many purposes. The Briggs-Haldane model has proven more useful:

$$E+S \rightleftarrows ES \rightleftarrows EP \rightarrow E+P$$

Under the Briggs-Haldane model, the graph of enzyme velocity vs. substrate looks the same as under the Michaelis-Menten model, but K_m is defined as a combination of all five of the rate constants in the model.

Assumptions of enzyme kinetic analyses

Standard analyses of enzyme kinetics (the only kind discussed here) assume:

- The production of product is linear with time during the time interval used.
- The concentration of substrate vastly exceeds the concentration of enzyme. This means that the free concentration of substrate is very close to the concentration you added, and that substrate concentration is constant throughout the assay.
- A single enzyme forms the product.
- There is negligible spontaneous creation of product without enzyme
- No cooperativity. Binding of substrate to one enzyme binding site doesn't influence the affinity or activity of an adjacent site.
- Neither substrate nor product acts as an allosteric modulator to alter the enzyme velocity.

How to determine V_{max} and K_m

To determine V_{max} and K_m:

1. Enter substrate concentrations into the X column and velocity into the Y column (entering replicates if you have them).
2. Enter this equation as a new equation into your program, and fit it to your data using nonlinear regression.

```
Y = (Vmax * X)/(Km + X)
```

Variable	Comment
X	Substrate concentration. Usually expressed in μM or mM.
Y	Enzyme velocity in units of concentration of product per time. It is sometimes normalized to enzyme concentration, so concentration of product per time per concentration of enzyme.
Vmax	The maximum enzyme velocity. A reasonable rule for choosing an initial value might be that V_{max} equals 1.0 times your maximal Y value. V_{max} is expressed in the same units as the Y values.
Km	The Michaelis-Menten constant. A reasonable rule for choosing an initial value might be 0.2 x your maximal X value.

Checklist for enzyme kinetics

When evaluating results of enzyme kinetics, ask yourself these questions:

Question	Comment
Was only a small fraction of the substrate converted to product?	The analysis assumes that the free concentration of substrate is almost identical to the concentration you added during the time course of the assay. You can test this by comparing the lowest concentration of substrate used in the assay with the concentration of product created at that concentration.
Is the production of product linear with time?	Check the concentration of product at several times to test this assumption.
Did you use high enough concentrations of substrate?	Calculate the ratio of the highest substrate concentration you used divided by the best-fit value of K_m (both in the same concentration units). Ideally, the highest concentration should be at least 10 times the K_m.
Are the standard errors too large? Are the confidence intervals too wide?	Divide the SE of the V_{max} by the V_{max}, and divide the SE of the K_m by the K_m. If either ratio is much larger than about 20%, look further to try to find out why.
Is product produced in the absence of enzyme?	The analysis assumes that all product formation is due to the enzyme. If some product is produced spontaneously, you'll need to do a fancier analysis to account for this.
Did you pick a time point at which enzyme velocity is constant?	Measure product formation at several time points straddling the time used for the assay. The graph of product concentration vs. time should be linear.
Is there any evidence of cooperativity?	The standard analysis assumes no cooperativity. This means that binding of substrate to one binding site does not alter binding of substrate to another binding pocket. Since many enzymes are multimeric, this assumption is often not true. If the graph of V vs. [S] looks sigmoidal, see "Allosteric enzymes" on page 251.

Comparison of enzyme kinetics with radioligand binding

The Michaelis-Menten equation for enzyme activity has a mathematical form similar to the equation describing equilibrium binding.

$$\text{Enzyme Velocity} = V = \frac{V_{max}[S]}{[S]+K_m}$$

$$\text{Specific Binding} = B = \frac{B_{max}[L]}{[L]+K_d}$$

However, the parameters mean different things. Note these differences between binding experiments and enzyme kinetics.

- It usually takes many minutes or hours for a receptor incubation to equilibrate. It is common (and informative) to measure the kinetics prior to equilibrium. Enzyme assays reach steady state (defined as constant rate of product accumulation) typically in a few seconds. It is uncommon to measure the kinetics of the transient phase before that, although you can learn a lot by studying those transient kinetics (see an advanced text of enzyme kinetics for details).
- The equation used to analyze binding data is valid at equilibrium - when the rate of receptor-ligand complex formation equals the rate of dissociation. The equation

used to analyze enzyme kinetic data is valid when the rate of product formation is constant, so product accumulates at a constant rate. But the overall system is not at equilibrium in enzyme reactions, as the concentration of product is continually increasing.

- K_d is a dissociation constant that measures the strength of binding between receptor and ligand. K_m is not a binding constant. Its value includes the affinity of substrate for enzyme, but also the kinetics by which the substrate bound to the enzyme is converted to product
- B_{max} is measured as the number of binding sites normalized to amount of tissue, often fmol per milligram, or sites/cell. V_{max} is measured as moles of product produced per minute.

Displaying enzyme kinetic data on a Lineweaver- Burk plot

The best way to analyze enzyme kinetic data is to fit the data directly to the Michaelis-Menten equation using nonlinear regression. Before nonlinear regression was available, investigators had to transform curved data into straight lines, so they could analyze with linear regression.

One way to do this is with a Lineweaver-Burk plot. Take the inverse of the Michaelis-Menten equation and simplify:

$$\frac{1}{V}=\frac{[S]+K_m}{V_{max}[S]}=\frac{[S]}{V_{max}[S]}+\frac{K_m}{V_{max}[S]}=\frac{1}{V_{max}}+\frac{K_m}{V_{max}}\cdot\frac{1}{[S]}$$

Ignoring experimental error, a plot of $1/V$ vs. $1/S$ will be linear, with a Y-intercept of $1/V_{max}$ and a slope equal to K_m/V_{max}. The X-intercept equals $-1/K_m$.

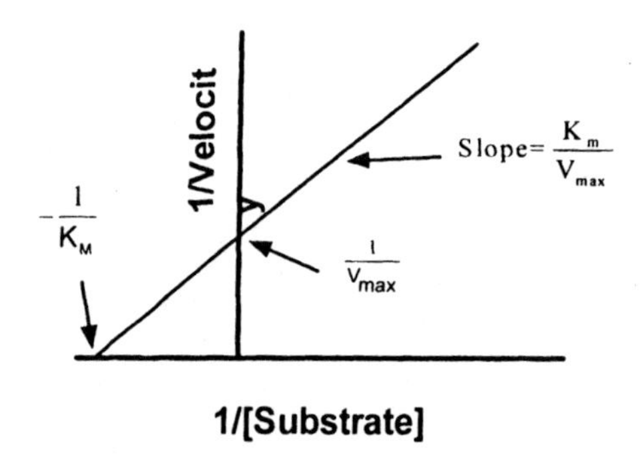

Use the Lineweaver-Burk plot only to display your data. Don't use the slope and intercept of a linear regression line to determine values for V_{max} and K_m. If you do this, you won't get the most accurate values for V_{max} and K_m. The problem is that the transformations (reciprocals) distort the experimental error, so the double-reciprocal plot does not obey the assumptions of linear regression. Use nonlinear regression to obtain the most accurate values of K_m and V_{max} (see page 19).

> Tip: You should analyze enzyme kinetic data with nonlinear regression, *not* with Lineweaver-Burk plots. Use Lineweaver-Burk plots to display data, not to analyze data.

Allosteric enzymes

One of the assumptions of Michaelis-Menten kinetics is that there is no cooperativity. If the enzyme is multimeric, then binding of a substrate to one site should have no effect on the activity of neighboring sites. This assumption is often not true.

If binding of substrate to one site increases the activity of neighboring sites, the term *positive cooperativity* is used. Activity is related to substrate concentration by this equation:

$$\text{Velocity}=V=\frac{V_{max}[S]^h}{[S]^h+K_{0.5}^h}$$

When the variable h equals 1.0, this equation is the same as the Michaelis-Menten equation. With positive cooperativity, h will have a value greater than 1.0. If there are two interacting binding sites, h will equal between one and two, depending on the strength of the cooperativity. If there are three interacting binding sites, h will equal between 1 and 3. Note that the denominator has the new parameter, $K_{0.5}$, instead of K_m.

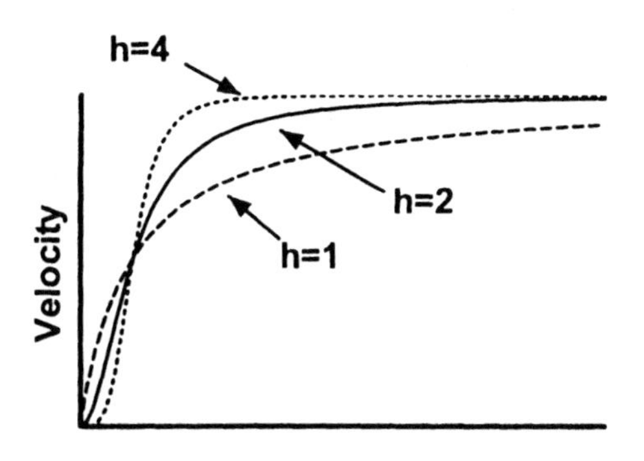

To fit data to the equation for enzyme activity with positive cooperativity, use the equation below. For initial values, try these rules: Vmax=1 x your maximal Y value, K=0.5 x your maximal X value, and h=1.0

```
Y=Vmax*X^h/(K^h + X^h)
```

The parameter h does not always equal the number of interacting binding sites (although h cannot exceed the number of interacting sites). Think of h as an empirical measure of the steepness of the curve and the presence of cooperativity.

Enzyme kinetics in the presence of an inhibitor

Competitive inhibitors

If an inhibitor binds reversibly to the same site as the substrate, the inhibition will be competitive. Competitive inhibitors are common in nature.

One way to measure the effect of an inhibitor is to measure enzyme velocity at a variety of substrate concentrations in the presence and absence of an inhibitor. As the graph below shows, the inhibitor substantially reduces enzyme velocity at low concentrations of substrate, but doesn't alter velocity very much at very high concentrations of substrate.

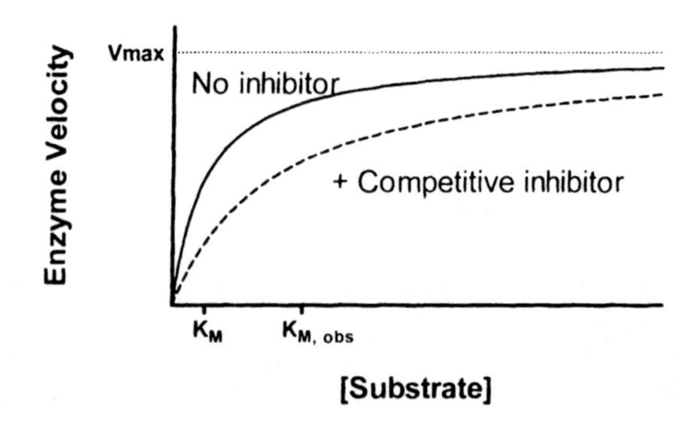

As the graph above shows, the inhibitor does not alter V_{max}, but it does increase the observed K_m (concentration of substrate that produces half-maximal velocity; some texts call this "apparent K_m"). The observed K_m is given by the following equation, where K_i is the dissociation constant for inhibitor binding (in the same concentration units as [Inhibitor]):

$$K_{M,obs} = K_m \cdot \left[1 + \frac{[\text{Inhibitor}]}{K_i}\right]$$

Using global fitting to fit K_i

The best way to find the best-fit value of K_i is to fit all your data at once using global nonlinear regression. Fit to this model.

```
KMobs=Km(1+[I]/Ki)
Y=Vmax*X/(KMobs+X)
```

Variable	Comment
X	Substrate concentration. Usually expressed in µM or mM.
Y	Enzyme velocity in units of concentration of product per time. It is sometimes normalized to enzyme concentration, so concentration of product per time per concentration of enzyme.
Vmax	The maximum enzyme velocity, in the absence of inhibitor. A reasonable rule for choosing an initial value might be that V_{max} equals 1.0 times your maximal Y value. V_{max} is expressed in the same units as the Y values. Share this parameter to get one best-fit value for the entire experiment.
Km	The Michaelis-Menten constant, in the same units as X. A reasonable rule for choosing an initial value might be 0.2 x your maximal X value. Share this parameter to get one best-fit value for the entire experiment.
Ki	The inhibition constant, in the same units as your X values and K_m. Share.
I	The concentration of inhibitor, in the same units as Ki. For each data set, this is a constant that you enter.

Fit the family of curves, setting up the global curve fitting to share the parameters V_{max}, K_m, and K_i. This means you'll get one best-fit value for each of these parameters from the entire set of experiments. Set I to be a constant whose value varies with each data set. The first data set is a control, where I equals 0.

> GraphPad note: In Prism, enter each inhibitor concentration as the column title for the appropriate data set. In the nonlinear regression dialog (constraints tab), define I (inhibitor concentration) to be a data set constant. This means that Prism will get its value from the heading of each column.

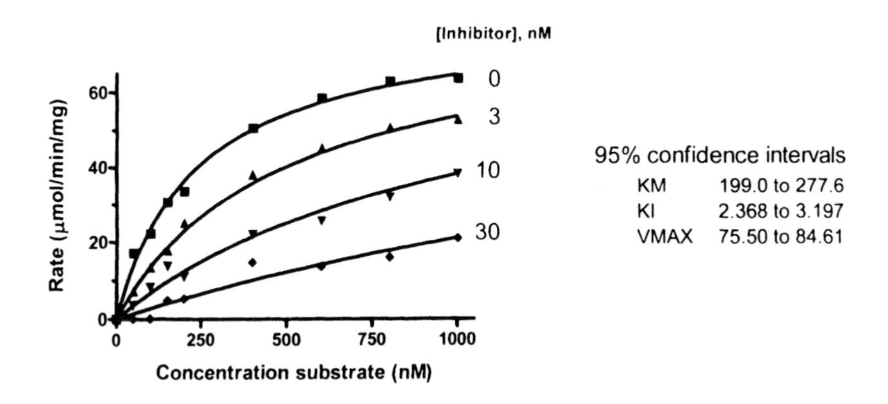

> GraphPad note: The companion step-by-step example book has detailed instructions for fitting a family of enzyme inhibition curves to determine the K_i. You'll also find this example at www.graphpad.com.

Finding K_i without global fitting

If you don't have a program that can fit data with global nonlinear regression, you can use a less accurate method.

If you have determined the K_m in the presence and absence of a single concentration of inhibitor, you can rearrange that equation to determine the K_i.

$$K_i = \frac{[\text{Inhibitor}]}{\frac{K_{M,obs}}{K_m} - 1.0}$$

You'll get a more reliable determination of K_i if you determine the observed K_m at a variety of concentrations of inhibitor. Fit each curve to determine the observed K_m. Enter the results onto a new table, where X is the concentration of inhibitor and Y is the observed K_m. If the inhibitor is competitive, the graph will be linear. Use linear regression to determine the X and Y intercepts. The Y-axis intercept equals the K_m and the X-axis intercept equals the negative K_i.

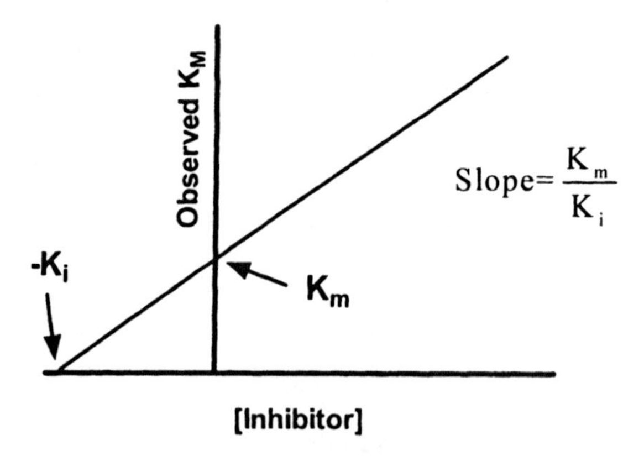

Another approach is to measure enzyme velocity at a single concentration of substrate with varying concentrations of a competitive inhibitor. The results will look like this.

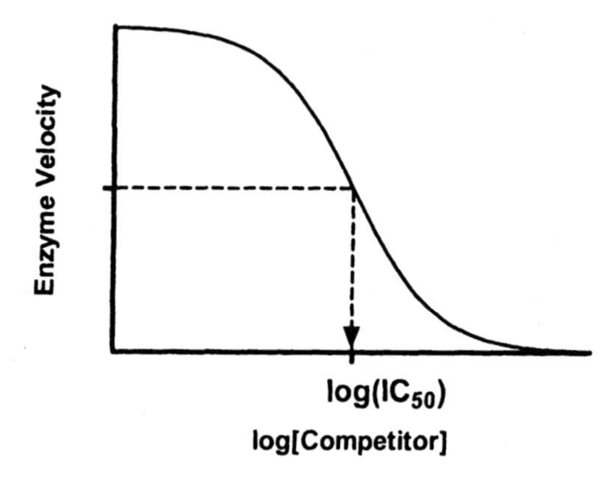

The concentration of competitor that reduces enzyme velocity by half is called the IC_{50}. Its value is determined by three factors:

- The dissociation constant for binding of inhibitor to enzyme, the K_i. If the K_i is low (the affinity is high), the IC_{50} will be low. The subscript i is used to indicate that the competitor inhibited enzyme activity. It is the concentration of the competitor that will bind to half the enzyme sites at equilibrium in the absence of substrate or other competitors.

- The concentration of the substrate. If you choose to use a higher concentration of substrate, it will take a larger concentration of inhibitor to compete for 50% of the activity.

- The K_m. It takes more inhibitor to compete for a substrate with a low K_m than for a substrate with a high K_m.

You can calculate the K_i, using the equation of Y. Cheng and W. H. Prusoff (*Biochem. Pharmacol.* 22: 3099-3108, 1973).

$$K_i = \frac{IC_{50}}{1+\frac{[\text{Substrate}]}{K_M}}$$

Inhibitors that are not competitive

Not all inhibitors are competitive. Some inhibitors decrease the observed V_{max}, with or without increasing the observed K_m. Consult an advanced text on enzyme kinetics for information about *non-competitive, uncompetitive,* and *mixed inhibition.*

Competitive and noncompetitive inhibitors bind reversibly. An inhibitor that binds covalently to irreversibly inactivate the enzyme is called an *irreversible inhibitor* or *inactivator.*

I. Fitting dose-response curves

41. Introduction to dose-response curves

What is a dose-response curve?

Dose-response curves can be used to plot the results of many kinds of experiments. The X axis plots concentration of a drug or hormone. The Y axis plots response, which could be almost any measure of biological function. For example, the response might be enzyme activity, accumulation of an intracellular second messenger, membrane potential, secretion of a hormone, change in heart rate, or contraction of a muscle.

The term "dose" is often used loosely. In its strictest sense, the term only applies to experiments performed with animals or people, where you administer various doses of drug. You don't know the actual concentration of drug at its site of action—you only know the total dose that you administered. However, the term "dose-response curve" is also used more loosely to describe *in vitro* experiments where you apply known concentrations of drugs. The term "concentration-response curve" is therefore a more precise label for the results of these types of experiments. The term "dose-response curve" is occasionally used even more loosely to refer to experiments where you vary levels of some other variable, such as temperature or voltage.

An *agonist* is a drug that binds to a receptor and causes a response. If you administer various concentrations of an agonist that causes a stimulatory response, the dose-response curve will go uphill as you go from left (low concentration) to right (high concentration). If the agonist causes an inhibitory response, the curve will go downhill with increasing agonist concentrations. A *full agonist* is a drug that appears able to produce the maximum cellular or tissue response. A *partial agonist* is a drug that provokes a response, but the maximum response is less than the maximum response to a full agonist in the same cell or tissue. An *inverse agonist* is a drug that reduces a pre-existing basal response, which is itself due to constitutive activation of a system in the absence of other ligands, e.g., perhaps due to an activating mutation in a receptor.

An *antagonist* is a drug that does not provoke a response itself, but blocks agonist-mediated responses. If you vary the concentration of antagonist (in the presence of a fixed concentration of agonist), the antagonist dose-response curve (also called an "antagonist inhibition curve") will run in the opposite direction to that of the agonist dose-response curve. It should be noted that the classification of drugs as full agonists, partial agonists, inverse agonists, and antagonists is highly dependent on the biological system in which they are tested. For example, if drug binding is strongly coupled to response in one system and only weakly coupled to response in another system, then a full agonist in the first system may appear as a partial agonist in the second system. Similarly, if a system is not constitutively active in the absence of ligands, then an inverse agonist in such a system would appear indistinguishable from a simple antagonist.

The shape of dose-response curves

Many steps can occur between the binding of the agonist to a receptor and the production of the response. So depending on which drug you use and which response you measure, dose-response curves can have almost any shape. However, in very many systems, dose-response curves follow a standard shape that is almost identical to that observed for the binding of a drug to a receptor. While a plot of response vs. the amount of drug is thus typically a rectangular hyperbola, the dose range for the full relationship may span several orders of magnitude, so it is more common to plot response vs. *logarithm* of the dose.

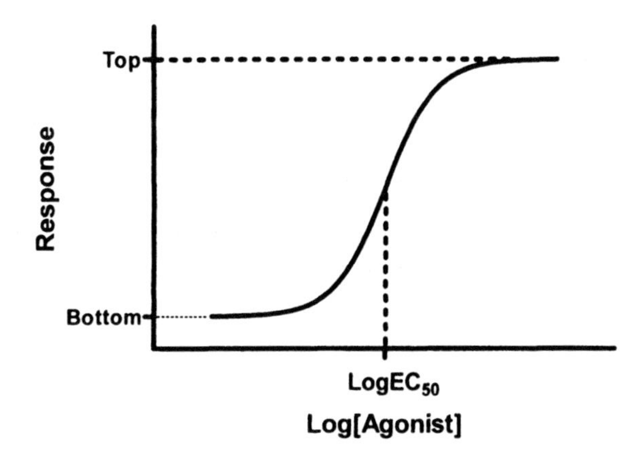

Dose-response experiments typically use around 5-10 doses of agonist, approximately equally spaced on a logarithmic scale. For example, doses might be 1, 3, 10, 30, 100, 300, 1000, 3000, and 10000 nM. When converted to logarithms, these values are equally spaced: 0.0, 0.5, 1.0, 1.5, 2.0, 2.5, 3.0, 3.5, and 4.0.

> Note: The logarithm of 3 is actually 0.4771, not 0.50. The antilog of 0.5 is 3.1623. So to make the doses truly equally spaced on a log scale, the concentrations ought to be 1.0, 3.1623, 10.0, 31.623, etc.

Since the linkage between agonist binding and response can be very complex, any shape is possible. It seems surprising, therefore, that so many dose-response curves have shapes almost identical to receptor binding curves. The simplest explanation is that the link between receptor binding and response is direct, so response is proportional to receptor binding. However, in most systems one or more second-messengers can link receptor binding to the final, measured, response. For example, the binding of some agonists can activate adenylyl cyclase, which creates the second-messenger cAMP. The second messenger then binds to an effector (such as a protein kinase) and initiates or propagates a response. Thus, direct proportionality between binding and response is not commonly observed in biological systems.

What do you expect a dose-response curve to look like if a second messenger mediates the response? If you assume that the production of second messenger is proportional to receptor occupancy, the graph of agonist concentration vs. second messenger concentration will have the same shape as receptor occupancy (a hyperbola if plotted on a linear scale, a sigmoid curve with a slope factor of 1.0 if plotted as a semilog graph). If the second messenger works by binding to an effector, and that binding step follows the law of mass action, then the graph of second messenger concentration vs. response will also have that same standard shape. It isn't obvious, but Black and Leff have shown that the graph of agonist concentration vs. response will also have that standard shape, provided that both binding steps follow the law of mass action (see Chapater 42). In fact, it doesn't matter how many steps intervene between agonist binding and response. So long as each messenger binds to a single binding site according to the law of mass action, the dose-response curve will follow the same hyperbolic/sigmoid shape as a receptor binding curve.

The EC_{50}

A standard dose-response curve is defined by four parameters: the baseline response (Bottom), the maximum response (Top), the slope (Hill slope), and the drug concentration that provokes a response halfway between baseline and maximum (EC_{50}).

It is easy to misunderstand the definition of EC_{50}. It is defined quite simply as the concentration of agonist that provokes a response halfway between the baseline (Bottom) and maximum response (Top). It is impossible to define the EC_{50} until you first define the baseline and maximum response. Depending on how you have normalized your data, this may not be the same as the concentration that provokes a response of Y=50. For instance, in the example below, the data are normalized to percentage of maximum response, without subtracting a baseline. The baseline is about 20%, and the maximum is 100%, so the EC_{50} is the concentration of agonist that evokes a response of about 60% (half way between 20% and 100%).

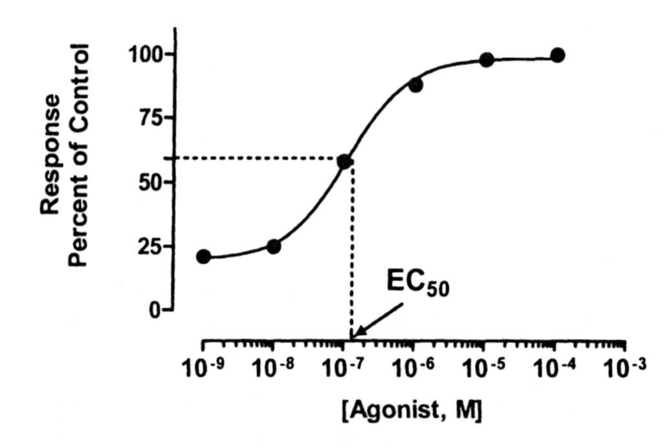

Don't overinterpret the EC_{50}. It is simply the concentration of agonist required to provoke a response halfway between the baseline and maximum responses. Because the EC_{50} defines the location of the dose-response curve for a particular drug, it is the most commonly used measure of an agonist's potency. However, the EC_{50} is usually not the same as the K_d for the binding of agonist to its receptor -- it is not a direct measure of drug affinity.

The steepness of a dose-response curve

Many dose-response curves follow the shape of a receptor binding curve. As shown below, 81 times more agonist is needed to achieve 90% response than a 10% response.

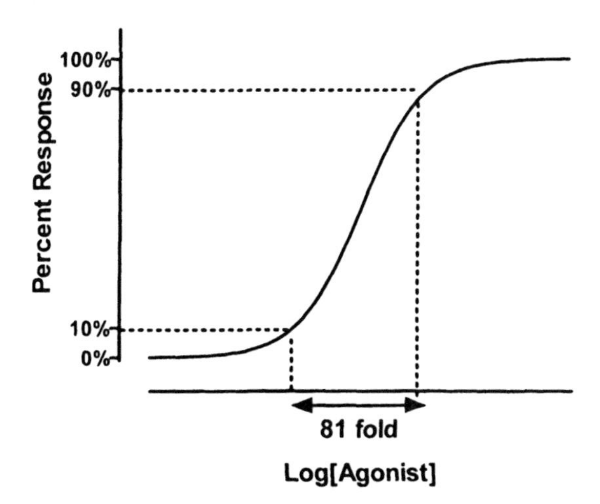

Some dose-response curves however, are steeper or shallower than the standard curve. The steepness is quantified by the Hill slope, also called a slope factor. A dose-response curve with a standard slope has a Hill slope of 1.0. A steeper curve has a higher slope factor, and a shallower curve has a lower slope factor. If you use a single concentration of agonist and varying concentrations of antagonist, the curve goes downhill and the slope factor is negative. The steeper the downhill slope, the more negative the Hill slope.

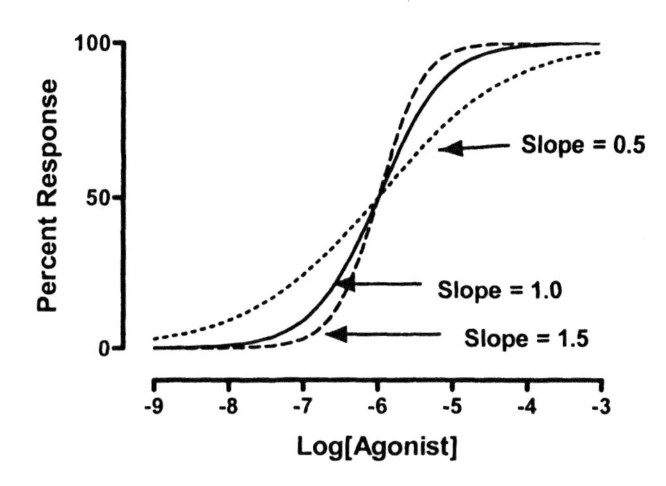

The equation for a dose-response curve

The general equation for a sigmoidal dose-response curve is also commonly referred to as the "Hill equation", the "four-parameter logistic equation", or the "variable slope sigmoid equation". One form of this equation is:

$$\text{Response}=\text{Bottom}+\frac{(\text{Top-Bottom})}{1+\frac{EC_{50}^{\text{HillSlope}}}{[\text{Drug}]^{\text{HillSlope}}}}$$

When you fit this equation, you want to find the best fit value of the LogEC_{50}, rather than the EC_{50} itself (see next section). Making that change, as well as defining Y to be the response and X to be the logarithm of [Drug] gives us:

$$\text{Response}=\text{Bottom}+\frac{(\text{Top-Bottom})}{1+\left(\frac{10^{\text{LogEC}_{50}}}{10^{X}}\right)^{\text{HillSlope}}}$$

$$=\text{Bottom}+\frac{(\text{Top-Bottom})}{1+10^{(\text{LogEC}_{50}-X)\text{HillSlope}}}$$

If you wish to manually enter this equation into your nonlinear regression program, the following syntax is standard for many different software packages:

```
Y=Bottom + (Top-Bottom)/(1+10^((LogEC50-X)*HillSlope))
```

Other measures of potency

The pEC_{50}

The pEC_{50} is defined as the negative logarithm of the EC_{50}. If the EC_{50} equals 1 micromolar (10^{-6} molar), the log EC_{50} is –6 and the pEC_{50} is 6. There is no particular advantage to expressing potency this way, but it is customary in some fields.

> Note: Expressing potency as the pEC_{50} is a similar practice to quantifying acidity with the pH, which is the negative logarithm of $[H^+]$.

If you want to fit the pEC_{50} directly rather than fitting the logEC_{50}, use the following equation syntax.

```
Y=Bottom + (Top-Bottom)/(1+10^((X - pEC50)*HillSlope))
```

Calculating any EC value from the EC_{50} and Hill slope

The potency of a drug is commonly quantified as the EC_{50} or the logarithm of the EC_{50}. But in some systems you might be more interested in the EC_{80} or the EC_{90} or some other value. You can compute the EC_{80} or EC_{90} (or any other EC value) from the EC_{50} and Hill slope. Or you can fit data to determine any EC value directly. If you express response as a percentage, a standard dose-response curve is described by this equation:

$$F=100\times\frac{[A]^{H}}{[A]^{H}+EC_{50}{}^{H}}$$

[A] is the agonist concentration, EC_{50} is the concentration that gives half-maximal effect, and H is the Hill coefficient or slope factor that defines the steepness of the curve. [A] and

EC_{50} are expressed in the same units of concentration, so the units cancel out. F is the fractional response, expressed as a percentage.

If you set F to any fractional response you want, and define EC_F as the agonist concentration necessary to achieve that response, then by substitution in the equation above,

$$F=100\times\frac{EC_F{}^H}{EC_F{}^H+EC_{50}{}^H}$$

and rearranging yields this equation:

$$EC_F=\left(\frac{F}{100-F}\right)^{1/H}\times EC_{50}$$

If you know the EC_{50} and Hill slope (H), you can easily compute the EC_{80} or EC_{10} or any other value you want. For example, if the Hill slope equals 1, the EC_{90} equals the EC_{50} times nine. If H equals 0.5, the curve is shallower and the EC_{90} equals the EC_{50} times 81.

Determining any EC value directly

You can also fit data directly to an equation written in terms of the EC_F. The advantage of this approach is that Prism will report the 95% confidence value for EC_F. Use the equation below, where X is the log of concentration and Y is response, which ranges from *Bottom* to *Top*. In the example below, F is set to a value of 80, but you can set it to be any desired value between 0 and 100.

```
F=80
logEC50=logECF - (1/HillSlope)*log(F/(100-F))
Y=Bottom + (Top-Bottom)/(1+10^((LogEC50-X)*HillSlope))
```

To fit data to this equation, you'll need to consider reasonable initial values for your parameters. We suggest setting *Top* equal to your maximal Y value and *Bottom* equal to your minimum Y value, as determined from your data points. For *HillSlope*, simply pick a value, probably +1.0 or −1.0. For *logEC*, enter the logarithm of your middle X value as a crude initial value, or enter a value based on the range of concentrations you use.

Here is a simplified equation for fitting the EC_{90}. Here, the response is expressed as a percentage ranging from zero to one hundred, so we dispense with the variables *Top* and *Bottom*.

```
logEC50=logEC90 - (1/HillSlope)*log(9)
Y=100/(1+10^((LogEC50-X)*HillSlope))
```

Dose-response curves where X is concentration, not log of concentration

Dose-response curves are generally performed with concentrations that are equally spaced on a log scale, and are usually fit to find the best-fit value of the $LogEC_{50}$ (see below). It is also possible to make the concentrations equally spaced on a linear scale, and fit to find the EC_{50}.

Start with the standard equation for the dose-response curve:

$$\text{Response}=\text{Bottom}+\frac{(\text{Top-Bottom})}{1+\dfrac{EC_{50}{}^{\text{HillSlope}}}{[\text{Drug}]^{\text{HillSlope}}}}$$

Define Y to be response, and X to be [Drug], and simplify.

$$\text{Response}=\text{Bottom}+\frac{(\text{Top-Bottom})}{1+\left(\dfrac{EC_{50}}{X}\right)^{\text{HillSlope}}}$$

Written as a user-defined equation for most nonlinear regression programs:

```
Y=Bottom + (Top-Bottom)/(1 + (EC50/X)^HillSlope)
```

When the Hill slope is set to 1.0, this is the same as the one-site binding hyperbola (except this equation adds a bottom baseline term).

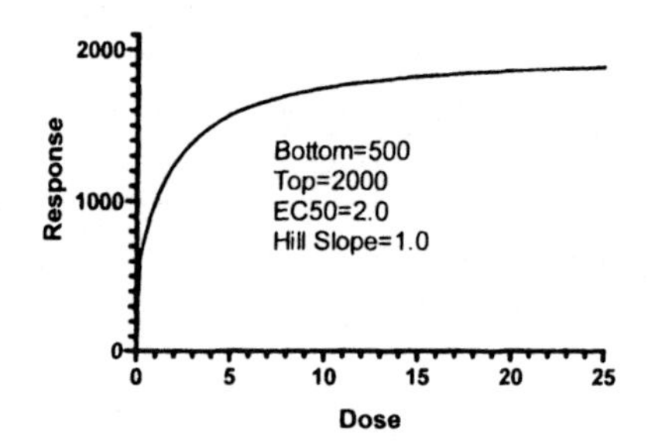

When the Hill slope is much greater than 1.0, the dose-response curve has a sigmoidal shape.

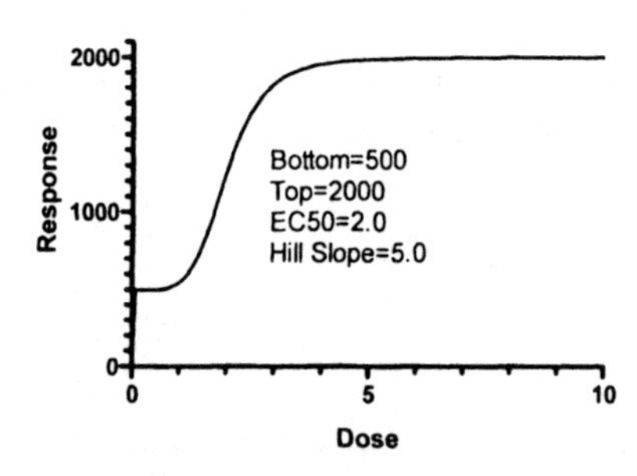

Note the confusing point here. A standard dose response curve, with a Hill Slope equal to 1.0, has a sigmoidal shape when X is the log of concentration or dose. The same standard dose response curve (with a Hill slope equal to 1.0) has a hyperbolic shape when X is concentration (or dose). It does not have a sigmoidal shape.

If the Hill slope is greater than 1.0, the curve has a sigmoidal shape either way – when X is concentration (or dose) or when X is the logarithm of concentration (or dose).

> Tip: When you see a sigmoidal dose-response curve, look carefully at the X axis to see if X is concentration (or dose) or the logarithm of concentration (or dose).

In general, you should avoid fitting dose-response curves on a linear scale, for two reasons. First, if the curve spans many orders of drug dose magnitude, then it becomes graphically difficult to present. Second, the error associated with the EC_{50} parameter (linear scale) of the standard dose-response model does not follow a Gaussian distribution and therefore cannot be used in standard statistical analyses that require the parameters follow a Gaussian distribution. This is discussed next.

Why you should fit the $logEC_{50}$ rather than EC_{50}

As shown above, you can write an equation for a dose-response curve either in terms of EC_{50} or log EC_{50}. Curve fitting finds the curve that minimizes the sum-of-squares of the vertical distance from the points. Rewriting the equation to change between EC_{50} and log EC_{50} isn't going to make a different curve fit better. All it does is change the way that the best-fit EC_{50} is reported.

However, rewriting the equation to change between EC_{50} and log EC_{50} has a major effect on standard error and confidence interval of the best-fit values. Consider these sample results:

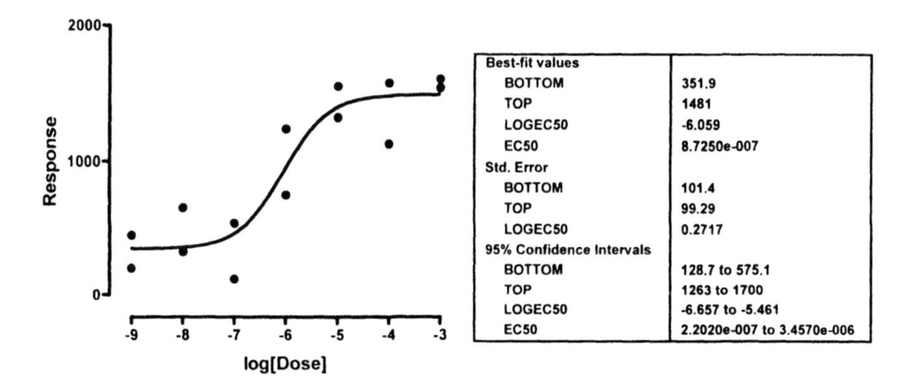

Best-fit values	
BOTTOM	351.9
TOP	1481
LOGEC50	-6.059
EC50	8.7250e-007
Std. Error	
BOTTOM	101.4
TOP	99.29
LOGEC50	0.2717
95% Confidence Intervals	
BOTTOM	128.7 to 575.1
TOP	1263 to 1700
LOGEC50	-6.657 to -5.461
EC50	2.2020e-007 to 3.4570e-006

These data were fit to a dose-response curve with a Hill slope of 1. The best-fit value for $logEC_{50}$ is -6.059. Converting to the EC_{50} is no problem – simply take the antilog. The EC_{50} is $10^{-6.059}$ M, about 0.87 μM.

The standard error of the $logEC_{50}$ is 0.2717. It is used as an intermediate result to calculate a confidence interval, which ranges from -6.657 to -5.461. This means that the 95%CI of the EC_{50} extends from $10^{-6.657}$ to $10^{-5.461}$ -- from 0.22 to 3.46 μM. Expressed as concentrations (rather than log of concentrations), the interval is not centered on the best-fit value (0.87 μM). Switching from linear to log scale turned a symmetrical confidence interval into a very asymmetrical interval, which you can report.

If you fit the same data to an equation describing a dose-response curve in terms of the EC_{50} rather than the $logEC_{50}$, the EC_{50}, remains 0.87 μM. But now the program computes the SE of the EC_{50} (0.5459 μM), and uses this to compute the 95% confidence interval of the EC_{50}, which ranges from -0.3290 to +2.074 μM. Note that the lower limit of the confidence interval is negative! Since the EC_{50} is a concentration, negative values are nonsense. Even setting aside the negative portion of the confidence interval, it includes all values from zero on up, which isn't terribly useful.

The problem is that the uncertainty of the EC_{50} really isn't symmetrical, especially when you space your doses equally on a log scale. Nonlinear regression (from Prism and most other programs) always reports a symmetrical confidence interval. In cases like this – fitting dose response data to a model written in terms of the EC_{50} – the confidence interval is not very helpful.

When some people see the SE of the $logEC_{50}$, they are tempted to convert this to the standard error of the EC_{50} by taking the antilog. In the example, the SE of the $logEC_{50}$ is 0.2717. The antilog of 0.2717 equals $10^{0.2717}$ or 1.869. What does this mean? It certainly is *not* the SE of the EC_{50}. The SE does *not* represent a point on the axis; rather it represents a distance along the axis. A distance along a log axis does not represent a consistent distance along a linear (standard) axis. For example, increasing the $logEC_{50}$ 1 unit from -9 to -8 increases the EC_{50} 9nM; increasing the $logEC_{50}$ 1 unit from -3 to -2 increases the EC_{50} by 9 mM (which equals 9,000,000 nM). So you cannot interpret the number 1.869 as a concentration. You can interpret it as a multiplier – a factor you multiply by or divide into the EC_{50}. To calculate the 95% CI, first multiply 1.869 by a constant from the t distribution for 95% confidence and the appropriate number of degrees of freedom (11 degrees of freedom in this example, so t is 2.201). The result is 4.113 Then compute the 95% CI of the EC_{50}. It extends from the best-fit EC_{50} divided by 4.113 to the best-fit EC_{50} times 4.113, from 0.21 µM to 3.58 µM.

Decisions when fitting sigmoid dose-response curves

From the preceding discussion, it is clear that most of the time you should enter your X values as logarithms of concentration or dose if you want to perform a standard sigmoidal dose-response curve fit. If you entered actual concentrations, most standard data analysis programs can transform those values to logarithms for you.

> Note: Since the logarithm of zero is undefined, you cannot enter a concentration of zero as a logarithm. If you enter a concentration of zero and then transform to logarithms, Prism will leave that result blank. Instead of entering a dose of zero, enter a low concentration, e.g., one log unit below your lowest non-zero concentration.

Before fitting a dose-response curve, you will need to make these decisions:

Decision	Discussion
Choose Hill slope of 1 or variable slope?	If you have plenty of data points, you should choose to fit the Hill Slope along with the other parameters. If data are scanty, you may wish to consider fixing the Hill slope to a value of 1.
Set *Top* to a constant value?	Ideally, the top part of the curve is defined by several data points. In this case, the nonlinear regression program will be able to fit the top plateau of the curve. If this plateau is not well defined by data, then you'll need to make the top plateau be a constant based on controls.

Decision	Discussion
Set *Bottom* to a constant value?	Ideally, the bottom part of the curve is defined by several data points. In this case, the nonlinear regression program will be able to fit the bottom plateau of the curve. If this plateau is not well defined by data, then you'll need to set the bottom plateau to a constant value based on controls. If you have subtracted a background value, then the bottom plateau of the curve must be 0. The program won't know this unless you tell it. Make *Bottom* a constant equal to zero in this case.
Absolute or relative weighting?	See "Weighting method" on page 307.
Fit each replicate or averages?	See "Replicates" on page 307.

Checklist: Interpreting a dose-response curve

After fitting a dose-response model to your data, ask yourself these questions:

Question	Comment
Is the logEC_{50} reasonable?	The EC_{50} should be near the middle of the curve, with at least several data points on either side of it.
Are the standard errors too large? Are the confidence intervals too wide?	The SE of the logEC_{50} should be less than 0.5 log unit (ideally a lot less).
Is the value of *Bottom* reasonable?	*Bottom* should be near the response you observed with zero drug. If the best-fit value of *Bottom* is negative, consider fixing it to a constant value equal to baseline response. If you know where the bottom of the curve should be, then set *Bottom* to that constant value. Or constrain Bottom to be greater than 0.
Is the value of *Top* reasonable?	*Top* should be near the response you observed with maximal concentration drug. If the best-fit value of *Top* is not reasonable, consider fixing it to a constant value. If you know where the top of the curve should be, then set *Top* that constant value.
If you used a variable slope model, are there enough data to define the slope?	If you asked Prism to find a best-fit value for slope, make sure there at least a few data points between 10 and 90% . If not, your data don't accurately define the slope factor. Consider fixing the slope to its standard value of 1.0
If you used a model with a Hill slope of 1, does the data appear to be steeper or shallower?	If the data appear to form a curve much steeper or shallower than the standard dose-response curve with a slope of 1, consider fitting to a model with a variable slope.
Does the curve appear to be biphasic or non-sigmoid?	The standard dose-response models assume that the curve is monotonic and sigmoid. If the curve goes up, and then down, you'll need a more complicated model (see next chapter).

42. The operational model of agonist action

Limitations of dose-response curves

Fitting a standard sigmoidal (logistic) equation to a dose-response curve to determine EC_{50} (and perhaps slope factor) doesn't tell you everything you want to know about an agonist. The problem is that the EC_{50} is determined by two properties of the agonist:

- How well it binds to the receptor, quantified by the *affinity* of the drug for binding to its receptor.
- How well it causes a response once bound. This property is known as the agonist's *efficacy*. Since efficacy depends on both agonist and tissue, a single drug acting on a single kind of receptor can have different efficacies, and thus different EC_{50} values, in different tissues.

A single dose-response experiment cannot untangle affinity from efficacy. Two very different drugs could have identical dose-response curves, with the *same* EC50 values and maximal responses (in the same tissue). One drug binds tightly with high affinity but has low efficacy, while the other binds with low affinity but has very high efficacy. Since the two dose-response curves are identical, there is no data analysis technique that can tell them apart. You need to analyze a *family* of curves, not an individual curve, to determine the affinity and efficacy. The rest of this chapter explains how.

Derivation of the operational model

Black and Leff (*Proc. R. Soc. Lond.* B, 220: 141-162, 1983) developed the *operational model of agonism* to help understand the action of agonists and partial agonists, and to develop experimental methods to determine the affinity of agonist binding and a systematic way to measure relative agonist efficacy based on an examination of the dose-response curves.

Start with a simple assumption: Agonists bind to receptors according to the law of mass action. At equilibrium, the relationship between agonist concentration ([A]) and agonist-occupied receptor ([AR]) is described by the following hyperbolic equation:

$$[AR]=\frac{[R_T]\times[A]}{[A]+K_A}$$

$[R_T]$ represents total receptor concentration and K_A represents the agonist-receptor equilibrium dissociation constant.

What is the relationship between agonist-occupied receptor (AR) and receptor action? We know biochemical details in some cases, but not in others. This lack of knowledge about all the steps between binding and final response prevents the formulation of explicit, mechanistic equations that completely describe a dose-response curve. However, Black and Leff derived a "practical" or "operational" equation that encompasses the behavior of all of these unknown biochemical cascades. They began with the observation that many dose-response curves have a sigmoidal shape with a Hill slope of 1.0, (the curves are hyperbolas when response is plotted against agonist concentration, and are sigmoidal when response is plotted against the log of agonist concentration). They then proved mathematically that if agonist binding is hyperbolic and the dose-response curve has a Hill slope of 1.0, the equation linking the concentration of agonist-occupied receptors to response must also be hyperbolic. This second equation, shown below, has been termed the "transducer function", because it is a mathematical representation of the transduction of receptor occupation into a response:

$$\text{Effect}=\frac{\text{Effect}_{\text{max}}\cdot[\text{AR}]}{[\text{AR}]+\text{K}_{\text{E}}}$$

The parameter Effect_{max} is the maximum response possible in the system. This may not be the same as the maximum response that a particular agonist actually produces. The parameter K_E is the concentration of AR that elicits half the maximal tissue response. The efficacy of an agonist is determined by both K_E and the total receptor density of the tissue ($[R_T]$). Black and Leff combined those two parameters into a ratio ($[R_T]/K_E$) and called this parameter tau (τ), the "transducer constant". Combining the hyperbolic occupancy equation with the hyperbolic transducer function yields an explicit equation describing the effect at any concentration of agonist:

$$\text{Effect}=\frac{\text{Effect}_{\text{max}}\cdot\tau\cdot[\text{A}]}{(\text{K}_{\text{A}}+[\text{A}])+\tau\cdot[\text{A}]}$$

This equation can be rewritten as follows, to make it easier to compare the operational model with the standard sigmoid equation for an agonist dose-response curve.

$$\text{Effect}=\frac{\text{Effect}_{\text{max}}\cdot\tau\cdot[\text{A}]}{\text{K}_{\text{A}}+[\text{A}]\cdot(\tau+1)}=\frac{[\text{A}]\cdot\left(\text{Effect}_{\text{max}}\cdot\frac{\tau}{\tau+1}\right)}{\frac{\text{K}_{\text{A}}}{\tau+1}+[\text{A}]}$$

This form of the equation makes it clear that the maximum effect in the dose-response relationship seen with a particular agonist is not Effect_{max}, but rather is Effect_{max} multiplied by $\tau/(\tau+1)$. Only a full agonist in a tissue with plenty of receptors (high values of τ) will yield a maximum response that approaches Effect_{max}.

The EC_{50} does not equal K_A (the equilibrium dissociation constant for agonist binding to the receptors) but rather $K_A/(1+\tau)$. With a strong agonist (large τ value), you'll get half-maximal response by binding fewer than half the receptors, so the EC_{50} will be much less than K_A.

The following figure shows a dose-response curve for a partial agonist, and shows the relationship between EC_{50} and maximum response to terms in the operational model.

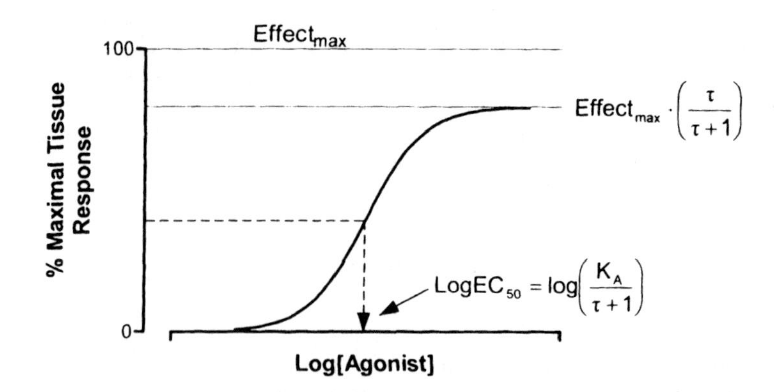

The parameter τ is a practical measure of efficacy. It equals the total concentration of receptors in the system divided by the concentration of receptors that need to be occupied by agonist to provoke a half-maximal tissue response. The τ value is the inverse of the fraction of receptors that must be occupied to obtain the half-maximal response. If τ equals 10, that means that occupation of only 10% of the receptors leads to a half-maximal response. If τ equals 1.0, that means that it requires occupation of all the receptors to give a half-maximal response. This would happen with a partial agonist or with a full agonist in a tissue where the receptors had been significantly depleted. Because τ is a property of *both* the tissue *and* receptor system, it is not a direct measure of *intrinsic efficacy*, which is commonly defined as a property belonging only to an agonist-receptor pair, irrespective of the assay system in which it is measured.

The equations above show agonist stimulated response, so the curves all begin at zero. It is easy to add a *Basal* parameter to model observed response in the absence of drug, so the response with no agonist equals *Basal* rather than zero.

Shallower and steeper dose-response curves

Some sigmoid dose-response curves are steeper or shallower than a curve with a standard slope factor of 1. The operational model can be extended to analyze these curves.

If you assume the initial binding of the agonist to the receptor follows the law of mass-action (Hill slope equals 1 for the binding step), then transduction step(s) between occupancy and final response must follow an equation that allows for variable slope. If the dose-response curve is still sigmoid, then the operational model can be extended fairly simply, by including a slope parameter, n. The extended form of the operational model is:

$$\text{Effect}=\frac{\text{Effect}_{max}\cdot\tau^n\cdot[A]^n}{(K_A+[A])^n+\tau^n\cdot[A]^n}$$

The relationship between this operational model and the variable slope sigmoid equation are as follows:

$$EC_{50}=\frac{K_A}{(2+\tau^n)^{1/n}-1}$$

$$E_{max}=\frac{\text{Effect}_{max}\cdot\tau^n}{\tau^n+1}$$

When n equals 1, the equation is the same as those shown earlier, describing dose-response curves with Hill slopes of 1.0. However, n is *not* the same as the Hill slope (but the two values will be very close for full agonists).

Designing experiments to fit to the operational model

A single dose-response curve does not define both the affinity and efficacy of an agonist. If you try to fit the operational model equation to a single dose-response curve, you'll run into a problem. Either the curve-fitting program will report an error message, or it will report best-fit values with enormously wide confidence intervals.

Any symmetrical dose-response curve is defined by four parameters: *Bottom* (response with no agonist), *Top* (response at very high concentrations), EC_{50} (concentration of agonist needed to provoke a response halfway between *Bottom* and *Top*) and the *Hill Slope*. However, the operational model equation has five parameters: *Basal* (response with no agonist), K_A (dissociation constant of agonist binding), $Effect_{max}$ (maximum possible effect with a full agonist and plenty of receptors), τ (a measure of agonist efficacy), and n (transducer slope).

Since the operational model has more parameters than are needed to describe a sigmoid dose-response curve, any curve can be defined by an infinite combination of operational model parameters. Even if a curve-fitting program could find best-fit values (rather than report an error message), the best-fit parameter estimates may not be correct.

To fit the operational model to data, therefore, you cannot analyze just a single dose-response curve. Instead you must fit a family of dose-response curves. Use one of these experimental approaches:

- One approach is to reduce the number of accessible receptors in a tissue or cell line to such an extent that a full agonist can no longer produce the maximal cellular response, no matter how high a concentration is used. A common method for reducing the number of functional receptors is to treat the tissue or cell line with a drug (e.g., alkylating agent) that binds irreversibly to the agonist binding site on the receptor, and thus permanently occludes that site. The agonist curve before alkylation is then compared to the curve after alkylation. This is the experimental method of choice for generating data that will allow affinity and efficacy estimates for drugs that are full agonists.

- A second approach that works only for partial agonist drugs is to directly compare the dose-response curve of a partial agonist with the dose-response curve of the full agonist. This method does not require receptor alkylation, but does require a known full agonist for the receptor of interest.

Note: All the dose-response curves should be obtained with the same tissue or cell line, in order to minimize variability in $Effect_{max}$ between preparations. This also applies to the use of recombinant expression systems (e.g., cell lines) with genetically engineered differences in receptor density; simultaneous analysis of curves obtained across different cell-lines will introduce between-tissue variability into the analysis, which can lead to problems with parameter estimation. In contrast, receptor depletion experiments using the same preparation of cells before and after treatment should only be subject to within-tissue variability.

Fitting the operational model to find the affinity and efficacy of a full agonist

Theory of fitting receptor depletion data to the operational model

To fit the operational model, one must account for data from two (or more) dose-response curves. For a full agonist, you must compare the dose-response curve in the absence or presence of receptor depletion. Experimentally, the key is to ensure that the receptors have been sufficiently alkylated such that the full agonist can no longer yield a maximal tissue response at saturating concentrations. These conditions are especially important to ensure that fitting the operational model to the data will yield a good estimate of the $Effect_{max}$ model parameter, which is crucial for successful estimation of the remaining model parameters. You can do this by globally fitting all the dose-response curves at one time, sharing model parameters across all the data sets.

Let's first consider an experiment where the dose-response curve to a full agonist is determined in the absence or presence of progressive depletion of accessible receptor binding sites (this is also referred to as reducing "receptor reserve"). Because $\tau = [R_T]/K_E$, irreversibly occluding agonist binding will reduce $[R_T]$ and thus reduce the value of τ. This will lower the maximal response and shift the dose-response curve to the right.

The operational model assumes that irreversibly occluding some receptors does not change the other three parameters. It assumes that the affinity of the agonist for the remaining receptors (K_A), the value of the transducer slope (n), and the value $Effect_{max}$ are properties of the tissue, not the drug, so have one value for all curves (note that $Effect_{max}$ refers to the maximum possible effect when no receptors are occluded, not the maximum effect attained in a particular dose-response curve). To fit the operational model, therefore, we want to globally fit all the data sets, sharing the value of K_A,, n, and $Effect_{max}$ but finding separate best-fit values of τ for each data set.

Fitting receptor depletion data to the operational model with Prism

Follow these steps:

1. Since concentrations are equally spaced on a log scale, enter data with X equal to the logarithm of the agonist concentration. Or transform your data to make X equal to the log of concentration if necessary.

2. Enter the operational model into your program. Here is one version:

```
operate= (((10^logKA)+(10^X))/(10^(logtau+X)))^n
Y=Basal + (Effectmax-Basal)/(1+operate)
```

3. If you have already subtracted off any basal activity, then constrain Basal to a constant value of zero.

4. Set up global fitting so $\log K_A$, n and $Effect_{max}$ are shared among data sets. If you didn't constrain Basal to be a constant also share it among data sets. Don't share logtau, as its value will be unique for each curve.

5. Consider the following recommendations for initial parameter values:

Parameter	Initial Values
Effectmax	1 x maximum Y value for the full agonist curve in the absence of receptor depletion
n	Set to 1 (initial value to be fit)
Basal	1 x minimum Y value for the full agonist curve. (If there is no basal response in the absence of agonist, then set this value as a constant of zero, or omit it from the equation).
logKA	1 x the X value corresponding to the response half way between the highest and lowest Y values for the agonist curve after receptor depletion.
logTau	Set to 0.0 (initial value to be fit). Since logtau starts at zero, this means that the initial value for τ is 1.0. This value of τ corresponds to a dose-response curve that plateaus at half $Effect_{max}$, and usually results in successful convergence.

Why fit the log of K_A and the log of τ? When writing any model for data analysis, you should arrange the parameters so that the uncertainty is symmetrical and Gaussian. If you fit to the logarithm of K_A and τ, the uncertainty is more symmetrical (and more Gaussian) than it would be if you fit to K_A and τ directly (see A. Christopoulos, *Trends Pharmacol. Sci*, 19:351-357, 1998).

Why fit *Basal*? You may measure a "response" even in the absence of agonist. So include a basal parameter in the model. *Basal* is the measured response in the absence of agonist. If there is no basal activity, or if you have subtracted away basal before analyzing your data, then constrain *Basal* to a constant value of zero.

Example of fitting receptor depletion data to the operational model

In this example, we fit the operational model to two data sets. One data set is the response of human M_1 muscarinic receptors, stably transfected into Chinese hamster ovary cells, to the agonist, acetylcholine in the absence of receptor alkylation, and the other shows the response to the same agonist in the same cells after receptor alkylation with the irreversible alkylating agent, phenoxybenzamine. The actual response being measured is agonist-mediatied [^{3}H]phosphoinositide hydrolysis (A. Christopoulos, University of Melbourne, unpublished). Shown below are the actual data (d.p.m.):

Log[Acetylcholine]	Vehicle	Alkylated
-8.000	516	241
-7.000	950	423
-6.523	2863	121
-6.000	7920	527
-5.523	11777	745
-5.000	14437	3257
-4.523	14627	3815
-4.000	14701	4984
-3.523	15860	5130

Analysis of these data according to the operational model yielded the curve fits shown below:

	Vehicle	Alkylated	Shared
Best-fit values			
$logK_A$	-4.981	-4.981	-4.981
$log\tau$	1.024	-0.2511	
n	1.279	1.279	1.279
Basal	218.0	218.0	218.0
E_{max}	15981	15981	15981
Std. Error			
$logK_A$	0.1349	0.1349	0.1349
$log\tau$	0.09797	0.02776	
n	0.1218	0.1218	0.1218
Basal	171.5	171.5	171.5
E_{max}	420.9	420.9	420.9
95% CI			
$logK_A$	-5.275 to -4.687	-5.275 to -4.687	-5.275 to -4.687
$log\tau$	0.8105 to 1.237	-0.3116 to -0.1906	
n	1.014 to 1.545	1.014 to 1.545	1.014 to 1.545
Basal	0 to 591.6	0 to 591.6	0 to 591.6
E_{max}	15064 to 16898	15064 to 16898	15064 to 16898

The table above shows some of the output from the Results page of the analysis (using GraphPad Prism). Note that the best-fit value for $logK_A$ is close to -5, a concentration that gives a full response in control conditions, and is a log unit away from the $logEC_{50}$ of the control dose-response curve (which is close to -6). Here is a graph of the curves.

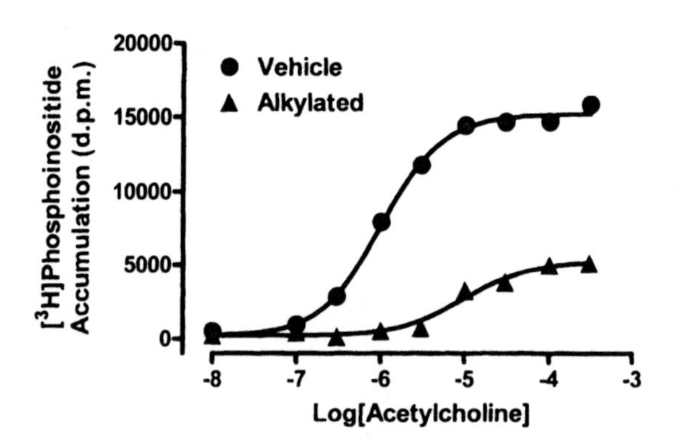

Fitting the operational model to find the affinity and efficacy of a partial agonist

Theory of fitting partial agonist data to the operational model

A second application of the operational model is to obtain affinity and efficacy estimates for one or more partial agonists by comparing their responses to a full agonist in the same tissue. The analysis is different from fitting to the operational model after receptor inactivation. With receptor inactivation, the goal is to get a single value of $\log K_A$ for the agonist. With partial agonists, we expect the get different values of $\log K_A$ for each partial agonist used.

It is impossible to determine the $\log K_A$ of a full agonist without inactivating receptors (see method above). However, for a full agonist ($\tau > 10$), the *Top* and *HillSlope* parameters obtained from the standard sigmoid dose-response equation are very good approximations of the $Effect_{max}$ and n parameters, respectively, of the operational model. This fact is exploited in the current method for obtaining operational model parameters for partial agonists. Specifically, the dose-response curve for the full agonist is fit to the standard sigmoidal dose-response equation, while the dose-response curves for the partial agonist(s) are fit to the operational model. The *Top* and *HillSlope* parameters of the full agonist curve are used by the operational model as *Effectmax* and *n*, respectively, when fitting the partial agonist curve.

> GraphPad note: Prism 4 lets you share parameters across data sets, even when different data sets are fit to different equations. This feature is also available in some other computer programs, but not in all of them.

Fitting partial agonist data to the operational model with Prism

Follow these steps:

1. Since concentrations are equally spaced on a log scale, enter data with X equal to the logarithm of the agonist concentration. Or transform your data to make X equal to the log of concentration.

2. Enter the data for the full agonist into your first data set column (e.g., column A) and the data for the partial agonist in column B.

3. Choose nonlinear regression, and enter a user-defined equation. The following example is specific for GraphPad Prism. If you use a different program, you will probably need to modify enter the model differently.

```
operate= (((10^logKA)+(10^X))/(10^(logtau+X)))^n
<A> Y = Basal + (Effectmax-Basal)/(1+10^((LogEC50-X)*n))
<~A> Y = Basal + (Effectmax-Basal)/(1+operate)
```

> GraphPad note: The second line in the equation is preceded by <A> so it only applies to data set A. It is a standard sigmoidal dose-response curve. The third line is preceded by <~A> so it applies to all data sets except A. It is the operational model and can be used for data in columns B and onward. This means that you can actually fit more than one partial agonist to the model at the same time, provided you have a full agonist curve in column A.

4. For *Basal*, *n* and *Effectmax*, choose to share the values for all data sets. Leave *logtau*, *logKA*, and *logEC50* to be individually fitted.

5. ***Consider the following recommendations for initial parameter values***:

Parameter	Initial Values
Effectmax	1 x maximum Y value for the full agonist curve in the absence of receptor depletion
n	Set to 1 (initial value to be fit)
Basal	1 x minimum Y value for the full agonist curve. (If there is no basal response in the absence of agonist, then set this value as a constant of zero, or omit it from the equation).
logKA	1 x the X value corresponding to the response half way between the highest and lowest Y values for the partial agonist curve.
logTau	Set to 0.0 (intial value to be fit). Since logtau starts at zero, this means that the intial value for τ is 1.0. This value of τ corresponds to a dose-response curve that plateaus at half $Effect_{max}$, and usually results in successful convergence.
logEC50	1 x the X value corresponding to the response half way between the highest and lowest Y values for the full agonist curve.

Example of fitting partial agonist data to the operational model

In this example, we wish to obtain affinity and efficacy estimates for the partial agonist, pilocarpine, by comparing its responses to those of the full agonist, oxotremorine-M, in Chinese hamster ovary cells transfected with the human M_3 muscarinic acetylcholine receptor (A. Christopoulos, University of Melbourne, unpublished). The response being measured is the same as that for the previous example. Note that the full agonist properties of oxotremorine-M were confirmed separately in receptor depletion experiments):

log[Agonist]	Oxotremorine-M	Pilocarpine
-7.70	280	
-7.10	1222	
-6.49	4086	
-5.89	6893	
-5.29	7838	
-4.69	8062	
-6.70		253
-6.17		502
-5.65		1263
-5.12		1879
-4.60		2467
-4.08		2818

	Oxotremorine-M	Pilocarpine	Shared
Best-fit values			
Basal	128.6	128.6	128.6
E_{max}	8076	8076	8076
$logEC_{50}$	-6.487	(not used)	
n	1.290	1.290	1.290
$logK_A$	(not used)	-5.446	
$log\tau$	(not used)	-0.2258	
Std. Error			
Basal	73.77	73.77	73.77
E_{max}	74.36	74.36	74.36
$logEC_{50}$	0.006110	(not used)	
n	0.06146	0.06146	0.06146
$logK_A$	(not used)	0.05676	
$log\tau$	(not used)	0.01655	
95% CI			
Basal	0.0 to 309.1	0.0 to 309.1	0.0 to 309.1
E_{max}	7894 to 8258	7894 to 8258	7894 to 8258
$logEC_{50}$	-6.502 to -6.472	(not used)	
n	1.140 to 1.441	1.140 to 1.441	1.140 to 1.441
$logK_A$	(not used)	-5.585 to -5.307	
$log\tau$	(not used)	-0.2663 to -0.1853	

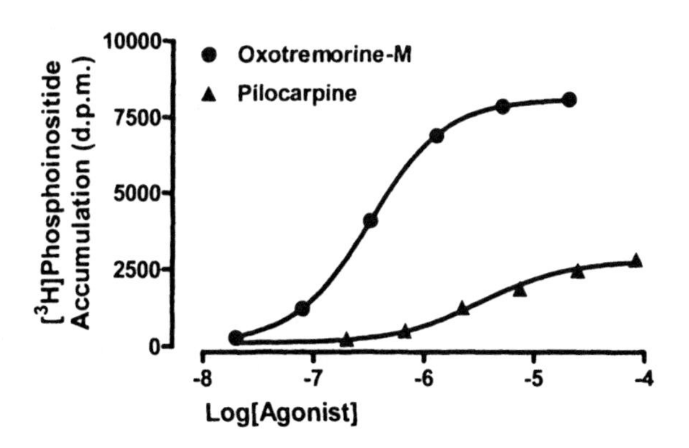

43. Dose-response curves in the presence of antagonists

Competitive antagonists

The term *antagonist* refers to any drug that will block, or partially block, a response. When investigating an antagonist, the first thing to check is whether the antagonism is surmountable by increasing the concentration of agonist. The next thing to ask is whether the antagonism is reversible. After washing away antagonist, does agonist regain response? If an antagonist is surmountable and reversible, it is likely to be competitive (see next paragraph). Investigations of antagonists that are not surmountable or reversible are beyond the scope of this manual.

A competitive antagonist binds reversibly to the same receptor as the agonist. A dose-response curve performed in the presence of a fixed concentration of antagonist will be shifted to the right, with the same maximum response and the same shape.

The dose ratio

Gaddum (*J. Physiol. Lond.*, 89, 7P-9P, 1936) derived the equation that describes receptor occupancy by agonist in the presence of a competitive antagonist. The agonist is drug A, its concentration is [A] and its dissociation constant is K_a. The antagonist is called drug B, so its concentration is [B] and dissociation constant is K_b. If the two drugs compete for the same receptors, fractional occupancy by agonist (*f*) equals:

$$f = \frac{[A]}{[A]+K_a\left(1+[B]/K_b\right)}$$

In the above equation, the occupancy of agonist [A] is determined by its K_a. It can therefore be seen that the presence of a competitive antagonist multiplies the K_a value by a factor equal to $1+[B]/K_b$. In other words, the only effect of a simple competitive antagonist on an agonist is to shift the occupancy of the agonist by this constant factor; it has no other effects on the properties of the agonist. This theoretical expectation forms the basis of all currently used methods for quantifying agonist-antagonist interactions, which therefore rely on the determination of agonist dose-response curve shifts in the presence of antagonists.

Because a competitive antagonist does not alter the relationship between agonist occupancy and final response, it is unnecessary for you to know this relationship for the Gaddum equation above to be useful in analyzing dose-response curves. Thus, the equation can just as easily be written in terms of an agonist's EC_{50} value in the dose-response curve, rather than its K_a value in the occupancy curve. You don't have to know what fraction of the receptors is occupied at the EC_{50} (and it doesn't have to be 50%). The key to the usefulness of the equation is that whatever the initial agonist occupancy, you'll get the same occupancy (and thus the same response) in the presence of antagonist when the agonist concentration is multiplied by $1+[B]/K_b$. Here is what the equation looks like when it is written in terms of the classic sigmoid dose-response curve relationship.

$$\text{Response} = \text{Bottom} + \frac{(\text{Top-Bottom})}{1+\left(\frac{EC_{50}\left[1+{[B]}/{K_b}\right]}{[A]}\right)^{\text{HillSlope}}}$$

The graph below illustrates this relationship. If concentration A of agonist gives a certain response in the absence of competitive antagonist, but concentration A' is needed to achieve the same response in the presence of a certain concentration of the antagonist, then A'/A represents the factor $1+[B]/K_b$. The ratio A'/A is called the "dose ratio" and is most conveniently (although not exclusively) determined using EC_{50} values. You'll get a different dose ratio if you use a different concentration of antagonist, but the shift will always reflect the constant ($1+[B]/K_b$) if the interaction is truly competitive. Thus, if you know [B] and can determine the dose ratio, you should be able to derive a value for K_b.

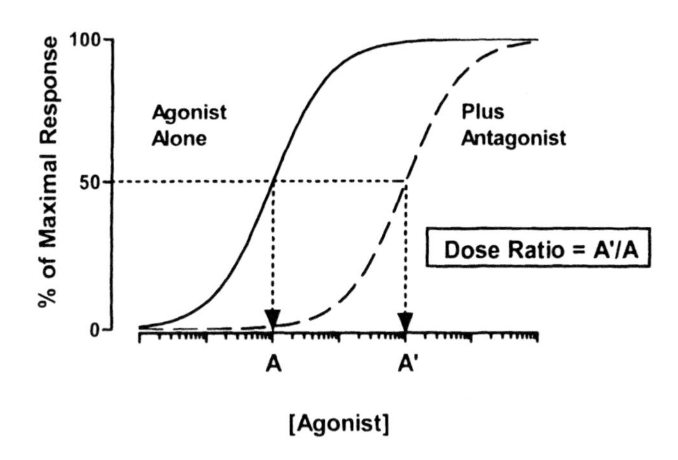

The Schild slope factor

In theory, the principal goal behind measuring agonist dose-response curve shifts in the presence of competitive antagonists is to use the relationship embodied in the dose ratio to obtain an estimate of the antagonist's K_b value, i.e., its equilibrium dissociation constant. In practice, this can only be accurately derived when you determine the effects of more than one concentration of antagonist on the dose-response curve of the agonist. If the interaction were truly competitive, then the shift of the agonist dose-response curve in the presence of antagonist will always correspond to $1+[B]/K_b$, irrespective of the value of [B]. Using different concentrations of B, therefore, allows you to check if the relationship holds. This procedure was first extensively developed by the pharmacologist, Heinz O. Schild (Arunlakshana and Schild, *Br. J. Pharmacol.*, 14: 48-57, 1959), and it is thus commonly associated with his name (i.e., "Schild analysis").

In his studies, Schild also asked the question: what happens if the relationship between antagonist concentration and agonist dose-response curve shift doesn't follow the factor $1+[B]/K_b$? For example, some noncompetitive antagonists can shift agonist dose-response curves to the right without changing agonist maximal response, minimal response, and slope, but the degree of the shift doesn't follow the competitive relationship of $1+ [B]/K_b$; in some cases the shift is greater than expected, whereas in others it is less than expected. Alternatively, an antagonist may be competitive, but the tissue or cellular preparation in which it is tested may contain more than one subtype of receptor with equal affinity and responsiveness to the agonist, but different affinities for the antagonist. Again, this latter situation may not result in changes of agonist dose-response curve shape in the presence of antagonist, but it can often result in agonist curve shifts that do not follow $1+[B]/K_b$. In order to accommodate agonist curve shifts in the presence of antagonists that were either greater than or less than expected for simple competition for a single receptor, Schild modified Gaddum's equation by introducing a slope factor, commonly referred to as the "Schild slope":

$$f = \frac{[A]}{[A]+K_a\left[1+\frac{[B]^S}{K}\right]}$$

In this equation, the antagonist term, [B], is now raised to the power S, where S denotes the Schild slope factor. Thus, if the antagonist shifts the agonist dose-response curve to the right in a parallel fashion, but greater than that predicted for simple competition, then the value of S will be greater than 1. In contrast, smaller-than expected agonist curve shifts can be accommodated by a value of S less than 1. Notice that we have also changed the "K_b" parameter from the previous equation to a "K" in the above equation. This is because a K_b value, i.e., an antagonist equilibrium dissociation constant, cannot be derived from the above equation if S does not equal 1, so by convention, we shouldn't call it the K_b in the above model. In practice, the K parameter should actually be estimated as a negative logarithm, so the equation can be re-written as follows:

$$f = \frac{[A]}{[A]+K_a\left[1+\frac{[B]^S}{10^{-pK}}\right]}$$

where pK is defined as the negative logarithm of K. Hence, the parameter, pK, represents a simple fitting constant that has no mechanistic meaning except when S=1, in which case $pK = pK_b$.

In Schild analysis, therefore, the determination of agonist dose-response curve shifts in the presence of different concentrations of antagonist allows you to first assess the *conformity* of the data to a model of simple competition, by determining whether the Schild slope is significantly different from 1 or not, and then *quantify* the antagonism (if S=1) by determining the pK_b value (the dissociation constant of the antagonist binding).

pK_b vs. pA_2

By convention, a pK_b value can only be derived when S =1. In this circumstance, the data are deemed to be consistent with a simple mechanism of one-to-one competition between agonist and antagonist for the receptor, and the pK_b is thus a mechanistic estimate of the negative logarithm of the antagonist's equilibrium dissociation constant. In practice, this is done by first fitting the Gaddum/Schild model to experimental data in order to obtain the estimate of S, and then performing a statistical test to determine whether this estimate of S is different from a value of 1. If S is not significantly different from 1, then the equation is refitted to the data with S fixed as a constant value of 1, and the resulting estimate of pK is the pK_b value.

What happens if S is significantly different from 1? In this case, the resulting estimate of pK is *not* the pK_b, and cannot be quoted as such. It is not the negative logarithm of the dissociation constant of antagonist binding. You will have to conclude that your experimental data are not consistent with a model of simple competition between agonist and antagonist. Nevertheless, you may still wish to quote an empirical estimate of the potency of your antagonist for the sake of comparison with other drugs. By convention, the most common estimate of antagonist potency that is independent of any particular mechanism is the "pA_2 value". The pA_2 is defined as the negative logarithm of the concentration of antagonist required to cause a twofold rightward shift of the agonist dose-response curve. It can readily be seen that for a competitive antagonist, $pA_2 = pK_b$, because a K_b concentration of a competitive antagonist will shift an agonist's curve by a factor of two ($1+[B]/K_b = 2$ when $[B]=K_b$). For a non-competitive antagonist, a pA_2 value is *not* a pK_b, but simply a measure of the potency of the antagonist to shift the curve of an agonist to the right by a factor of two. Historically, the pA_2 was determined from the X-intercept of the Schild plot (see page 285), when S was not fixed to a value of 1, but it can easily be calculated from the following relationship between Gaddum/Schild model parameters:

$$pA_2 = \frac{pK}{S}$$

Thus, even if an antagonist is not competitive, a pA_2 value can be quoted as an empirical estimate of the antagonist's potency.

An alternative to the classic Schild slope factor

Although the Gaddum/Schild equation in its original form (see above) is still the most commonly used model for fitting agonist-antagonist dose-response data, we and others have noted that the two most relevant parameters of interest, the slope, S, and the pK, are highly correlated with one another. That is, when the nonlinear regression algorithm changes one parameter while trying to find its best-fit value, the other parameter also changes to compensate. Parameters are always somewhat correlated, but these two are especially correlated, making the results less adequate. We need to find a way to minimize this problem, while still allowing for the derivation of appropriate estimates of the Schild slope and pK. One modification of the Gaddum/Schild equation that overcomes this problem is shown below (see D. R. Waud et al., *Life Sci.*, 22, 1275-1286, 1978; Lazareno and Birdsall, *Br. J. Pharmacol.*, 109, 1110-1119, 1993):

$$f = \frac{[A]}{[A] + K_a\left[1 + \left(\frac{[B]}{10^{-pK}}\right)^S\right]}$$

It can be seen that the difference between this modified equation and the original Gaddum/Schild equation is that the entire [B]/K term is now raised to the power S, rather than just the [B] term in the original equation.

What effect does this have on the model and its ability to fit agonist-antagonist interaction data? We have performed many simulations to investigate the properties of the modified Gaddum/Schild model, and have found the following results. First, the value of S is the same if we use the modified equation compared to the original form of the equation, so the S parameter can still be quoted as an estimate of the Schild slope. Second, and most importantly, the parameters S and pK are far less correlated in the modified equation. Third, if the value of S is significantly different from 1 in the modified equation, then the estimate of the pK is not valid as an estimate of the antagonist's pK_b, but it *is* a valid estimate of the pA_2. In contrast, if the original form of the Gaddum/Schild equation were to be used, then the estimate of pK when the value of S is not 1 is meaningless; it *cannot* be used as an estimate of the pA_2 unless it was first divided by the value of S (see above).

Obviously, it is far better to use a model that allows separate estimates of S and pA_2 to be obtained directly from the curve-fitting process, rather than having to indirectly calculate a pA_2 value from two previously estimated parameters that are each associated with their own standard errors. Based on these findings, therefore, we can rewrite the modified Gaddum/Schild model as follows:

$$f=\frac{[A]}{[A]+K_a\left[1+\left(\frac{[B]}{10^{-pA_2}}\right)^S\right]}$$

The remaining sections of this chapter describe analyses based on this modified Gaddum/Schild model of agonist-antagonist interactions.

Using global fitting to fit a family of dose-response curves to the competitive interaction model

The most rigorous method for quantifying agonist-antagonist interactions is to globally fit the modified Gaddum/Schild model to all the agonist dose-response curves that were constructed in the absence or presence of different concentrations of antagonist. For this approach to work, the nonlinear regression program must be able perform global fitting (i.e., share model parameters between all the data sets).

If we combine the standard sigmoid dose-response equation with the modified Gaddum/Schild equation describe above, we obtain the following equation:

$$\text{Response}=\text{Bottom}+\frac{(\text{Top-Bottom})}{1+\left(\frac{10^{\text{LogEC}_{50}}\left[1+\left(\frac{[B]}{10^{-pA_2}}\right)^S\right]}{[A]}\right)^{\text{HillSlope}}}$$

This equation defines a dose-response curve to an agonist, A, in the presence of increasing concentrations of antagonist, B. When [B]=0, the equation becomes the standard sigmoid dose-response model described elsewhere in this book. Using a program that allows global parameter sharing across multiple data sets, you can fit this equation directly to all your dose-response curves for a particular agonist determined in the absence or presence of different concentrations of antagonist by choosing to share the values of each parameter across all the data sets. However, parameter-sharing alone won't work because you are using an equation with two independent variables, agonist concentration [A] *and* antagonist concentration, [B]. You will also need to use a program that allows you to have parameter-sharing feature in conjunction with two independent variables.

> GraphPad note: You can use two independent variables in Prism by assigning one of the variables to be in your X column, and the other variable as a Column title for its respective data sets. Other programs have different rules for multiple independent variables, or don't allow you to use them at all.

The worked example below illustrates data for the inhibition of acetylcholine-mediated $[^3H]$phosphoinositide hydrolysis (dpm) by the antagonist, N-methylscopolamine, in Chinese hamster ovary cells stably transfected with the human M_1 muscarinic receptor (A. Christopoulos, University of Melbourne, unpublished). Note that the concentrations of N-methylscopolamine (M) that were used in the experiment were entered as the column titles for the appropriate acetylcholine dose-response data set.

log[Acetylcholine]	0	3e-10	1e-9	3e-9	1e-8
-8.00	688	162	310		
-7.00	3306	478	3209		
-6.52	12029	4663	564		
-6.00	29865	15009	9769	1501	462
-5.52	35802	31041	25158	7833	1531
-5.00	38300	36406	29282	23995	9463
-4.52				35642	22583
-4.00	36291	34412	36245	40341	31046
-3.52				35573	33407

The following syntax can be used to specify the model:

```
EC50=10^LogEC50
Antag=1+(B/(10^(-1*pA2)))^SchildSlope
LogEC=Log(EC50*Antag)
Y=Bottom + (Top-Bottom)/(1+10^((LogEC-X)*HillSlope))
```

Here are some comments about the variables in the equation.

Variable	Comments
X	The logarithm of agonist concentration in molar. The independent variable.
Y	The response (the dependent variable). In this example, the response is measured in dpm, but it could be in almost any units.
B	The concentration of antagonist in molar. In Prism, set this to be a data set constant, so its value comes from the title of each column. For any given data set (column) this is a constant. But its value varies from column to column, so you can also think of this as being a second independent variable.
Bottom	The bottom plateau of the dose-response curves, in the same units as Y. In most cases, this is a parameter to fit, and you can set its initial value to the minimum Y value for the agonist curves. Set this parameter to be shared, so you get one best-fit value for the family of curves, and not one value for each curve. If there is no basal response in the absence of agonist, then constrain this value as to be a constant equal to zero, or omit it from the equation.
Top	The top plateau of the dose-response curves, in the same units as Y. This is a parameter to fit, and you can set its initial value to the maximum Y value for the agonist curves. Set this parameter to be shared, so you get one best-fit value for the family of curves, and not one value for each curve.
logEC50	The logarithm of the EC50 of the agonist alone, in log molar. This is a parameter to fit. Set is initial value to the X value corresponding to the response halfway between the highest and lowest Y values for the full agonist curve. Set this parameter to be shared, so you get one best-fit value for the family of curves, and not one value for each curve. Its value only makes sense for the control curve, but it mathematically enters the equation for each curve so you must share its value.

Variable	Comments
HillSlope	The slope of the agonist curve. Unitless. Set its initial value to 1.0. In some cases, you might want to constrain this to a constant value of 1.0. If you don't hold it constant, you must set this parameter to be shared among all your data sets. You want a single value of the Hill slope for the experiment.
SchildSlope	The Schild slope, which will be close to 1.0 if the drug is a competitive antagonist. Set its initial value to 1.0. In some cases, you might want to constrain this to a constant value of 1.0. If you don't hold it constant, you must set this parameter to be shared among all your data sets. You want a single value of the Schild slope for the experiment, not one for each concentration of antagonist.
pA2	The negative logarithm of the concentration (in molar) of antagonist that shifts the agonist EC50 by a factor of 2. This is a parameter to fit (the whole point of this analysis). Enter a rough estimate of the negative logarithm of the antagonist's K_b, based on the minimum concentration of antagonist that you observed experimentally to cause a discernible shift in the control agonist dose-response curve. It is essential that you set this parameter to be shared among all your data sets. You want a single value of pA2 for the experiment, not one for each concentration of antagonist.

Note that EC50, LogEC, and Antag are intermediate variables used to write the model, and are not parameters that are to be fit.

Ideally, the most appropriate approach for analyzing the entire family of dose-response curves is to fit them to two versions of the above equation, one where the SchildSlope parameter is set as a constant equal to 1, and the other where it is a shared value for all data sets, and then compare the two different forms of the equation using the F-test. If the simpler model (Schild slope=1) is the better fit, then the estimate of pA_2 is in fact the pK_b, and may be quoted as such. If the equation where the Schild slope does not equal 1 is the better fit, then the estimate of pA_2 is not the pK_b.

Shown below are the results of this analysis for the interaction between acetylcholine and N-methylscopolamine at the M_1 muscarinic receptor, based on the table above. The equation where the Schild slope is allowed to vary does not fit the data significantly better than the model where the Schild slope is set to its conventional value 1.0. So we fixed the Schild slope to 1.0 in the results below.

Comparison of Fits	
Null hypothesis	SCHILDSLOPE = 1.0
Alternative hypothesis	SCHILDSLOPE unconstrained
P value	0.5594
Conclusion (alpha = 0.05)	Do not reject null hypothesis
Preferred model	SCHILDSLOPE = 1.0
F (DFn, DFd)	0.3494 (1,27)

Parameter	Best-fit value	95% CI
LogEC50	-6.366	-6.470 to -6.261
PA2	9.678	9.552 to 9.803
Bottom	362.9	-1194 to 1920
Top	36500	35050 to 37951
HillSlope	1.559	1.252 to 1.867

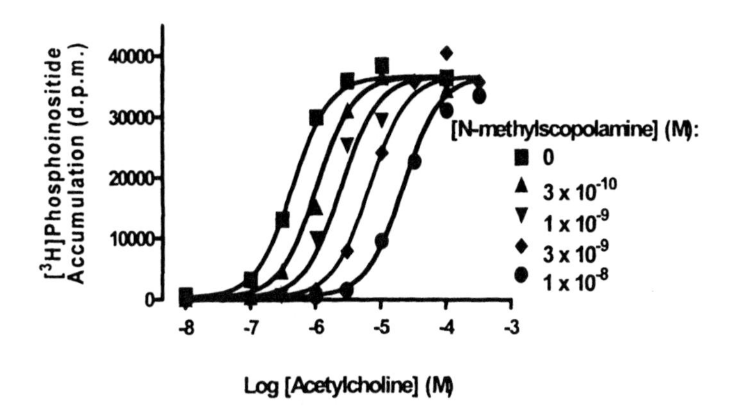

The whole point of the analysis is to find the K_b of the antagonist. Since the data are consistent with a Schild slope of 1.0 (and we fixed Schild slope to 1.0 in the results shown), the PA2 value can be interpreted as the pK_b. Therefore the K_b value equals 10^{-pkb}, which equals 2.09×10^{-10}M, or 0.209 nM. We can also transform each end of the confidence interval of the PA2 to obtain the 95% confidence interval for the pK_b, which ranges from 0.157 nM to 0.281 nM.

Fitting agonist EC_{50} values to the competitive interaction model

Although the procedure described in the previous section is the preferred method for analyzing agonist-antagonist interaction data, there are some situations where you are not able to use this method, or simply don't need the level of rigor associated with it. For instance, you may be doing experiments where only a few agonist concentrations are tested, such that only the linear portion of the sigmoid dose-response curve (on a logarithmic scale) is determined in the absence and presence of each antagonist concentration. Alternatively, you may be using a nonlinear regression program that doesn't allow you to use global parameter-sharing and/or two independent variables. In these instances, you can't fit the complete sigmoid model presented above, but you can still determine equieffective agonist concentrations, perhaps as EC_{50} values. Lew and Angus (*Trends Pharmacol. Sci.*, 16: 328-337, 1995) have presented a simple method for analyzing agonist-antagonist interactions using nonlinear regression of agonist EC_{50} values obtained in the absence or presence of antagonist.

Start with the Gaddum equation for occupancy as a function of agonist and antagonist concentrations:

$$f=\frac{[A]}{[A]+K_a\left[1+{[B]}/{K_b}\right]}$$

Simple algebra expresses the equation this way:

$$f=\frac{1}{1+\frac{K_a}{K_b}\left[\frac{[B]+K_b}{[A]}\right]}$$

Thus you can obtain any particular occupancy f, with any concentration of antagonist [B] so long as you adjust A to keep the quantity in the parentheses constant (C).

$$\frac{[B]+K_b}{[A]}=C$$

Rearrange to show how you must change the agonist concentration to have the same response in the presence of an antagonist.

$$\frac{[B]+K_b}{C}=[A]$$

The EC_{50} is the concentration needed to obtain 50% of the maximal response. You don't know the fraction of receptors occupied at that concentration of agonist, but you can assume that the same fractional occupancy by agonist leads to the same response regardless of the presence of antagonist. So you can express the equation above to define EC_{50} as a function of the antagonist concentration [B].

$$EC_{50}=\frac{[B]+K_b}{C}$$

You determined the EC_{50} at several concentrations of antagonist (including 0), so you could fit this equation to your data to determine a best-fit value of K_b (and C, which you don't really care about). But it is better to write the equation in terms of the logarithm of EC_{50}, because the uncertainty is more symmetrical on a log scale. See "Why you should fit the $\log EC_{50}$ rather than EC_{50}" on page 263. By tradition, we use the negative logarithm of EC_{50}, called the pEC_{50}. For similar reasons, you want to determine the best-fit value of log K_b (logarithm of the dissociation constant of the antagonist) rather than K_b itself.

$$pEC_{50} = -\log\left([B]+10^{-pK_b}\right)-\log(C)$$

Define Y to be the pEC_{50}, X to be the antagonist concentration [B], and a new constant P to be log(C). Now you have an equation you can use to fit data:

$$Y = -\log\left(X+10^{-pK_b}\right)-P$$

Determining the K_b using nonlinear regression of agonist pEC_{50} values

1. Determine the EC_{50} of the antagonist in the presence of several concentrations of antagonist, including zero concentration. Enter these values into a data table as follows: Into the X column, enter the antagonist concentrations in molar. Into the Y column, enter the negative logarithm of the EC_{50} values.

2. Use nonlinear regression to fit this equation.

```
Y=-1*log(X +(10^(-1*pKb)))-P
```

Note: This equation is written so Y is the pEC50. It won't work if you enter EC50 or logEC50 values instead.

Dealing with Schild slopes that do not equal 1

Here is how the equation above can be recast to find the Schild slope and the pA_2 using nonlinear regression analysis of EC_{50} values. The equation is again based on the modified Gaddum/Schild model presented earlier in this chapter:

$$pEC_{50}=-\log\left([B]^S+10^{-pA_2 \times S}\right)-\log(c)$$

Note that the parameter pK_b in the original Lew and Angus equation has been replaced with pA_2 in the above equation. This is because if the value for S is significantly different from 1, then the antagonist fitting parameter is *not* the pK_b, but will be the pA_2.

Enter the following user-defined equation:

```
K=-1*pA2
Y=-1*log(X^S+(10^(K*S)))-P
```

When performing this analysis, it is a good idea to fit the data to both equations at the same time and use the F-test to decide which one is the more appropriate equation. If the simpler equation is the better equation, then the pK_b estimate may be quoted. Otherwise, you must conclude that your data are not consistent with a model of simple competition; you can still quote the pA_2, however, as an empirical estimate of antagonist potency.

Parameter	Initial Values
pA2	Enter a rough estimate of the negative logarithm of the antagonist's K_b, based on the minimum concentration of antagonist that you observed experimentally to cause a discernible shift in the control agonist dose-response curve.
S	Set to 1.
P	An initial value of P = 0 usually results in a succesful convergence.

The Schild plot

The oldest method for analyzing agonist-antagonist interactions from functional experiments is the original linear regression method developed by Schild. This method relies explicitly on the determination of agonist dose ratios in the absence and presence of antagonist. If you perform experiments with several concentrations of antagonist, you can create a graph with log(Antagonist) on the X-axis and log(Dose ratio –1) on the Y-axis; this is commonly referred to as the Schild plot. If the antagonist is competitive, you expect a slope of 1.0 and an X-intercept of log K_b for the antagonist.

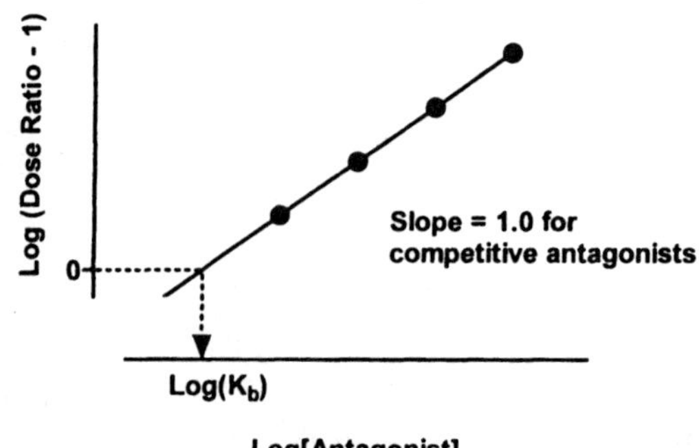

In comparison to the nonlinear regression methods outlined above, the linear regression method of the Schild plot is potentially flawed. The problem is that the EC_{50} of the control agonist dose-response curve is used to compute dose ratios for all other curves. Any error in that control value shows up in all the data points. The Schild plot was developed in an era when nonlinear regression was unavailable, so it was necessary to transform data to a linear form. This is no longer an advantage, and Schild plots can be thought of in the same category as Scatchard plots. That is, they are useful for graphical representations of agonist-antagonist interaction data, but for analytical purposes the nonlinear regression methods outlined above are superior.

Antagonist inhibition curves

There are often instances where complete agonist dose-response curves in the absence or presence of antagonist cannot be readily determined to fully define the effects of the antagonist over more than one or two orders of magnitude of antagonist concentrations. For example, there may be solubility problems with the agonist, or it may only be available in very small quantities such that large concentrations cannot be prepared, or it may rapidly desensitize the preparation when used at high, but not low, concentrations. These practical difficulties with the agonist, in turn, limit the investigator's ability to accurately discriminate whether the antagonist is competitive or noncompetitive, because noncompetitive antagonists may appear competitive when tested at low concentrations, but reveal their noncompetitive nature when tested at high concentrations.

One approach to overcoming these limitations that has become increasingly popular is to test the effects of increasing, graded, concentrations of antagonist on a single, fixed, concentration of agonist. This kind of experimental design is referred to as the "antagonist inhibition curve" design, and can readily test the effects of antagonist concentrations that span many orders of magnitude. This method is particularly widespread in the measurement of biochemical responses using cell-based or tissue extract-based assays. Shown below is an example of an agonist dose-response curve as well as the corresponding antagonist inhibition curve determined in the presence of a fixed agonist concentration (3×10^{-8} M) that produces the response denoted by the dotted line.

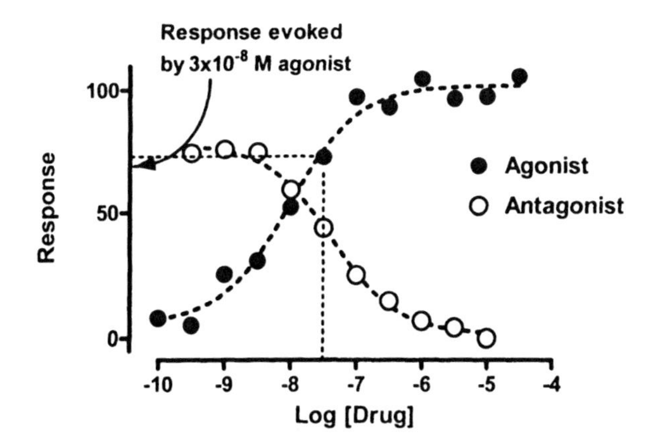

It can be seen that the shape of the antagonist inhibition curve appears similar to that of an antagonist competition binding curve obtained from a standard radioligand binding assay. Indeed, the concentration of antagonist that reduces the initial level of agonist response by 50% is usually called the IC_{50}, just like the concentration of antagonist that reduces specific radioligand binding by 50% in a binding assay. However, this is where the similarities end. Although it is relatively straightforward to obtain the $LogK_d$ of a competitive antagonist from a competition binding assay using the IC_{50} and the Cheng-Prusoff equation, you generally *cannot* obtain an equivalent estimate of $LogK_b$ from a functional antagonist inhibition curve using the same method (see Leff and Dougall, *Trends Pharmacol. Sci.*, 14, 110-112, 1993). This is because the shape of the antagonist inhibition curve in a functional assay is dependent on the shape of the agonist dose-response curve. If an agonist produces steep dose-response curves in a given tissue or cell line, then the resulting antagonist inhibition curve will be very different from if the agonist produces shallow curves, or curves with a slope of 1.

In order to properly analyze functional antagonist inhibition curve experiments, you need to include information about the control agonist dose-response curve in the analysis. The appropriate experimental design requires that you construct a control agonist curve and the antagonist inhibition curve in the same tissue or cell line. You can then analyze your data as follows:

1. Enter your agonist dose-response data into the first column of a new Data table.

2. Enter your antagonist inhibition curve data into the second column of your Data table.

3. Analyze your data according to the following user-defined equation. The syntax is specific for GraphPad Prism, and will need to be slightly modified if you are using a different program:

```
Control=Bottom + (Top-Bottom)/(1+10^((LogEC50-X)*HillSlope))
Antag=(10^LogEC50)*(1+((10^X)/(10^(-1*pA2)))^SchildSlope)
WithAntag=Bottom+(Top-Bottom)/(1+(Antag/FixedAg)^HillSlope)
<A>Y=Control
<B>Y=WithAntag
```

GraphPad note: The fourth line in the equation is preceded by <A> so it only applies to data set A. It is a standard sigmoidal dose-response curve. The last line is preceded by <B> so it applies to data set B. In this equation, therefore, it matters which data set you enter into column A, and which data set is entered into column B, so make sure that the control agonist dose-response data go into the first column and the antagonist inhibition curve data go into the second column.

Globally share the values of all the parameters across all data sets, except for the parameter, *FixedAg*, which represents the initial fixed concentration of agonist used in the determination of the antagonist inhibition curve. Choose to set *FixedAg* as a constant value equal to the fixed agonist concentration (Molar) used in your antagonist inhibition curve assay; for the above example, this value would be set as 3e-8. The nonlinear regression algorithm will then work its way through the equation. The desired parameters determined by the algorithm will then reflect the best-fit values that describe both agonist and antagonist curves. Because the Schild slope=1 in our example, the estimate of pA_2 is the pK_b. As with the previous examples, however, you should also fit this model with the Schild slope fixed to 1 and compare it using the F test with the model where the Schild slope is shared (but allowed to vary) by both data sets.

Shown below is the same example data set from above fitted to the model, as well as some of the output from the GraphPad Prism Results page.

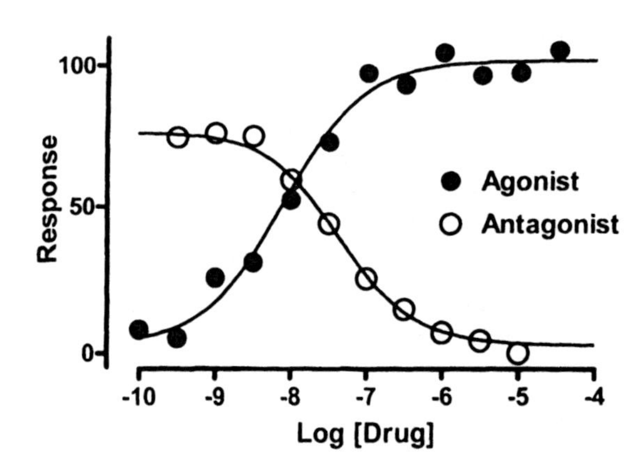

	Control	With Antagonist	Shared
Bottom	2.544	2.544	2.544
Top	101.9	101.9	101.9
logEC50	-8.078	-8.078	-8.078
HillSlope	0.8206	0.8206	0.8206
PA2	(not used)	8.135	
SchildSlope	(not used)	1.00	
FixedAg	(not used)	3.0000e-008	

An important consideration with fitting antagonist inhibition data is with respect to your initial parameter values. Because this is a relatively complicated model, it is probably best to enter the initial values for each data set manually, based on reasonable first guesses. For example, the *HillSlope* and *SchildSlope* values can each be initially set to 1. The *logEC50* and *pA2* parameters can be assigned the X values corresponding to the approximate midpoints of the control agonist curve and the antagonist inhibition curve, respectively. Note that the *Top* and *Bottom* parameters *must* be assigned according to the basal and maximal responses, respectively, of the control agonist dose-response curve.

44. Complex dose-response curves

Asymmetric dose-response curves

The standard (Hill) sigmoidal dose-response model is based on the assumption that the log(dose) vs. response curve is *symmetrical* around its midpoint.

But some dose-response curves are not symmetrical. In a recent study, Van der Graaf and Schoemaker (*J. Pharmacol. Toxicol. Meth.*, 41: 107-115, 1999) showed that the application of the Hill equation to asymmetric dose-response data can lead to quite erroneous estimates of drug potency (EC_{50}). They suggested an alternative model, known as the *Richards equation,* which could provide a more adequate fit to asymmetric dose-response data. Here is the Richards model shown both as an equation and as computer code.

$$\text{Response}=\text{Bottom}+\frac{(\text{Top-Bottom})}{\left[1+10^{(\text{LogX}_b-\text{X})\text{HillSlope}}\right]^S}$$

```
Numerator = Top - Bottom
Denominator=(1+10^((LogXb-X)*HillSlope))^S
Y=Basal+Numerator/Denominator
```

The Hill equation (variable slope sigmoidal dose-response curve) is sometimes called the *four-parameter logistic equation*. It fits the bottom and top of the curve, the EC50, and the slope factor (Hill slope). The Richards equation adds an additional parameter, S, which quantifies the asymmetry. Accordingly, this equation is sometimes referred to as a *five-parameter logistic equation*.

If S=1, the Richards equation is identical to the Hill equation (the four-parameter logistic equation) and the curve is symmetrical. However, if S does not equal 1.0, then the curve is asymmetric. The figure below shows a series of curves with various values of the symmetry parameter, S. Only the value of S differs between the three curves.

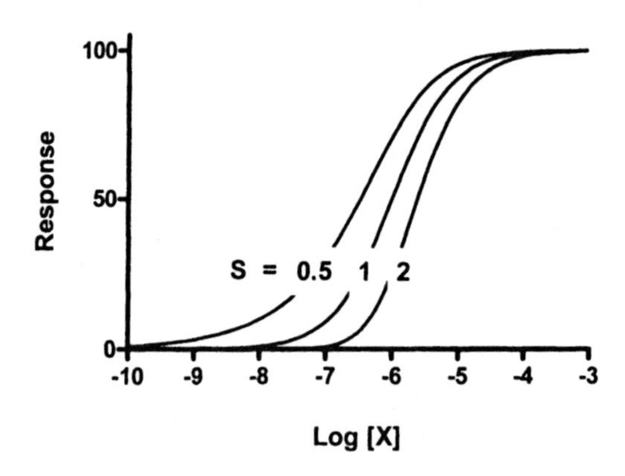

The parameter $LogX_b$, is not equivalent to the $logEC_{50}$. In the graph above, all three curves have $LogX_b$ = -6 yet the three curves have very different EC_{50}s. If you fit your dose-response data to the asymmetric Richards model, you can compute the LogEC50 using this equation:

$$LogEC_{50}=LogX_b-\left(\frac{1}{HillSlope}\right)\cdot Log\left(2^{1/S}-1\right)$$

You can also rewrite the model, as shown below, to fit the logEC50 directly (instead of fitting $LogX_b$.

```
LogXb = LogEC50 + (1/HillSlope)*Log((2^(1/S))-1)
Numerator = Top - Bottom
Denominator = (1+10^((LogXb-X)*HillSlope))^S
Y = Bottom + (Numerator/Denominator)
```

Unless the curve is symmetrical (S=1), the LogEC50 is not the same as the inflection point ($LogX_i$) which equals:

$$LogX_i=LogX_b+\left(\frac{1}{HillSlope}\right)\cdot Log(S)$$

The Hill (four-parameter variable slope dose-response curve) equation is a simpler case of the Richards (five-parameter asymmetric dose-response curve) equation. As discussed previously in Chapter 21, this means that the two models are nested. You can fit data to both models and then compare the two fits with the extra sum-of-squares F test (or the AIC_c approach). Use the results of model comparison to decide whether the data are better described by a symmetric or an asymmetric model.

When you fit the Richards five-parameter model to your data, you'll probably find that the standard errors are large (so the confidence intervals are wide). This is because the Hill slope and asymmetry parameter (S) in the Richards equation tend to be highly correlated with one another. You'll needs lots of data (with little scatter) to reliably fit both the Hill slope and the asymmetry factor. It is common to fix the Hill slope to 1.0 when fitting the Richards equation.

For more information on asymmetrical dose-response curves, see the review by Girlado et al. (*Pharmacol. Ther.*, 95, 21-45, 2002).

Bell-shaped dose-response curves

When plotted on a logarithmic axis, dose-response curves usually have a sigmoidal shape, as discussed in the previous chapter. However, some drugs may cause an inhibitory response at low concentrations, and a stimulatory response at high concentrations, or vice-versa. The net result is a bell-shaped dose-response curve.

Bell-shaped dose-response curves have been observed experimentally, for example, for many receptors that couple to both stimulation and inhibition of the enzyme, adenylyl cyclase (see S. Tucek et al., *Trends Pharmacol. Sci.*, 23: 171-176, 2002).

Tip: Unless you know that the reason for a non-standard dose-response curve shape in your experiment is due to an experimental error, avoid excluding data points simply to make a non-sigmoid curve fit a sigmoid model. Instead, it is relatively easy to extend the standard model of dose-response curves to accommodate different nonlinear and saturating curve shapes.

Combining two sigmoid equations

The following equation combines two sigmoid dose-response relationships to describe a bell-shaped dose-response curve. In the figure below, the curve begins at *Plateau1*, turns over at the *Dip* and then approaches *Plateau2*. The two different values for the $LogEC_{50}$ and n_H parameters denote the midpoint potency and the slope factors, respectively, of each phase of the curve. The variable, [A], denotes the agonist concentration.

$$Y=Dip+\frac{(Plateau_1\text{-}Dip)}{1+10^{(LogEC_{50}1-Log[A])\cdot n_H1}}+\frac{(Plateau_2\text{-}Dip)}{1+10^{(Log[A]-LogEC_{50}2)\cdot n_H2}}$$

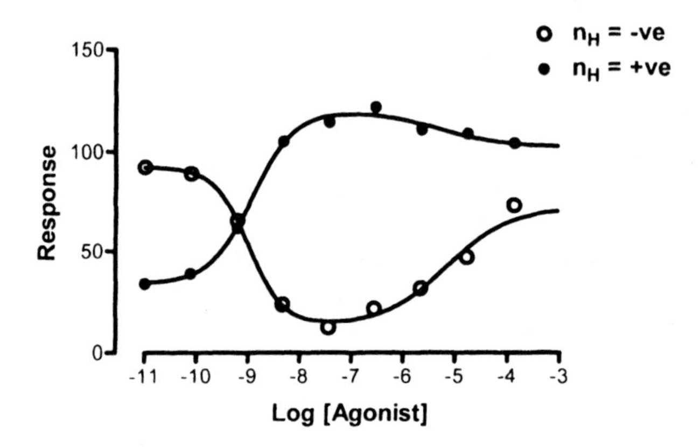

Here is one way the equation can be typed directly into a computer program:

```
Span1=Plateau1-Dip
Span2=Plateau2-Dip
Section1=Span1/(1+10^((LogEC50_1-X)*nH1))
Section2=Span2/(1+10^((X-LogEC50_2)*nH2))
Y=Dip+Section1+Section2
```

Because this model is more complicated than the standard monotonic sigmoid dose-response curve, there are a number of practical considerations when it comes to using the model to fit data. First, it is important that there are sufficient data points to adequately define both phases of the response; otherwise the model will fail to converge because it will have too many parameters relative to the number of points. Second, it can be seen from the graph that there are two general types of bell-shaped relationships possible, one where the dip occurs at the highest level of response, and one where the dip occurs at the lowest level of response. In order for the model to converge successfully, you need to be careful with your choice of initial parameter values. Of particular importance is the sign of the slope parameter, n_H. As can be seen in the graph, the slope factors are positive for one kind of curve, but negative for the other.

Using the Gaussian distribution equation

Sometimes dose-response curves exhibit a dip after reaching a maximal response level, but the response after this dip is not sufficiently defined for the investigator to conclude whether the curve is truly bell-shaped or not. This is often observed, for instance, when agonists cause desensitization of the tissue at the highest concentrations of drug used. Another common cause of these kinds of curve shapes is insufficient drug to fully define the entire dose-response relationship, perhaps due to solubility issues.

When the data don't follow a standard sigmoid shape, you should fit the data to a model that more closely approximates the shape of the curve yet still gives you measures of agonist potency, maximal response range, and slope factor.

> Tip: If your dose-response data show this kind of dip, beware of fitting your data to the standard sigmoid dose-response curve. The best-fit values for the maximal response and the EC_{50} will not be very accurate.

One possibility for fitting these kinds of data is combining two sigmoidal shape curves, as described above. However, this approach is only useful when you have sufficient data points to fully define both phases of a curve.

An alternative is to fit the data to the Gaussian distribution. While this distribution is rarely used to fit dose-response data, the figure below shows that a portion of the Gaussian distribution (solid) looks like a dose-response curve with a dip at the top. This is not a mechanistic model, but it is a way to empirically fit your data and get parameters that you can compare between treatments.

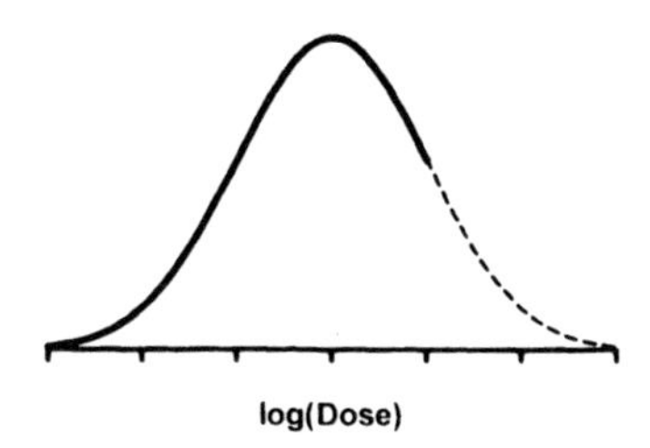

This Gaussian distribution has been used successfully to fit dose-response data (see A. Christopoulos et al., *J. Pharmacol. Exp. Ther.*, 298: 1260-1268, 2001). Shown below is how this equation can be rewritten to define a bell-shaped dose-response curve.

$$E=Basal+Range\times e^{-\left[\frac{10^{Log[A]}-midA}{slope}\right]^2}$$

where

$$midA=LogEC_{50}+slope\sqrt{-ln(0.5)}$$

In the original formulation of the Gaussian equation, the "midA" parameter would normally define the mean of the distribution, i.e., the x-axis value corresponding to the midpoint of the distribution – the peak. For fitting dose-response curves, this is not useful because the x-axis value we want is the logarithm of drug causing the response halfway between the Basal and the top of the dip, i.e., the $LogEC_{50}$. Thus, in the revised equation shown above, midA is corrected to allow for the estimation of the $LogEC_{50}$.

The reparameterized Gaussian dose-response equation defines a bell-shaped dose-response curve in terms of only four parameters, the *Basal* response, the *$LogEC_{50}$*, the maximal response *Range* (which is the maximal response range from Basal to the dip in the curve) and the *Slope*. Because the Gaussian equation requires fewer parameters than the bell-shaped equation described at the start of this section (4 vs. 7), it can be used to fit dose-response curves with fewer data points.

Here is how you type the equation into a computer program.

```
midA=LogEC50+Slope*SQRT(-ln(0.5))
Y=Basal+(Range*exp(-1*((X-midA)/Slope)^2))
```

The figure below shows a curve fit based on the modified Gaussian equation. The dotted line shows the fit of the standard sigmoid equation to the same data set. Neither the EC_{50}, nor the maximum response, would be correct if the data were fit to the standard sigmoidal dose response curve.

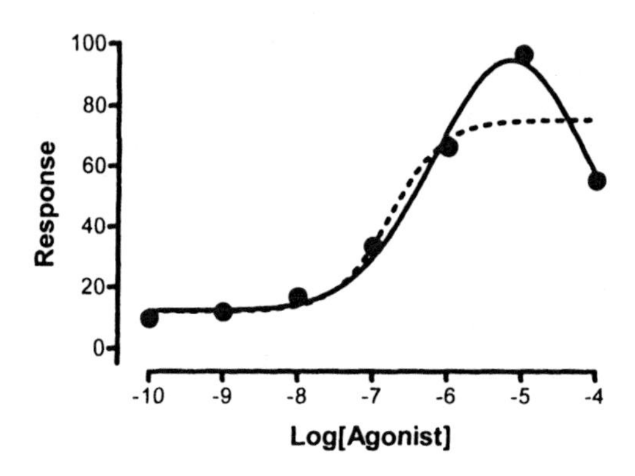

> Troubleshooting tip: What does it mean if the curve fits the data well, but the $LogEC_{50}$ is obviously way too high (beyond the range of your data)? Because of its symmetric nature, the Gaussian equation actually has *two* $LogEC_{50}$ values, a $LogEC_{50}$ for responses to the left of the dip, and a higher $LogEC_{50}$ for responses to the right of the dip. If the best-fit $LogEC_{50}$ is too high, your program probably fit the wrong one. Enter a smaller initial estimate of the $LogEC_{50}$, and the program will fit the $LogEC_{50}$ for the part of the curve where you actually have data.

What are the advantages and disadvantages of using this equation to fit bell-shaped data compared to the previous equation (combining two sigmoidal curves)? The main advantage of using the Gaussian is that you are dealing with a model containing fewer parameters, and thus increase your chances of obtaining a satisfactory fit with fewer data points. The main disadvantage of the model is in its symmetric nature. In the Gaussian, the "down" phase of the bell-shape is a mirror image of the "up" phase of the bell-shape. If you have a complete data set that fully defines both phases of the bell-shape, then the Gaussian will only provide a satisfactory fit if the two phases are practically mirror images, which is not that common. In this latter instance, you are better off using the more complicated bell-shaped model described earlier, which accommodates different slopes and plateaus for the two different phases. The Gaussian is best reserved for those data sets where one of the phases of the curve is well-defined, but the other is not, as shown in the figure above.

Note that the slope parameter is *not* equivalent to the Hill slope (n_H) found in the sigmoid dose-response equations. Although the slope parameter of the Gaussian allows for curves of varying degrees of steepness, its actual value changes *opposite* to that of the Hill slope in a sigmoid fit. That is, for steep curves, the value of the Gaussian slope gets smaller, whereas for shallow curves it gets larger, in contrast to the Hill slope. As with the other bell-shaped equation, therefore, you need to be careful when entering the initial values for the Gaussian equation.

Biphasic dose-response curves

Another common deviation from the standard monotonic sigmoid shape is the biphasic sigmoid shape. An example of an equation for a biphasic dose-response curve is shown below.

$$Y = Bottom + \frac{(Top-Bottom)\cdot Frac}{1+10^{(LogEC_{50}1-Log[A])\cdot n_H 1}} + \frac{(Top-Bottom)\cdot(1-Frac)}{1+10^{(LogEC_{50}2-Log[A])\cdot n_H 2}}$$

Here *Top* and *Bottom* are the maximal and minimal responses, respectively, $LogEC50_1$ and $LogEC50_2$ are the midpoint potency parameters for the two different phases, respectively, nH_1 and nH_2 are their corresponding Hill slopes, and *Frac* is the fraction of the curve comprising the more potent phase. The equation syntax is shown below, as is a figure illustrating a fit of the equation to a simulated (with random error) data set.

```
Span=Top-Bottom
Section1=Span*Frac/(1+10^((LogEC50_1-X)*nH1))
Section2=Span* (1-Frac)/(1+10^((LogEC50_2-X)*nH2))
Y=Bottom + Section1 +Section2
```

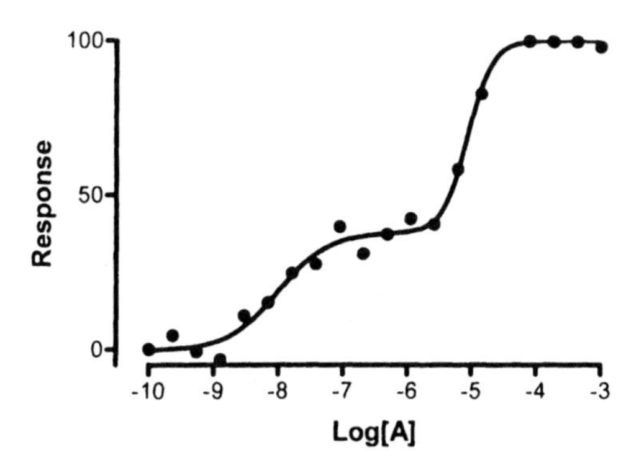

As with the preceding equations, successful curve-fitting with this model relies on the number of data points, the quality of the data, and your initial values. Our experience with this model is that it is especially sensitive to changes in the slope parameters, which often turn out to be significantly different from 1. In the example above, for instance, the value for nH_2 (for the less potent, right-most phase) was greater than 2, and the model had difficulty converging unless an estimate greater than 1 was entered as the initial value for that parameter. For the *Frac* parameter, you can use the Constraints feature in the nonlinear regression dialog box to constrain this value to always be between 0 and 1.

J. Fitting curves with GraphPad Prism

45. Nonlinear regression with Prism

Using Prism to fit a curve

To perform nonlinear regression with Prism, follow these steps. Later chapters explain your options in detail.

1. Enter or import your data. It is essential that the table be formatted to have a numerical X column. This will be the first column. You can enter one column of data for Y into column A, or several columns of Y data for different data sets (in columns B, C, etc.).

2. You can start nonlinear regression from a data table, from a results table (often a transform of the data), or from a graph. Press the *Analyze* button and choose to do a built-in analysis. Then select *Nonlinear regression (curve fit)* from the list of curves and regressions. This will bring up the nonlinear regression dialog.

3. On the first (Equation) tab of the dialog, pick an equation (model). You can choose a built-in (classic) model or enter your own. On this same tab, select the additional calculations you wish to perform.

4. The dialog has six more tabs where you choose analysis options.

5. Prism displays the results of nonlinear regression on a results sheet, which may have several subsheets (pages). It also superimposes the best-fit curve on a graph of those data.

> Note: The companion book of step-by-step examples includes several examples of nonlinear regression. These tutorials are also available on www.graphpad.com.

Which choices are most fundamental when fitting curves?

When you first learn about nonlinear regression, it is easy to feel overwhelmed by the many options, and be tempted to just skip over most of them. However, two choices are really important, and you must spend enough time to make sensible decisions:

- Which model? Nonlinear regression fits a model to your data, so it is important that you pick a model appropriate for your experimental situation. If you pick the wrong model, the results simply won't be helpful.

- Which parameters (if any) to hold constant? A model defines Y as a function of X and one or more parameters. You won't always want to ask Prism to find the best-fit value of all the parameters. For example, if you have subtracted away any nonspecific or background signal, a dose-response or kinetic curve must plateau at zero. In this case, you should tell Prism to constrain that parameter to have a constant value of zero. Setting constraints in this way can have a huge impact on the results.

Tip: Failing to set constraints is probably the most common error in curve fitting. Don't rush to fit a curve without stopping to consider whether some parameters should be constrained to have a constant value.

Prism's nonlinear regression error messages

In some cases, nonlinear regression is simply not able to find a best-fit curve, so Prism reports an error message instead. Prism reports error messages at the top of the tabular results. Each error messages refers to a fit of a particular data set (column) so you might get results for some data sets and different error messages for others.

Message	Meaning
Interrupted	Either you clicked "Cancel" on the progress dialog, or Prism exceeded the maximum allowed number of iterations specified in the Weighting tab of the Nonlinear regression dialog.
Bad initial values	A math error, such as division by zero or taking the log of a negative number, occurred when Prism first evaluated the equation. The problem occurred before Prism began to fit the equation to your data. You may have picked the wrong equation or have picked wildly bad initial values for one or more parameters.
Incorrect model	A math error occurred when Prism first began the fitting process. This means that your model, with your initial values, doesn't come close enough to your data for the fitting process to proceed. Check that you picked the right model and chose sensible initial values.
Does not converge	The fitting process began ok, but was unable to converge on best-fit values. Usually this means that one or more parameters were taking on unrealistic values. You may have chosen a model that is too complicated to describe your data. Try simplifying the model (for example, fitting a one-phase model rather than a two-phase model).
Floating point error	During the fitting process Prism tried to do a mathematically impossible operation (divide by zero, take the log of a negative number) so the fitting procedure had to terminate.

Note that these messages only tell you *when* the error occurred in the fitting process. In most cases, this does not tell you why the error occurred. For help in troubleshooting, see Chapter 6.

46. Constraining and sharing parameters

The Constraints tab of the nonlinear regression parameters dialog

The Constraints tab of the nonlinear regression dialog is very versatile. For each parameter, you can choose to fix it to a constant value, constrain it to a range of values, or share its value between data sets.

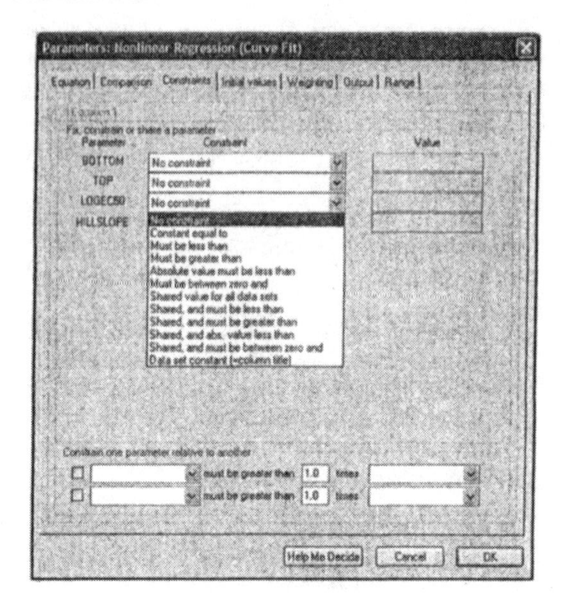

Constraining to a constant value

In many cases, it makes sense to constrain one (or more) of the parameters to a constant value. For example, even though a dose-response curve is defined by four parameters (bottom, top, $\log EC_{50}$ and Hill slope), you don't have to ask Prism to find best-fit values for all the parameters. If the data represent a "specific" signal (with any background or nonspecific signal subtracted), it can make sense for you to constrain the bottom of the dose-response curve to equal zero. In some situations, it can make sense to constrain the Hill slope to equal a standard value of 1.0.

If you constrain a parameter to a constant value, this same value applies for all the data sets you are fitting.

> Tip: Think carefully about which parameters, if any, you want to fix to a constant value. This can have a big impact on the results. Failing to fix a parameter to a constant value is a common error in curve fitting.

You don't have to enter the constant value in the dialog. Instead, you can link to a value you entered on an Info sheet. For example, here is an Info sheet with two values entered at the bottom.

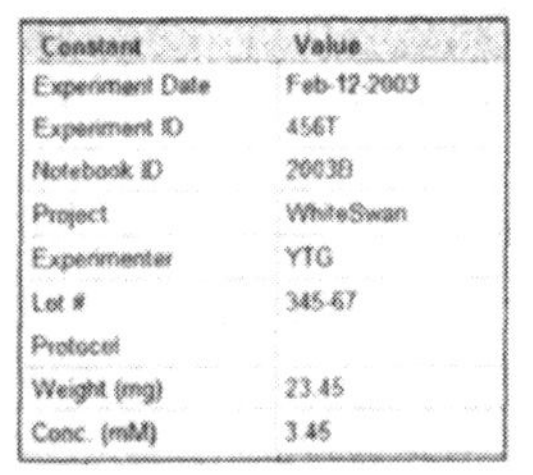

Constant	Value
Experiment Date	Feb-12-2003
Experiment ID	456T
Notebook ID	2003B
Project	WhiteSwan
Experimenter	YTG
Lot #	345-67
Protocol	
Weight (mg)	23.45
Conc. (mM)	3.45

Using an Info constant is easy. On the constraints tab, you'll choose to make a parameter a constant value. Then when you click to enter that value ...

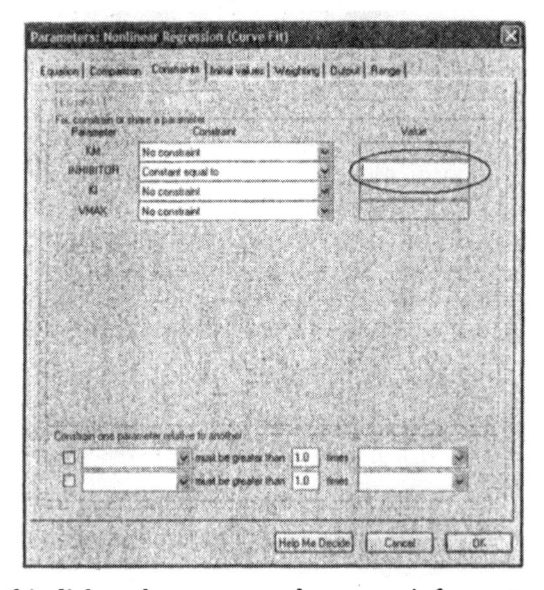

... Prism will pop up this dialog where you can choose any info constant on a linked info sheet.

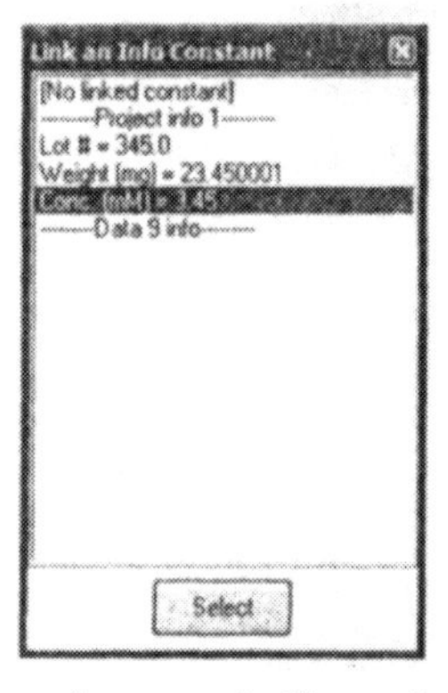

Prism shows linked constants in parentheses to indicate that they are linked and will be automatically updated when you edit the Info sheet.

INHIBITOR	Constant equal to	(3.45)

Data set constants

Define a parameter to be a *Data set Constant* and it will have a different constant value for each data set. For example, set the parameter I to be a data set constant:

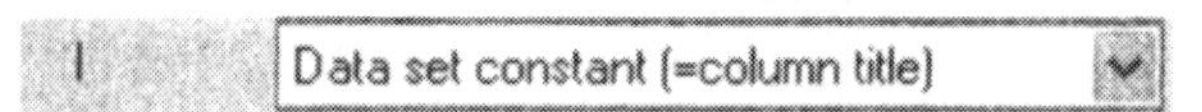

The values are not set in the Parameters dialog. Instead, Prism reads the values from the column headings of the data table. For example, each column may represent a different concentration of antagonist. In this case, use the concentrations as column (data set) titles, and set the parameter to be a data set constant. Prism will read the column titles, and set the parameter to the appropriate value for each data set.

X Values	A	B	C
Time	1 nM	2 nM	5nM
X	Y	Y	Y

It is ok to enter column titles like "10.3 nM" or "Time= 5 seconds". Prism will find the numbers within the column title and use that as the parameter value. But Prism extracts only the number. It doesn't attempt to interpret any units you enter. So in this example, the parameter will equal 1, 2 or 5. Prism doesn't attempt to interpret what the "nM" means.

> Note: Since data set constants are defined by the column title, and each column has only one title, it never makes sense to define more than one parameter to be a data set constant.

You can think of the data set constant as being a second independent variable, and so it provides a way to perform multiple nonlinear regression. For example, you can look at how response varies with both time and concentration. Enter the time values in the X column, so each row represents a different time point. Enter the concentration values as column titles so each column represents a different concentration. Now fit a model that uses both time and concentration, setting concentration to be a data set constant.

For an example of how data constants are useful, see pages 236 and 280. As those examples show, data set constants are most useful when combined with global curve fitting.

> Tip: Don't confuse the two ways of setting a parameter to a constant value. If you choose *constant equal to* you will enter a single value in the dialog, and this becomes the value of the parameter for all data sets. If you choose *data set constant*, you don't enter a value into the dialog. Prism gets the value from the column titles of the data table you are analyzing. The parameter can have a different value for each data set.

Constrain to a range of values

Use the Constraints tab of the nonlinear regression parameters dialog to constrain any parameter to be greater than, or less than, some value you enter, or to be between zero and a value you enter. For example, rate constants can't be negative, so next to K, choose "must be less than" and enter 0. If a parameter must be a fraction, constrain it to be between 0.0 and 1.0.

You can also constrain the relationship between two parameters. For example, if you are fitting a two-phase dose-response model, constrain the first logEC50 to be greater than the second.

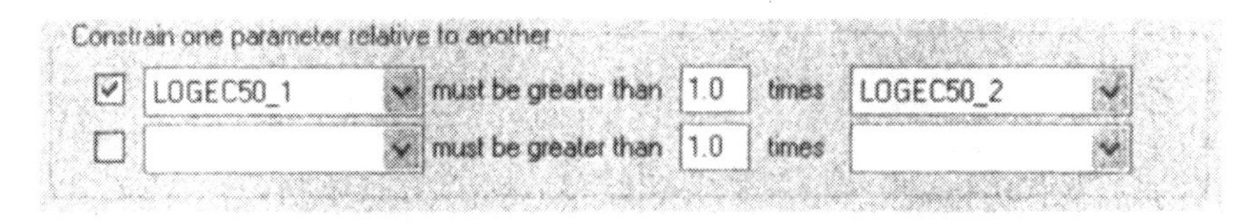

Shared parameters (global fitting)

When Prism fits a family of data sets at once, it usually fits each data set independently of the rest. It is convenient to analyze all the data sets on a table at once, and helpful to have all the results organized in a single table. But the results will be the same as they would be if each fit were done separately.

Prism 4 makes it possible to perform global fitting, where you specify that one or more parameters are to be shared among data sets. A shared parameter has a single best-fit value for all the data sets. Prism does a global fit of all the data sets, finding individual best-fit values for some parameters and a single, shared, best-fit value for others.

To specify that you want to share a value, choose *Shared value for all data sets* in the Constraints tab. Or choose one of the other Shared choices that also includes a constraint (for example, *Shared and must be less than*).

47. Prism's nonlinear regression dialog

The equation tab

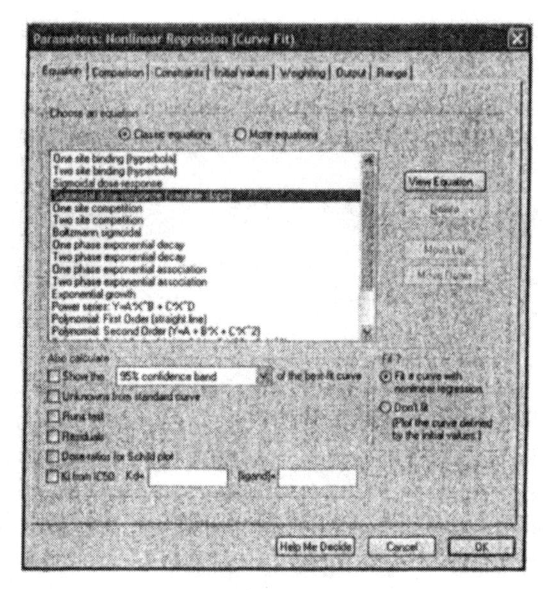

Choose an equation

Choose a classic equation (Chapter 48) , choose a library equation (Chapter 49), or enter your own equation (Chapter 50).

Optional calculations

Option	Discussion
Confidence or prediction bands of the best-fit curve	Plot the best-fit curve as well as an envelope around the curve denoting the 95% confidence or prediction interval. See page 32.
Interpolate unknowns from a standard curve	Enter the unknowns at the bottom of the same data table as the standard curve. Enter Y values only, leaving X on those rows blank. Or enter X values only, leaving Y on those rows blank. See page 97.
Runs test	The runs test can help you determine whether your data systematically deviate from the model you chose. See page 36.
Residuals	Viewing a table and graph of residuals can help you decide whether your data follow the assumptions of nonlinear regression. See page 35.
Dose-ratios for Schild plot	See Chapter 43 for a detailed discussion and alternative (better) ways to analyze Schild experiments.
K_i from IC_{50}	For competitive radioligand binding experiment, compute the dissociation constant (K_i) from the IC_{50}.

To fit or not to fit?

Choose whether you want to fit your model with nonlinear regression. The alternative is to plot the curve defined by the initial values. This is a very useful choice, as it lets you be sure that the initial values are sensible and generate a curve that comes near your data.

Comparison tab

Use the Comparison tab of the Nonlinear Regression Parameters dialog to specify one of three kinds of comparisons.

Compare the fit of each data set to two equations (models)

You've already chosen the first model (equation) on the first (equation) tab. Select the other equation at the bottom of the Comparison tab. For details, see Chapter 25.

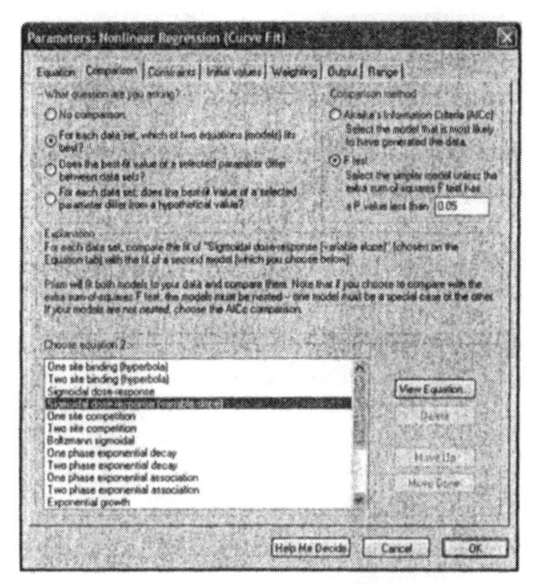

Does the best-fit value of a parameter (or several parameters) differ among data sets?

Select the parameter or parameters to compare at the bottom of the dialog. If you select all the parameters, the comparison will ask whether the curves differ at all among data sets. If one data set is control and one is treated, this will ask whether the treatment changes the curve at all. If you choose only one parameter, the comparison will ask whether the best-fit value of that parameter is changed by the treatment. If you select two or more parameters, but not all, the comparison asks if those selected parameters differ among the data sets. You get one P value comparing the curves with those parameters shared vs. a fit with those parameters fit separately. Prism won't compute separate comparisons of each parameter. See Chapter 27.

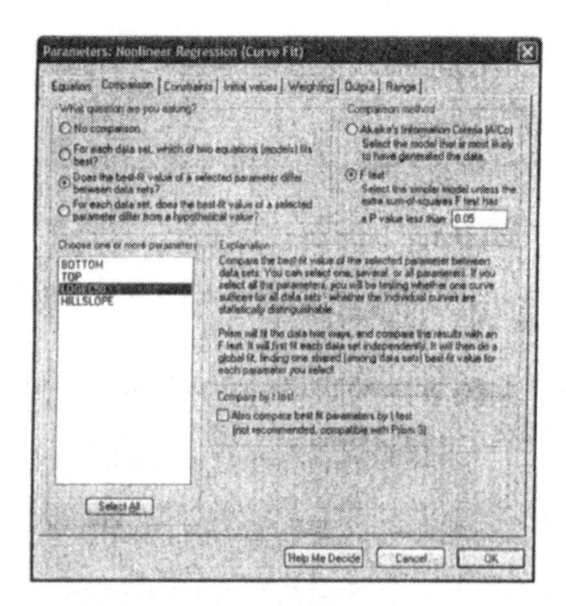

Test whether the best-fit value of a specified parameter differs significantly from a hypothetical value you propose

Choose the parameter and enter the value at the bottom of the dialog. For example, you might test if a Hill Slope differs from 1.0 (a standard value). Or you might test if a baseline or intercept differs from 0.0. For details, see Chapter 26.

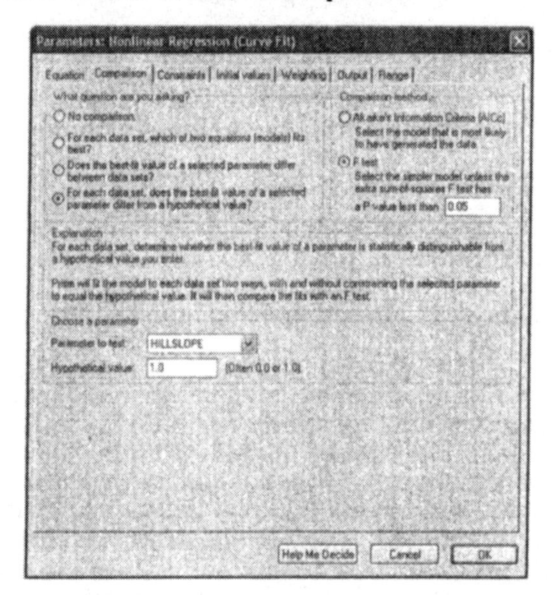

F test or AIC_c?

Prism offers two methods to compare models and data sets, the extra sum-of-squares F test and Akaike's Information Criteria (AIC_c). Choose in the upper right of the dialog. The extra sum-of-squares F test is used more commonly by biologists, but the AIC_c method has some advantages. For a detailed discussion, see Chapter 24.

Initial values tab

Nonlinear regression is an iterative procedure. The program must start with estimated initial values for each parameter. It then adjusts these initial values to improve the fit.

Prism automatically provides initial values for each parameter, calculated from the range of your data. If you select a classic equation, the rules for calculating the initial values are built-in to the program. If you enter a user-defined equation, you define the rules. See page 331.

The initial values tab of the Nonlinear Regression Parameters dialog lets you view and alter the initial values computed by the rules.

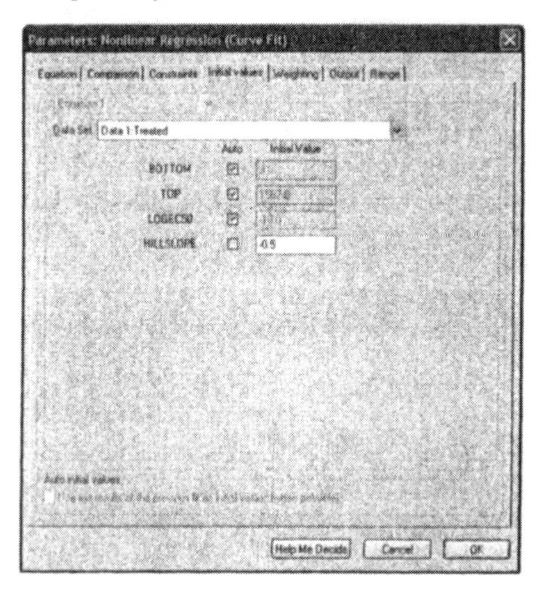

How much difference do initial values make?

When fitting a simple model to clean data, it won't matter much if the initial values are fairly far from the correct values. You'll get the same best-fit curve no matter what initial values you use, unless the initial values are very far from correct. Initial values matter more when your data have a lot of scatter or your model has many parameters.

Viewing the curve generated by the initial values

You'll find it easy to estimate initial values if you have looked at a graph of the data, understand the model, and understand the meaning of all the parameters in the equation. Remember that you just need an estimate. It doesn't have to be very accurate.

If you aren't sure whether the initial values are reasonable, check *Don't fit. (Plot the curve generated by the initial values.)* on the first tab of the nonlinear regression parameters dialog. When you click OK from the nonlinear regression dialog, Prism will not fit a curve but will instead generate a curve based on your initial values. If this curve is not generally in the vicinity of the data points, change the initial values before running nonlinear regression.

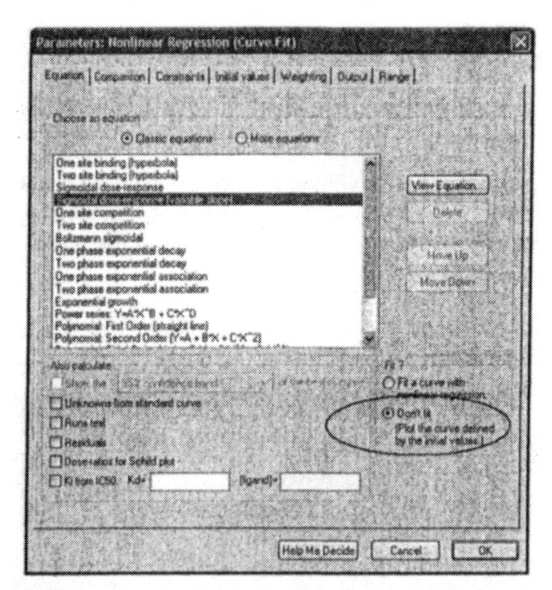

If you are having problems estimating initial values, set aside your data and simulate a family of curves using the analysis "Create a family of theoretical curves" (don't use the Nonlinear Regression analysis). Once you have a better feel for how the parameters influence the curve, you might find it easier to go back to nonlinear regression and estimate initial values.

Constraints for nonlinear regression

The constraints tab is very useful. Besides setting constraints, you can also define a parameter to be a data set constant or to be shared among data sets. These choices are discussed in depth in Chapter 46.

Weighting tab

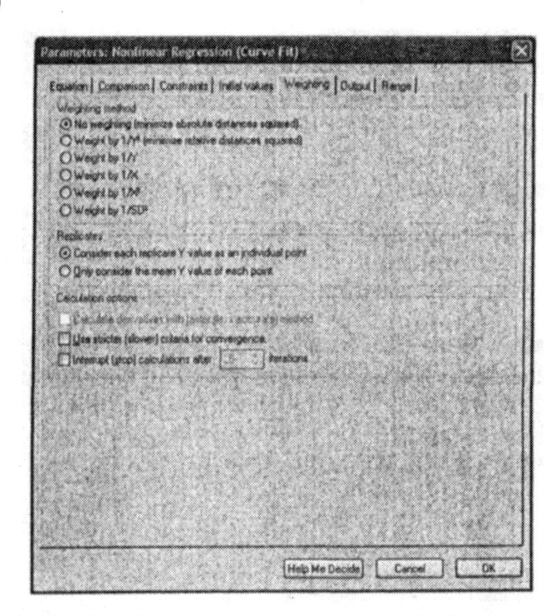

Weighting method

Chapter 14 explained the concept of differential weighting of data. Prism offers five ways to weight your data points.

The last choice weights by the inverse of the square of the SD. If you enter replicates, Prism computes the SD of the replicates automatically. Or you can format the data table to enter SD directly. In this case, you can enter any values you want in the "SD" subcolumn. This lets you set up any kind of weighting scheme you want.

If the weighting scheme you chose would result in a division by zero for any value, Prism does not fit the data set and reports "Weighting impossible" at the top of the results page.

Prism also considers sample size when weighting. If you entered individual replicates, and chose to treat the replicates separately, then no special calculations are needed. If you entered replicate values but chose to fit to the mean of the replicates, then Prism can multiply the weighting factor by N. The means computed from a large number of replicates get more weight than means computed from a few replicates. Similarly, if you enter mean, SD (or SEM) and N, Prism can multiply the weighting factor by N.

> Tip: If you are confused by weighting, just stick to the default choice (no weighting; minimize sum-of-squares). Changing to a different weighting method rarely has a huge impact on the results.

Replicates

In most experiments, you collect replicate Y values at every value of X. You should enter these into Prism into side-by-side subcolumns, from which Prism can automatically calculate error bars. When you fit a curve to these data, there are two ways Prism can fit a model to the data. It can treat each replicate as a separate point, or it can average the replicate Y values and treat the mean as a single point. Choose on the Weighting tab of the nonlinear regression parameters dialog. See a more detailed discussion on page 87.

> Tip: If you are confused by the choice of fitting to individual replicates vs. the mean, choose to fit individual replicates (which is the default). This choice rarely has a huge impact on the results.

Calculation options

Derivatives

While performing nonlinear regression, Prism repeatedly evaluates the partial derivative of your equation with respect to each parameter. This is the most time consuming part of nonlinear regression.

If you choose a built-in equation, Prism uses analytical derivatives built into the program. In other words, our programmers did the necessary calculus to define each derivative as an equation. There is no choice for you to make. If you enter your own equation (or use a library equation), Prism evaluates the derivatives numerically, and you can choose the method Prism uses.

Ordinarily, Prism uses Richardson's method to evaluate the derivatives. This method calculates the derivative by determining Y after both increasing and decreasing the value of the parameter a bit. Check the option box to use a faster, but potentially less accurate, method (which only evaluates the equation after increasing the value).

In most cases, the results will be identical regardless of the method. We recommend that you use the slow but accurate method to validate results with a new equation. You can then switch to the quick method for routine analyses if the speed of nonlinear regression matters to you (only an issue with huge files). With small data sets and fast computers, nonlinear regression will seem instantaneous even if you pick the slower method.

Convergence criteria

Prism stops iterating and declares the results to have converged when two iterations in a row change the sum-of-squares by less than 0.01%. If you check the box for strict convergence criteria, Prism will continue the iterations until five consecutive iterations each reduce the sum-of-squares by less than 0.000001%.

We recommend that you use the slow method only when you are having difficulty fitting an equation, or to validate your results. Use the standard method for routine analyses. If you select the standard method, Prism will automatically switch to the stricter criteria if the R^2 is less than 0.3.

Selecting the stricter criteria rarely affects the results but slows the calculations a bit (only noticeable with huge data sets or slow computers).

Stop calculating after a certain number of iterations

If this option is checked, Prism will stop nonlinear regression after the number of iterations you specify. In most cases, nonlinear regression converges in fewer than a few dozen iterations. If the iterations continue on and on and on, it may be because you've picked an inappropriate equation, picked unhelpful initial values, or have very scattered data. This option ensures that Prism won't spend a long time on calculations that won't be helpful. It is especially useful when you use a Prism script to fit many data sets.

> Tip: If you are curious to see how nonlinear regression works, set this to option to stop after one iteration. Then you can look at the graph after a single iteration. If you want to view the curve after another iteration, check the option in the initial values tab to *Use the results of the previous fit as initial values.* Otherwise, you'll just run the first iteration repeatedly.

Output tab

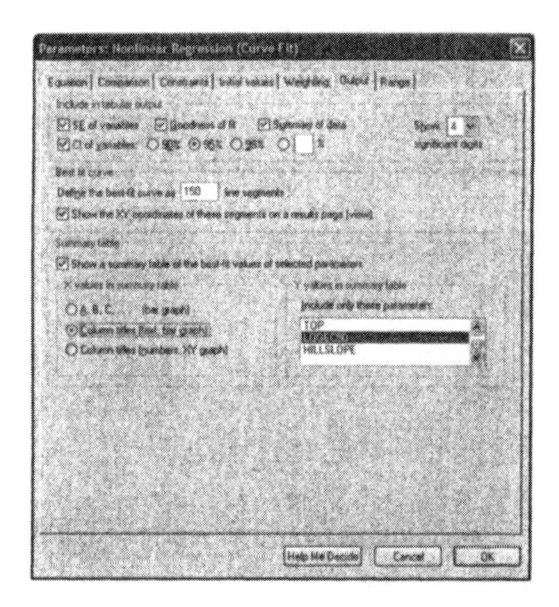

> Tip: You'll probably only need to change the output options from their default setting on rare occasions. The default settings will suffice almost all the time.

Include in tabular output

Choose which parts of the output you wish to see. We recommend that you leave all the boxes checked to get the most complete output. Also choose the number of significant digits used to report results. This is especially useful if you embed the results table on a graph or layout.

Table of XY values

Curves are defined by many short line segments. You decide how many segments Prism will create. Prism initially creates all curves with 150 line segments. Increasing the number may improve the accuracy of standard curve calculations and make the curve appear smoother (especially if it has many inflection points).

Normally, Prism hides this table of XY values of the line segments used to plot curves. Check the option box if you want to see this table as a results subpage.

Summary table and graph

When analyzing several data sets, the results table is rather lengthy. To display key results on a summary table, check the option box to create a summary table and select the variable you wish to summarize. Prism creates a summary table (as an additional results view) that shows the best-fit value of that variable for each data set, and graphs this table. Depending on your choices in the dialog, this may be a bar graph or an XY graph. It shows the best-fit value of a selected variable for each data set on the table. In some cases, you may analyze the summary table with linear or nonlinear regression. For example, the summary graph may show the best-fit value of a rate constant as a function of concentration (obtained from the column titles of the original data). You can fit a line or curve to that graph.

Note: When Prism compares the fits of two equations, it creates a summary table only from the results with the second equation. Since this may not be helpful, we suggest that you only make summary tables when fitting a single equation. When you choose other comparisons, the summary table is not available at all.

Range tab

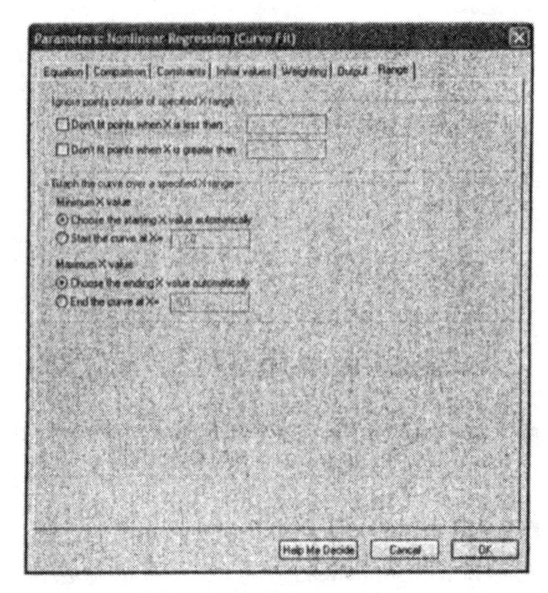

You don't have to fit a curve to all the data. Choose to ignore points with X values less than a limit you specify or greater than another limit you specify.

After determining the best-fit values of the variables in the equation, Prism calculates and plots the best-fit curve. It automatically plots the curve starting at the X position of the first (lowest X) data point and ends at the last data point. You may enter different limits.

Notice that the two range choices are very different. The first set of choices affects which data are analyzed, so affects the results. The second set of choices affects only how the curve is graphed, but does not affect the results.

Tip: Range options will be useful only occasionally. In most cases, you'll be happy with the default choices.

Default preferences for nonlinear regression

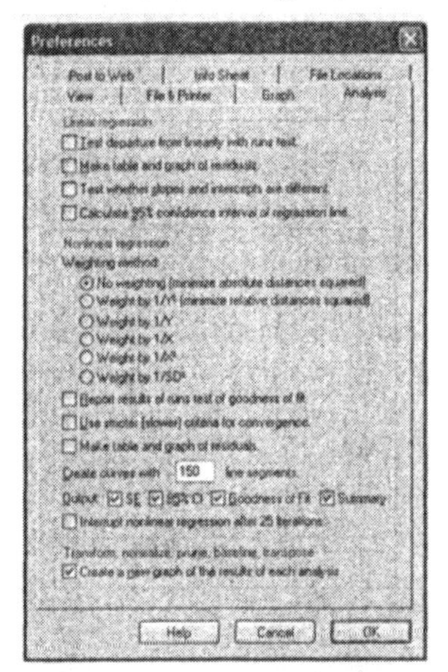

The nonlinear regression parameters dialog affects one particular nonlinear regression analysis. Change settings on the Preferences dialog to change default settings for future nonlinear regression analyses. To open this dialog, pull down the *Edit* menu and choose *Preferences* and go to the *Analysis* tab. You can change these default settings:

- Minimize sum-of-square of absolute distances or relative distances?
- Report results of runs test of goodness-of-fit?
- Use stricter (slower) criteria for convergence?
- Make table and graph of residuals?
- Number of line segments to generate curves.

> Note: Changing the analysis options changes the default settings for *future* nonlinear regression analyses. It will not change analyses you have *already* performed. When you do an analysis in the future, of course you can override any of the default settings.

48. Classic nonlinear models built into Prism

Prism comes with 19 built-in classic equations that fill the needs of many biologists.

Don't try too hard to use a classic equation. If your needs are not quite filled by the classic equations, don't hesitate to enter a user-defined equation, which is quite easy.

Equilibrium binding

One-site binding (hyperbola)

$$Y = \frac{B_{max} \cdot X}{K_d + X}$$

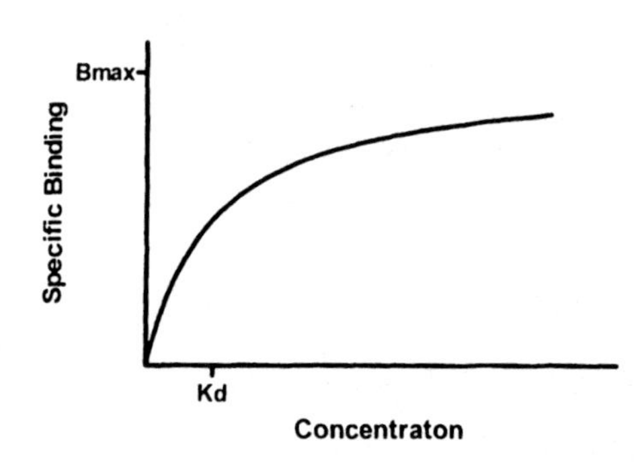

This curve is known as a *rectangular hyperbola, binding isotherm, or saturation binding curve*. Y is zero initially, and increases to a maximum plateau value B_{max}.

This equation describes the equilibrium binding of a ligand to a receptor as a function of increasing ligand concentration. X is the concentration of the ligand, and Y is the specific binding. *Bmax* is the maximum number of binding sites, expressed in the same units as the Y-axis (usually radioactive counts per minute, sites per cell, or fmol of receptor per mg of tissue). *Kd* is the equilibrium dissociation constant, expressed in the same units as the X-axis (concentration). When the drug concentration equals K_d, half the binding sites are occupied at equilibrium.

> Note: In this equation, Y should be the *specific* binding, not the total binding. To learn how Prism analyzes saturation binding curves, see "Analyzing saturation radioligand binding data" on page 199.

This equation also describes the activity of an enzyme as a function of substrate concentration. In this case, the variable labeled *Bmax* is really V_{max}, the maximum enzyme activity, and the parameter labeled K_d is really K_m, the Michaelis-Menten constant.

See also "Analyzing saturation radioligand binding data" on page 199, and "How to determine V_{max} and K" on page 248.

Two-site binding

$$Y=\frac{B_{max1}\cdot X}{K_{d1}+X}+\frac{B_{max2}\cdot X}{K_{d2}+X}$$

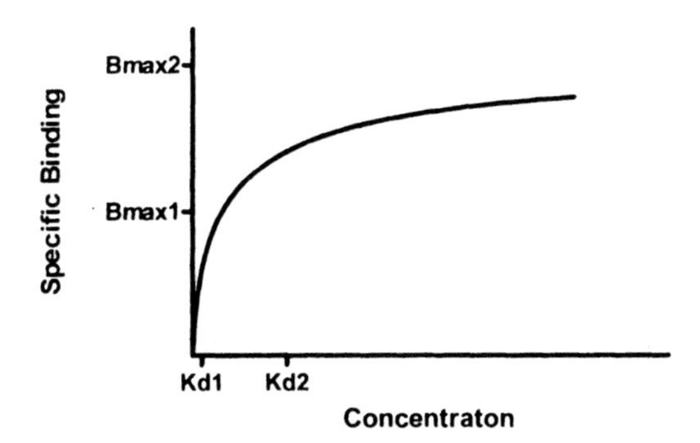

This equation is an extension of the one-site binding curve. It shows the binding of a ligand to two receptors with different affinities (different K_d values). It also describes the enzyme activity as a function of substrate concentration when two isozymes are present. The curve in the example has K_d values that differ by a factor of ten, with equal B_{max} values. Even with such a large difference between K_d values, the curve is not obviously biphasic.

See "Determining K_d and B_{max} for two classes of binding sites" on page 204.

One-site competition

$$Y=Bottom+\frac{(Top-Bottom)}{1+10^{X-LogEC50}}$$

This equation describes the competition of a ligand for receptor binding. It is identical to the sigmoid dose-response curve with a Hill slope of -1.

The parameter *LogEC50* is the concentration of the competitor required to compete for half the specific binding. This is also referred to as LogIC50, the "I" denoting that the curve is inhibitory.

Usually the Y values are total binding. If you enter specific binding instead, fix *Bottom* to have a constant value of zero. If you enter percent specific binding, also set *Top* to be a constant equal to 100.

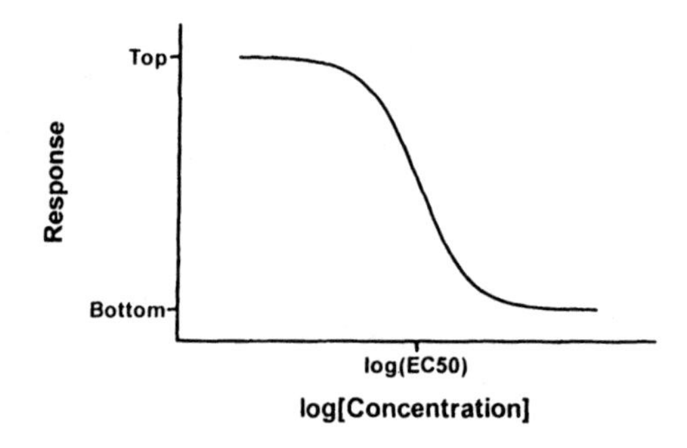

See "Competitive binding data with one class of receptors" on page 213.

Tip: If you have subtracted away any nonspecific binding, then you know the curve must plateau at zero. In this case, be sure to set the parameter Bottom to be a constant value equal to zero. If you transformed your data to percentage of control binding, then also set the parameter Top to be a constant equal to 1.0.

Two site competition

$$Y=Bottom+(Top-Bottom)\left[\frac{Fraction1}{1+10^{X-LogEC50_1}}+\frac{1-Fraction1}{1+10^{X-LogEC50_2}}\right]$$

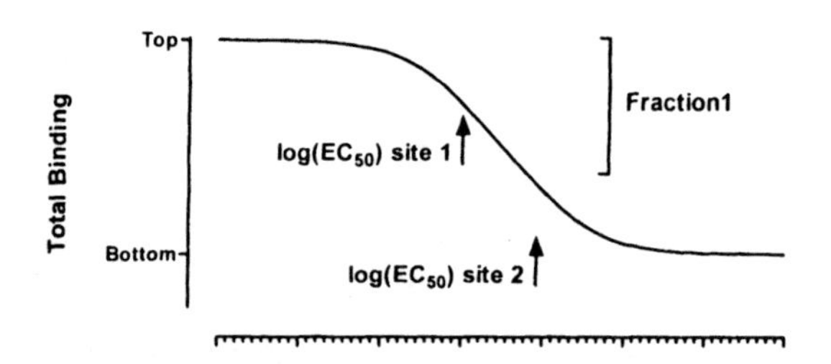

This equation describes the competition of a ligand for two types of receptors. The radioligand has identical affinities for both receptors, but the competitor has a different affinity for each.

Y is binding (total or specific) and X is the logarithm of the concentration of the unlabeled ligand. *Fraction1* is the fraction of the receptors that have an affinity described by *LogEC501*. The remainder of the receptors have an affinity described by *LogEC502*. If *LogEC501* is smaller than *LogEC502*, then *Fraction1* is the fraction of high affinity sites. If *LogEC501* is larger than *LogEC502*, then *Fraction1* is the fraction of low affinity sites.

Tip: If you have subtracted away any nonspecific binding, then you know the curve must plateau at zero. In this case, be sure to set the parameter Bottom to be a constant value equal to zero. If you transformed your data to percentage of control binding, then also set the parameter Top to be a constant equal to 1.0.

Dose-response

Sigmoidal dose-response

$$Y=\text{Bottom}+\frac{(\text{Top-Bottom})}{1+10^{\text{LogEC50-X}}}$$

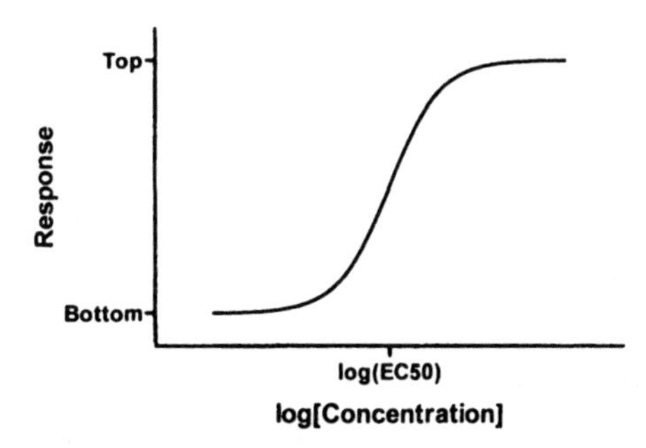

This is a general equation for a dose-response curve. It shows response as a function of the logarithm of concentration. X is the logarithm of agonist concentration and Y is the response. This equation is also called a *three-parameter logistic equation.*

The parameter *Bottom* is the Y value at the bottom plateau; *Top* is the Y value at the top plateau, and *LogEC50* is the X value when the response is halfway between *Bottom* and *Top. LogEC50* is the logarithm of the EC_{50} (effective concentration, 50%). This parameter is sometimes called ED_{50} (effective dose, 50%), or IC_{50} (inhibitory concentration, 50%, used when the curve goes downhill).

This equation assumes a standard midpoint slope (i.e., 1 for a rectangular hyperbola), where the response goes from 10% to 90% of maximal as X increases over about two log units. The next equation allows for a variable slope.

This book has four chapters on analyzing dose-response curves. See "Introduction to dose-response curves" on page 256.

> Tip: If you have subtracted away any nonspecific or blank signal, then you know the curve must plateau at zero. In this case, be sure to set the parameter Bottom to be a constant value equal to zero. If you transformed your data to percentage of control response, then also set the parameter Top to be a constant equal to 1.0.

Sigmoidal dose-response (variable slope)

$$Y=\text{Bottom}+\frac{(\text{Top-Bottom})}{1+10^{(\text{LogEC50-X})\cdot\text{HillSlope}}}$$

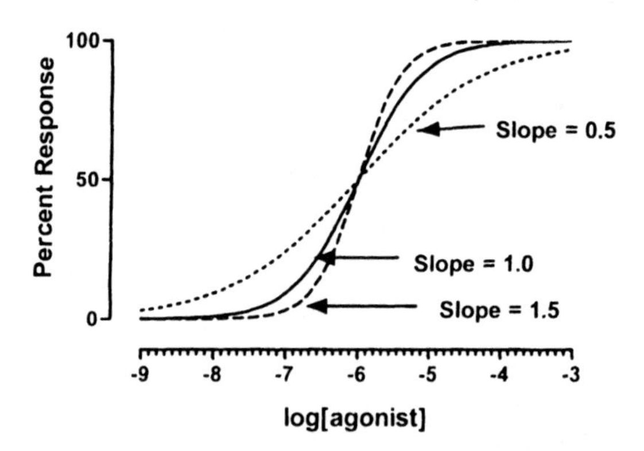

This equation extends the previous equation, but allows for a variable slope. This equation is also called a *four-parameter logistic equation* or the *Hill equation*.

The parameter *Bottom* is the Y value at the bottom plateau; *Top* is the Y value at the top plateau, and *LogEC50* is the X value when the response is halfway between *Bottom* and *Top*. With different kinds of variables, this variable is sometimes called ED_{50} (effective dose, 50%), or IC_{50} (inhibitory concentration, 50%, used when the curve goes downhill).

The parameter *Hillslope* (called the Hill slope, the slope factor, or the Hill coefficient) describes the steepness of the curve. If it is positive, the curve rises as X increases. If it is negative, the curve falls as X increases. A standard sigmoid dose-response curve (previous equation) has a Hill slope of 1.0. When the Hill slope is less than 1.0, the curve is more shallow. When the Hill slope is greater than 1.0, the curve is steeper. The Hill slope has no units.

This book has four chapters on analyzing dose-response curves. See "Introduction to dose-response curves" on page 256.

> Tip: If you have subtracted away any nonspecific or blank signal, then you know the curve must plateau at zero. In this case, be sure to set the parameter Bottom to be a constant value equal to zero. If you transformed your data to percent of control response, then also set the parameter Top to be a constant equal to 1.0.

Boltzmann sigmoidal

$$\text{Y=Bottom}+\frac{(\text{Top-Bottom})}{1+\exp\left(\frac{\text{V50-X}}{\text{Slope}}\right)}$$

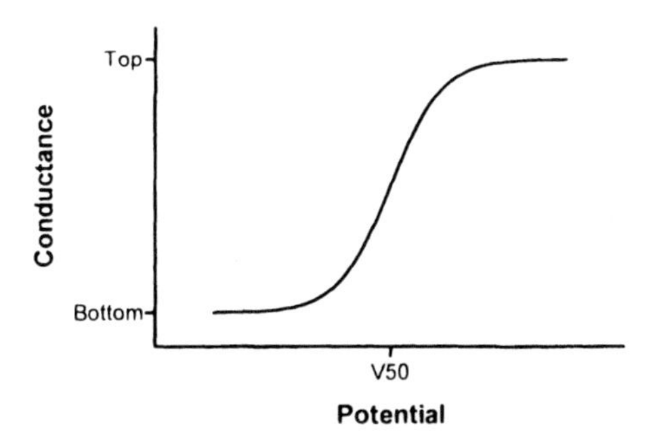

This equation describes voltage-dependent activation of ion channels. It describes conductance (Y) as a function of the membrane potential (X). Conductance varies from *Bottom* to *Top*. *V50* is the potential at which conductance is halfway between *Bottom* and *Top*. *Slope* describes the steepness of the curve, with a larger value denoting a shallow curve. The slope is expressed in units of potential, usually mV, and is positive for channels that activate upon depolarization.

Under appropriate experimental conditions, you can use the slope to calculate the valence (charge) of the ion moving across the channel. The slope equals RT/zF where R is the universal gas constant, T is temperature in °K, F is the Faraday constant, and z is the valence. Since $RT/F \approx 26$ mV at 25°C, z = -26/slope.

Bottom is commonly made a constant equal to 0.0. If you also make *Top* a constant equal to 1.0, then Y can be viewed as the fraction of channels that are activated.

Exponential

One-phase exponential decay

$$Y = \text{Span} \cdot e^{-K \cdot X} + \text{Plateau}$$

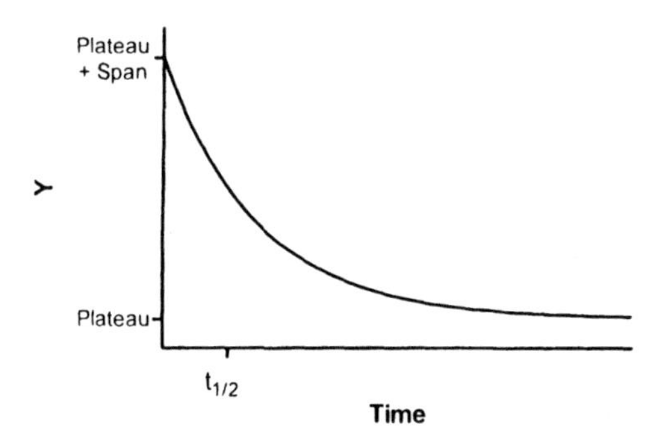

This equation describes the kinetics such as the decay of radioactive isotopes, the elimination of drugs, and the dissociation of a ligand from a receptor.

X is time, and Y may be concentration, binding, or response. Y starts out equal to *Span* + *Plateau* and decreases to *Plateau* with a rate constant *K*. The half-life of the decay is 0.693/*K*. *Span* and *Plateau* are expressed in the same units as the Y axis. *K* is expressed in the inverse of the units used by the X axis. In many circumstances, the plateau equals zero. When fitting data to this equation, consider fixing the plateau to zero.

> Tip: If you have subtracted away any nonspecific binding, then you know the curve must plateau at zero. In this case, be sure to set the parameter Plateau to be a constant value equal to zero.

Two-phase exponential decay

$$Y = Span1 \cdot e^{-K_1 \cdot X} + Span2 \cdot e^{-K_2 \cdot X} + Plateau$$

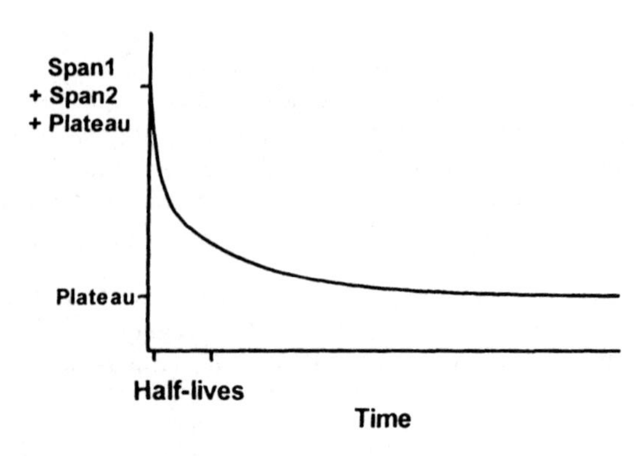

This equation describes a two phase exponential decay. Y starts out equal to *Span1 + Span2 + Plateau* and decays to *Plateau* with fast and slow components. The two half-lives are 0.693/*K1* and 0.693/*K2*. In the figure, the two rate constants differ tenfold, but the spans were equal. The curve is not obviously biphasic, and it takes a very practiced eye to see that the curve does not follow a single phase model.

> Tip: If you have subtracted away any nonspecific binding, then you know the curve must plateau at zero. In this case, be sure to set the parameter Plateau to be a constant value equal to zero.

One-phase exponential association

$$Y = Y_{max} \cdot \left(1 - e^{-K \cdot X}\right)$$

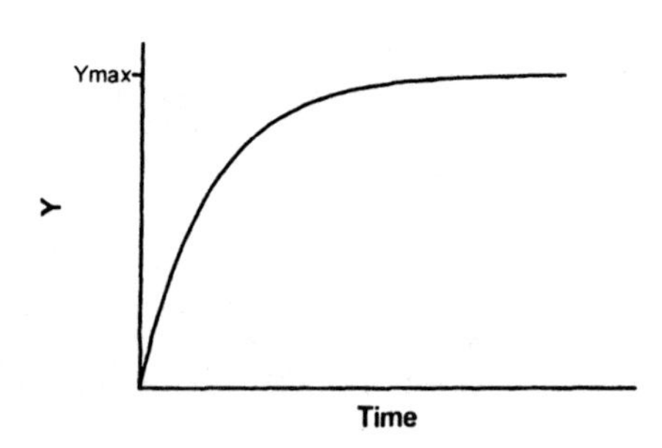

This equation describes the pseudo-first order association kinetics of the interaction between a ligand and its receptor, or a substrate and an enzyme. Y is either binding or enzyme activity. X is time. See "Association binding experiments" on page 234.

Y starts out equal to zero and increases to a maximum plateau (at equilibrium) equal to *Ymax*. When X equals 0.693/*K*, Y equals 0.5***Ymax*.

Two phase exponential association

$$Y=Y_{max1}\cdot\left(1-e^{-K_1\cdot X}\right)+Y_{max2}\cdot\left(1-e^{-K_2\cdot X}\right)$$

This is an extension of the exponential association to two phases, corresponding to a radioligand binding to two independent sites.

Exponential growth

$$Y=Start\cdot e^{K\cdot X}$$

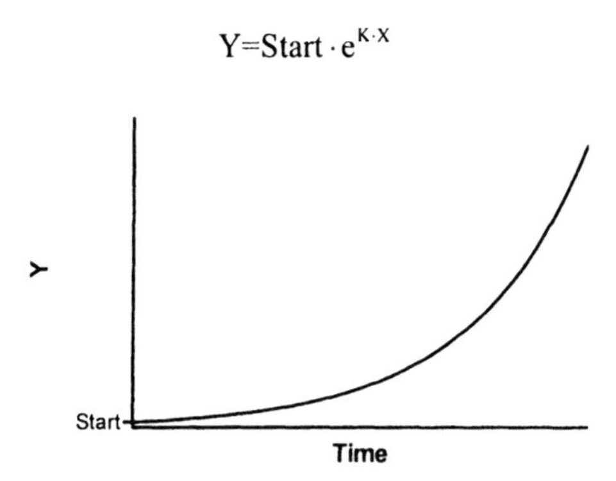

This describes an exponential growth curve. Y is population size (perhaps cell number) and X is time. At X=0, Y equals *Start*. Y increases geometrically with a doubling time equal to 0.693/*K*.

It is often difficult to fit this equation with nonlinear regression, because a tiny change in the initial values will drastically alter the sum-of-squares.

Before fitting data to the exponential growth equation, consider converting your Y values to logarithms. With exponential growth data, the experimental error sometimes increases dramatically as Y gets bigger. If you transform to logarithms, the resulting values may have a more uniform and Gaussian error than the raw data. If so, fit the transformed data with linear regression. The Y intercept represents the Y value at time zero (Start) and the slope is the rate constant (k).

Other classic equations

Power series

This versatile equation has many uses.

$$Y=A\times X^B+C\times X^D$$

Fitting data to a power series model can be difficult. The initial values generated automatically by Prism are not always helpful (A and B are set to 1.0; C and D to 1.1). You'll probably need to enter better initial values in order to fit this equation to data. The initial values of B and D are important, because small changes in those values can make a huge change in Y.

The equation is not defined and so leads to a floating point error, if X equals zero and B or D are negative numbers, or if X is negative and B or D are between 0.0 and 1.0.

Polynomial (linear) equations

Prism offers first-, second-, third- and fourth-order polynomial equations. Although few chemical or pharmacological models are described by polynomial equations, these equations are often used to fit standard curves. The higher order equations have more inflection points.

Unlike all other equations, you don't have to worry about initial values when fitting data to polynomial equations. You will get exactly the same answer no matter what the initial values are.

The "order" of a polynomial equation tells you how many terms are in the equation.

Order	Equation
First	$Y=A+B\cdot X$
Second	$Y=A+B\cdot X+C\cdot X^2$
Third	$Y=A+B\cdot X+C\cdot X^2+D\cdot X^3$
Fourth	$Y=A+B\cdot X+C\cdot X^2+D\cdot X^3+E\cdot X^4$

You can enter a higher-order equation (up to 14th order) as a user-defined equation (or select one from the equation library).

Note: The parameters A, B and C in the second order polynomial equation are in a different order than you usually see in algebra texts.

Sine wave

$$Y = \text{Baseline} + \text{Amplitude}\cdot\sin\left(\text{Frequency}\cdot X + \text{PhaseShift}\right)$$

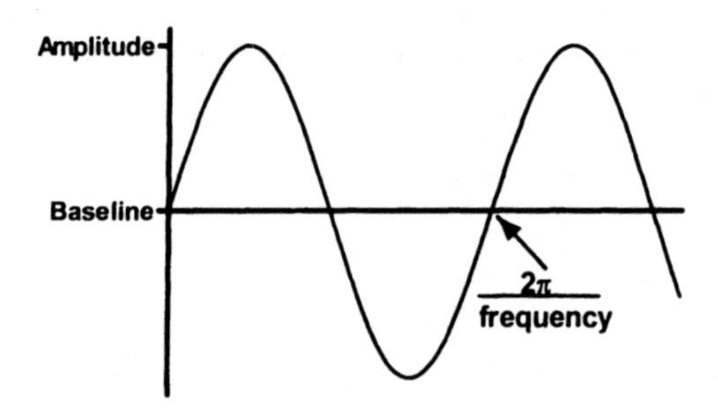

X is in radians. In most cases, you'll want to fix Baseline to a constant value of zero. Amplitude is the maximum height of the curve away from the baseline. Frequency is the number of complete oscillations per 1 X unit.

Gaussian distribution

$$Y = \frac{AREA}{SD\sqrt{2\pi}} \cdot e^{-\frac{1}{2}\left[\frac{X-Mean}{SD}\right]^2}$$

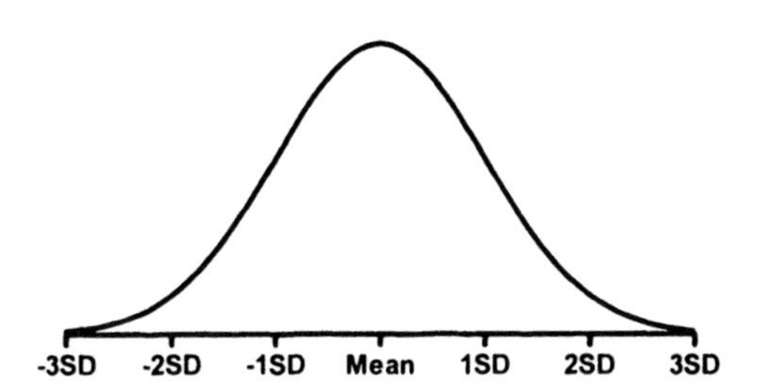

This equation defines the cumulative probability distribution of a Gaussian bell-shaped distribution with specified mean and SD. The area under the entire curve is *Area*. A standard probability distribution is scaled so that *Area* equals 1.0. The units of the Y-axis are arbitrary, determined by your choice of *Area*.

49. Importing equations and equation libraries

Selecting from the equation library

If the equation you want is not one of the classic equations built-in to Prism, look in Prism's equation library. To choose an equation from a library, click *More equations* (from the parameters dialog for nonlinear regression), then choose to select from the equation library.

On the *Equation Selection* dialog, choose a library file (left panel) then choose an equation (right panel). Confirm your choice with the preview on the bottom of the dialog.

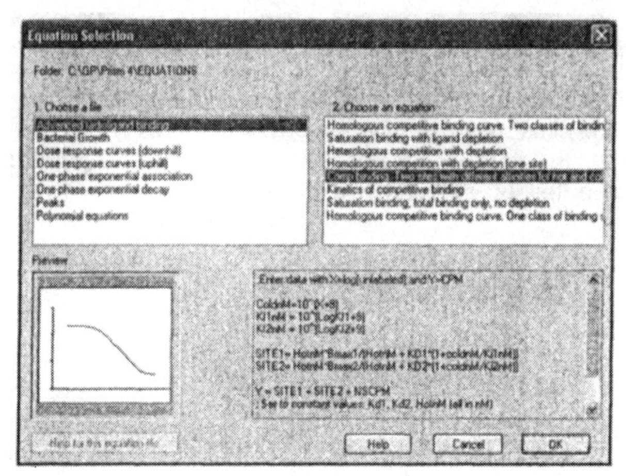

When you select an equation from the library, you transfer it from the library file to your own list of equations. You can then edit that copy of the equation without affecting the library. Prism stores the equation with every file that uses it, and also places the equation in your list of user-defined equations (the list you see when you choose "more equations"). Prism does not remember that the equation came from a library.

Adding equations to the equation library

Add to the library by following these steps:

1. Create a new Prism project. Use the *Create a family of theoretical curves* analysis and enter a new equation. Pick reasonable values for the minimum and maximum X values of the curve.

2. Customize the graph so it will be clear when seen as a preview (test this by looking at the graph gallery).

3. Repeat with any number of related equations that you want to store in one file.

4. Save the file into the "Equations" folder within the Prism program folder.

5. If you are creating an equation file that will be used by others, consider creating a help file that explains the equations. Using any HTML (web) editor, create a help

file with the same name as the file containing the equations, but with the extension "htm". Place the help file in the same folder with the equation file.

Importing equations

You can import any equation from any project. From the parameters dialog for nonlinear regression, click *More equations.* Then choose to import an equation, select a file and select the equation.

When you import an equation, you transfer it to your own list of equations (the list you see when you choose *More equations*). Prism does not store any sort of link back to the file the equation was imported from.

50. Writing user-defined models in Prism

What kinds of equations can you enter?

You are not limited to the equations (models) that we provide with Prism. You can enter your own equations, subject to these limitations:

Limitation	Explanation
No implicit equations.	Y must be defined as a function of X and one or more parameters. The variable Y can only appear once, on the left side of the last line of the equation. If Y also appears on the right side of the equation, you have an implicit equation, which Prism cannot handle. In many cases, you'll be able to algebraically rearrange the equation.
No differential equations.	You must define Y as a function of X and one or more variables. It is not sufficient to define the derivatives.
No equations with more than one X variable.	Prism does not calculate multiple regression, so it cannot fit models with two or more independent (X) variables. But note that you can define a parameter to be a column constant, in which case its value comes from the column titles. In some cases, you can think of these column constants as being a second independent variable. See page 301.
No discontinuous equations.	If you enter a discontinuous equation (where an infinitesimal change in X can create a huge change in Y) the results of nonlinear regression may not be reliable.
The equation must define Y as a function of X.	The independent variable must be X. The dependent variable must be Y. For example, if you measure a voltage as a function of time, you cannot enter an equation that defines V as a function of t. It must define Y as a function of X.

Equation syntax

At the top of the nonlinear regression (or create a family of theoretical curves) parameters dialog, select *More equations*. Then select *Enter your own equation* to bring up the *User-defined Equation* dialog.

First enter a name for the equation, which will then appear on the *More equations* list in the nonlinear regression dialog.

Then enter the equation itself, following these guidelines:

- Variable and parameter names must not be longer than 13 characters. If you want to use two words to name a variable, separate with the underscore character, for example Half_Life. Don't use a space, hyphen or period.

- Prism does not distinguish between upper and lower case letters in variable, parameter, or function names.
- Use an asterisk (*) to indicate multiplication. Prism does not always recognize implied multiplication. To multiply A times B, enter "A*B" and not "AB".
- Use a caret (^) to indicate power. For example, "A^B" is A to the B power.
- Use parentheses as necessary to show the order of operations. To increase readability, substitute brackets [like this] or braces {like this}. Prism interprets parentheses, brackets, and braces identically.
- Use a single equals sign to assign a value to a variable.
- You don't need any special punctuation at the end of a statement.
- To enter a long line, type a backslash (\) at the end of the first line, then press ***Return*** and continue. Prism treats the two lines as one.
- To enter a comment, type a semicolon (;) and then the text. Comments can begin anywhere on a line.

Here are three examples of one-line equations:

```
Y=Bmax*X/(Kd + X)
```

```
Y=A*(X+1)
```

```
Y=Start*exp[(-0.693*X/Half_Life)*K]
```

You don't have to write your equation on one line. Use intermediate variables to simplify longer equations. Prism automatically distinguishes between *intermediate variables* and *equation parameters* that you can fit. If a variable is used first on the left side of an equals sign, then it is an intermediate variable. If a variable is used first on the right side of an equals sign, then it is an equation parameter.

Below is an example of a longer equation. Because K is used first on the left of the equals sign, Prism recognizes that it is an intermediate variable rather than a parameter to be fit by nonlinear regression. When you fit data to this equation, you'll find the best-fit value of the parameter HalfLife, not K. Note two comments, one on a line by itself and the other on the same line with equation code.

```
; One-phase exponential decay
K=0.693/HalfLife ;rate constant
Y=Start*exp(-K*X)
```

Available functions for user-defined equations

When you enter your equations, you can use any of the functions listed below.

Function	Explanation	Excel equivalent

Function	Explanation	Excel equivalent
abs(k)	Absolute value. If k is negative, multiply by -1.	abs(k)
arccos(k)	Arccosine. Result is in radians.	acos(k)
arccosh(k)	Hyperbolic arc cosine.	acosh(k)
arcsin(k)	Arcsine. Result is in radians.	asin(k)
arcsinh(k)	Hyperbolic arcsin. Result is in radians.	asinh(k)
arctan(k)	Arctangent. Result is in radians.	atan(k)
arctanh(k)	Hyperbolic tangent. K is in radians.	atanh(k)
arctan2(x,y)	Arctangent of y/x. Result is in radians.	atan2(x,y)
besselj(n,x)	Integer Order J Bessel, N=0,1,2...	besselj(x,n)
bessely(n,x)	Integer Order Y Bessel, N=0,1,2...	bessely(x,n)
besseli(n,x)	Integer Order I Modified Bessel, N=0, 1, 2...	besseli(x,n)
besselk(n,x)	Integer Order K Modified Bessel, N=0, 1 ,2...	besselk(x,n)
beta(j,k)	Beta function.	exp(gammaln(j) +gammaln(k) -gammaln(j+k))
binomial(k,n,p)	Binomial. Probability of k or more "successes" in n trials, when each trial has a probability p of "success".	binomdist(k,n,p,false)
chidist(x2,v)	P value for chi square equals x2 with v degrees of freedom	chidist(x2,v)
ceil(k)	Nearest integer not smaller than k. Ceil (2.5) = 3.0. Ceil(-2.5 = -2.0.	(no equivalent)
cos(k)	Cosine. K is in radians.	cos(k)
cosh(k)	Hyperbolic cosine. K is in radians.	cosh(k)
deg(k)	Converts k radians to degrees.	degrees(k)
erf(k)	Error function.	2*normsdist(k*sqrt(2))-1
erfc(k)	Error function, complement.	2-2*normsdist(k*sqrt(2))
exp(k)	e to the kth power.	exp(k)
floor(k)	Next integer below k. Floor(2.5)=2.0. Floor(-2.5)=-3.0.	(no equivalent)
fdist(f,v1,v2)	P value for F distribution with V1 degrees of freedom in the numerator and V2 in the denominator.	fdist(f,v1,v2)
gamma(k)	Gamma function.	exp(gammaln(k))
gammaln(k)	Natural log of gamma function.	gammaln(k)

Function	Explanation	Excel equivalent
hypgeometricm(a,b,x)	Hypergeometric M.	(no equivalent)
hypgeometricu(a,b,x)	Hypergeometric U.	(no equivalent)
hypgeometricf(a,b,c,x)	Hypergeometric F.	(no equivalent)
ibeta(j,k,m)	Incomplete beta.	(no equivalent)
if(condition, j, k)	If the condition is true, then the result is j. Otherwise the result is k. See next section below.	(similar in excel)
igamma(j,k)	Incomplete gamma.	(no equivalent)
igammac(j,k)	Incomplete gamma, complement	(no equivalent)
int(k)	Truncate fraction. INT(3.5)=3 INT(-2.3) = -2	trunc()
ln(k)	Natural logarithm.	ln(k)
log(k)	Log base 10.	log10(k)
max(j,k)	Maximum of two values.	max(j,k)
min(j,k)	Minimum of two values.	min(j,k)
j mod k	The remainder (modulus) after dividing j by k.	mod(j,k)
psi(k)	Psi (digamma) function. Derivative of the gamma function.	(no equivalent)
rad(k)	Converts k degrees to radians.	radians(k)
sgn(k)	Sign of k. If k>0, sgn(k)=1. If k<0, sgn(k)= -1. If k=0, sgn(k)=0.	sign(k)
sin(k)	Sine. K is in radians.	sin(k)
sinh(k)	Hyperbolic sine. K is in radians.	sinh(k)
sqr(k)	Square.	k*k
sqrt(k)	Square root.	sqrt(k)
tan(k)	Tangent. K is in radians.	tan(k)
tanh(k)	Hyperbolic tangent. K is n radians	tanh(k)
tdist(t,v)	P value (one-tailed) corresponding to specified value of t with v degrees of freedom. T distribution.	tdist(t,v,1)
zdist(z)	P value (one-tailed) corresponding to specified value of z. Gaussian distribution.	normsdist(z)

Tip: Don't use any of these function names for your variables or parameters.

Using the IF function

Prism allows you to introduce some branching logic through use of the *IF* function. The syntax is:

IF (conditional expression, value if true, value if false)

You can precede a conditional expression with NOT, and can connect two conditional expressions with AND or OR. Examples of conditional expressions:

```
MAX>100
Ymax=Constraint
(A<B or A<C)
NOT(A<B AND A<C)
FRACTION<>1.0
X<=A and X>=B
```

Note: "<>" means not equal to, "<=" means less than or equal to, and ">=" means greater than or equal to.

Here is an example.

```
Y= IF (X<X0, Plateau, Plateau*exp(-K*X))
```

If X is less than X0, then Y is set equal to *Plateau*. Otherwise Y is computed as *Plateau**exp(-*K**X).

You may also insert a conditional expression anywhere in an equation, apart from an *IF* function. A conditional expression evaluates as 1.0 if true and 0.0 if false. Example:

```
Y=(X<4)*1 + (X>=4)*10
```

When X is less than 4, this evaluates to 1*1 + 0*10=1. When X is greater than 4, this evaluates to 0*1+1*10=10.

Here is a function that returns Y if Y is positive, but otherwise leaves the results blank. In other words, it removes all negative values. The way to leave a result blank is to do an impossible mathematical transform such as dividing by zero.

```
Y = IF (Y<0, Y/0, Y)
```

How to fit different portions of the data to different equations

In some situations you may wish to fit different models to different portions of your data. This often occurs in kinetic experiments where you add a drug or otherwise perform some sort of intervention while recording data. The values collected before the intervention follow a different model than those collected afterward.

Although Prism has no built-in way to fit different equations to different portions of the data, you can achieve that effect using a user-defined equation containing the IF function.

Example 1. Plateau followed by exponential association

In this example, you collected data that established a baseline early in the experiment, up to "Start". You then added a drug, and followed the outcome (Y) as it increased toward a plateau. Prior to the injection the data followed a horizontal line; after the injection the data formed an exponential association curve.

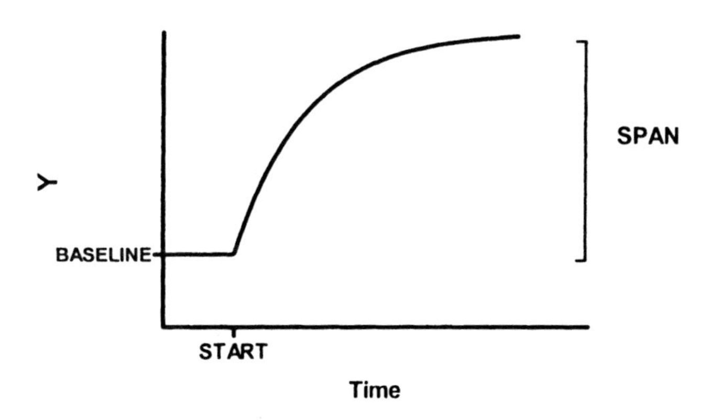


```
Y1=BASELINE
Y2=BASELINE + SPAN*(1-exp(-K*(X-START)))
Y=IF[(X<START),Y1,Y2)
```

It is easiest to understand this equation by reading the bottom line first. For X values less than *START*, Y equals *Y1*, which is the baseline. Otherwise, Y equals *Y2*, which is defined by the exponential association equation.

This equation has two intermediate variables (*Y1* and *Y2*). Prism can fit the four true parameters: *START*, *SPAN*, *K*, and *BASELINE*.

In many cases, you'll make START a constant equal to the time of the experimental intervention.

Example 2. Two linear regression segments

This equation fits two linear regression lines, ensuring that they intersect at $X=X_0$.

```
Y1 = intercept1 + slope1*X
YatX0 = slope1*X0 + intercept1
Y2 = YatX0 + slope2*(X - X0)
Y = IF(X<X0, Y1, Y2)
```

The first line of the equation defines the first line segment from its intercept and slope.

The second line of the equation computes the Y value of the first regression at the right end of that segment, when $X=X_0$.

The third line of the equation computes the second regression segment. Since we want a continuous line, the Y value at the left end of the second segment must equal the Y value at the right end of the first segment (*YatXo*). The Y value at any other position along the second segment equals *YatXo* plus the increase due to the second regression line. That increase equals the slope of the second segment (*slope2*) times the distance from X to *Xo*.

The final line defines Y for all values of X. If X is less than *Xo* then Y is set equal to *Y1*. Otherwise Y is set equal to *Y2*.

Here are the results with sample data. The program found that the best-fit value of *Xo* was 5.00, and the two lines meet at that X value.

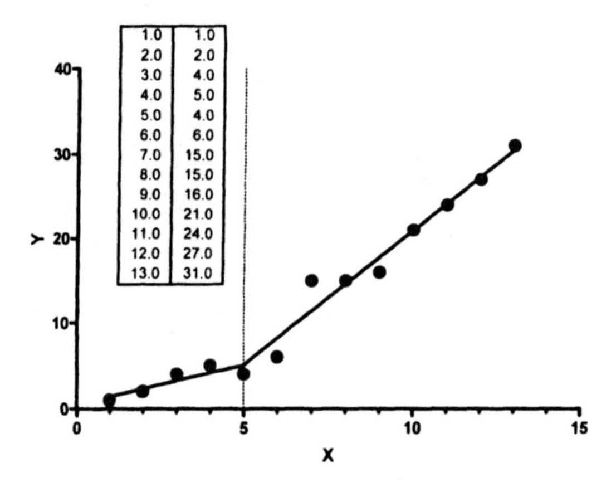

Use segmental linear regression cautiously

Segmental linear regression is useful when something happens at X_o to change the slope of the line. For example, use segmental linear regression if X is time, and you added a drug (or changed a voltage) at time X_o.

If you didn't perform an intervention at X_o, segmental linear regression is probably not the analysis of choice. Instead, you probably want to fit some sort of curve.

> Tip: Do not use segmental linear regression to analyze a biphasic Scatchard plot. A biphasic Scatchard plot does not have an abrupt break point. You should fit the original data to a two-site binding curve instead.

How to define different models for different data sets

You can define an equation so data set A is fit to one model and data set B to another.

To specify that a line in an equation should only be used for a particular data set, precede the line with the data set identifier in angled brackets. So a line preceded by <A> would apply only to data set A, while a line preceded by <AC> would only apply to data set AC. The letters correspond to the labels Prism shows on the data table.

For example, use this equation to fit a table where column A is nonspecific binding and column B is total binding (nonspecific plus specific). To make this work, you also need to set the parameter NS to be shared in the Constraints tab. This means Prism will fit a single value of NS for both data sets.

```
; total and nonspecific binding
; Define NS to be shared in the constraints tab
Nonspecific=NS*X
Specific = Bmax*X(Kd+X)
<A>Y=Nonspecific
<B>Y=Nonspecific + Specific
```

Use the syntax "<~A>" To specify that a line pertains to all data sets except one. For example, use this equation to fit a table where column A is nonspecific binding, and columns B and C (and maybe others) represent total binding. To make this work, you'd also need to define NS to be a shared parameter.

```
; total and nonspecific binding
; Define NS to be shared in the constraints tab
Nonspecific=NS*X
Specific = Bmax*X(Kd+X)
<A>Y=Nonspecific
<~A>Y=Nonspecific + Specific
```

Defining rules for initial values and constraints

Rules for initial values

Before it can perform nonlinear regression, Prism must have initial values for each parameter. You can define rules for generating the initial values at the time you enter a new equation. Then Prism will calculate the initial values automatically. If you don't enter rules for initial values, you will need to enter the initial values for every parameter, for every data set, every time you fit data.

To define rules for initial values for user-defined equations:

1. While entering or editing a user-defined equation, click on the tab labeled *Rules for Initial Values.*

2. Enter the rule for finding the initial value of each parameter. Enter a number in the first column and select a multiplier or divisor from the drop-down list in the second column.

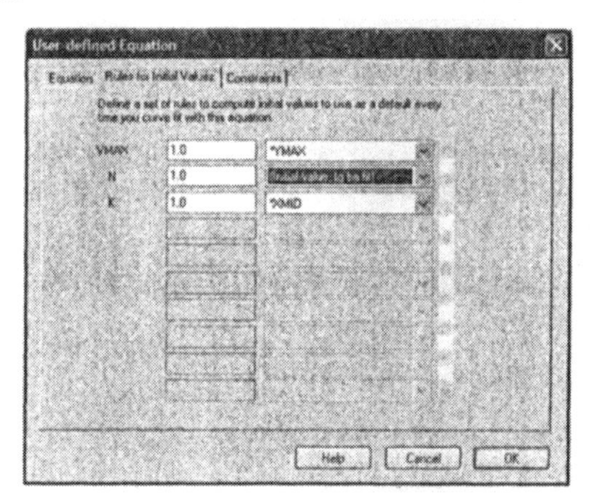

All but two choices on the drop-down list are used to multiply or divide the number you entered by a value determined from the range of the data (e.g., YMIN, XMAX, XMID, or

YMAX-YMIN). The abbreviation YMIN is the minimum value of Y; XMAX is the maximum value of X, and XMID is the average of XMIN and XMAX. For example, if you enter "0.5" in the first column and select "YMAX" in the second column, Prism sets the initial value to half of YMAX (which may differ for each data set).

The first choice on the drop-down list is *(Initial value, to be fit)*. The value you entered will be the initial value for all data sets.

> Note: You won't always be able to find rules that generate useful initial values. For some models, you'll need to enter initial values manually.

Constraints

Click the *Default Constraints* tab on the User-defined equation dialog (page 306) to define constraints that you will want to use almost every time you select the equation. These constraints will become the defaults whenever you choose this equation in the future.

> Tip: Use the Default Constraints tab on the User-defined equation dialog to set constraints that are part of the model so will be useful almost every time you use the equation. Use the *Constraints* tab on the *Nonlinear regression parameters* dialog to set constraints for a particular fit.

Managing your list of equations

When you choose *More equations* from the nonlinear regression parameters dialog, Prism shows you a list of equations you have entered or imported. If you don't plan to use an equation again, select it and click *Delete* to erase it from the list. That won't affect any files that use the equation you erased. If you open one of these files, and change the parameters of the nonlinear regression, Prism will automatically add the equation back to your list.

You can change the order of equations in your list by selecting an equation and then clicking *Move Up* or *Move Down*.

Modifying equations

You can edit any equation you entered yourself or imported (or chose from the equation library). From the nonlinear regression parameters dialog, select the equation from the list of "more equations" and then click *Edit Equation*.

Classic equations cannot be modified. But you can create a new user-defined equation based on a classic equation.

To copy and paste a built-in equation:

1. Start from the *Parameters: Nonlinear regression* or *Simulate Theoretical Curve* dialog.

2. Select a built-in classic equation, and click *View Equation*.

3. Press *Copy All.*

4. Cancel from that dialog.

5. Select *More equations*, then *Enter your own equation.*

6. Enter an equation name. Then move the insertion point to the Equation block and press Paste.

51. Linear regression with Prism

Entering data for linear regression

From the Welcome or New Table dialog, choose any XY graph. Prism will create a data table formatted with numbers for the X column. You can choose one of several formats for Y columns.

- If you enter Y values for several groups (into columns A, B, ...) Prism will report the results of linear regression of X with each of the Y variables. Prism, however, cannot calculate multiple regression.
- If you format the Y columns for replicates (for example, triplicates), Prism can average these and perform all calculations with the means. Or it can treat each replicate as a separate value to fit.
- If you format the Y columns for entry of SD or SEM, Prism analyzes only the means and ignores the SD or SEM values.
- If you format the table to have a subcolumn for X error bars, these will be ignored by linear regression.

Choosing a linear regression analysis

Start from a data table or graph (see "Entering data for linear regression" on page 334). Click on the *Analyze* button and choose to do built-in analysis. Then select *Linear regression* from the list of *curves and regressions.*

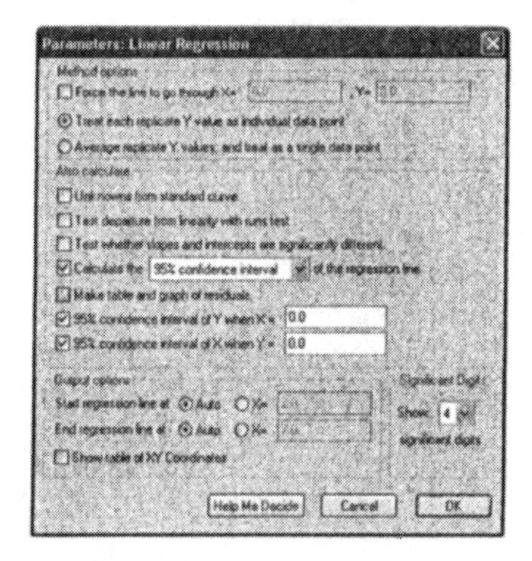

Force a regression line through the origin (or some other point)?

You may force the regression line to go through a particular point such as the origin. In this case, Prism will determine only the best-fit slope, as the intercept will be fixed. Use this option when scientific theory tells you that the line must go through a particular point (usually the origin, X=0, Y=0) and you only want to know the slope. This situation arises rarely, and you should use common sense when making your decision. For example, consider a protein assay. You measure optical density (Y) for several known concentrations of protein in order to create a standard curve. You then want to interpolate unknown protein concentrations from that standard curve. When performing the assay, you adjusted the spectrophotometer so that it reads zero with zero protein. Therefore you might be tempted to force the regression line through the origin. But this constraint may result in a line that doesn't fit the data well. Since you really care that the line fits the standards very well near the unknowns, you will probably get a better fit by not constraining the line.

Tip: Most often, you should let Prism find the best-fit line without any constraints.

Fit linear regression to individual replicates or means?

If you collected replicate Y values at every value of X, there are two ways Prism can calculate linear regression. It can treat each replicate as a separate point, or average the replicate Y values and treat the mean as a single point.

You should choose *Treat each replicate Y value as individual data point* when the sources of experimental error are the same for each data point. If one value happens to be a bit high, there is no reason to expect the other replicates to be high as well. The errors are independent.

Choose *Average replicate Y values and treat and treat as a single data point* when the replicates are not independent. For examples, the replicates would not be independent if they represent triplicate measurements from the same animal, with a different animal used at each value of X (dose). If one animal happens to respond more than the others, that will affect all the replicates. The replicates are not independent.

This choice will affect the best-fit values of the slope and intercept only if the number of replicates varies among data points. However, this choice will alter the SE of the slope and intercept and therefore the width of the confidence intervals.

Additional calculations with linear regression

Prism offers five optional calculations with linear regression.

Calculation	Description
Calculate unknowns from a standard curve	After fitting a regression line, Prism will interpolate unknown values from that curve.
Runs test	Tests whether the line deviates systematically from your data.
Residuals	Helps you test the assumptions of linear regression.
Compare whether slopes and intercepts differ between data sets.	If you fit two data sets, compare the best-fit values of slope and intercept. The results will be in a separate results subpage (also called a view).
Confidence or prediction intervals of the regression line.	These show you graphically how certain you can be of the best-fit line.

Changing where the regression lines starts and stops

By default Prism draws the regression line from the smallest X value in your data to the largest. To change these limits, uncheck *Auto* **(under** *Output options)* **and** enter a new starting or stopping X value. This affects how the line is graphed, but does not affect any of the numerical results.

Default preferences for linear regression

The linear regression parameters dialog affects one particular linear regression analysis. Change settings on the Analysis tab of the Preferences dialog to change default settings for future linear regression analyses. To open this dialog, pull down the *Edit* menu and choose *Preferences* and then go to the *Analysis* tab. You can change these settings:

- Report results of runs test of goodness-of-fit.
- Make table and graph of residuals.
- Test whether slopes and intercepts differ between data sets.
- Plot the 95% confidence interval of the best-fit line.

> Note: Changing the analysis preferences changes the default settings for *future* linear regression analyses. It will not change analyses you have *already* performed.

Using nonlinear regression to fit linear data

Since Prism's nonlinear regression analysis is more versatile than its linear regression analysis, it can make sense to fit linear data using a nonlinear regression program. Here are several such situations.

> GraphPad note: To fit linear regression using Prism's nonlinear regression analysis, choose the built-in equation called *Polynomial: First Order (straight line).*

Comparing models

You might wish to compare an unconstrained linear regression line with one forced through the origin (or some other point). Or you might compare a linear model with a second-order (quadratic) polynomial model.

Weighting

Standard linear (and nonlinear) regression assumes that the scatter of the points around the line is (on average) the same all the way along the line. But in many experimental situations, the scatter goes up as Y goes up. What is consistent is the scatter as a fraction of Y. For example, you might see about a 5% error all the way along the line.

Prism's linear regression analysis does not let you choose a weighting scheme. But Prism's nonlinear regression analysis does let you choose weighting schemes. See Chapter 14. By weighting the points to minimize the sum of the relative distance squared, rather than the distance squared, you prevent the points with the highest Y value (and thus the highest scatter) from having undue influence on the line.

To choose relative weighting, use Prism's nonlinear regression analysis and choose relative weighting in the Weighting tab.

Sharing parameters between data sets

Prism's nonlinear regression analysis (but not linear regression) lets you fit a model to several data sets, sharing one or more parameters between data sets. This means you could fit a line to several data sets, sharing the slope (so you get one best-fit slope for all data sets) but not sharing the intercept.

To share parameters, use Prism's nonlinear regression analysis and choose to share a parameter on the Constraints tab.

Deming (Model II) linear regression

Standard linear regression assumes that you know the X values perfectly, and all the uncertainty is in Y. If both X and Y variables are subject to error, fit linear regression using a method known as Deming, or Model II, regression. For background, see Chapter 7.

To do the analysis with Prism, click the analyze button, and then choose Deming (Model II) regression from the list of clinical lab analyses. Most often, your X and Y are subject to the same average error.

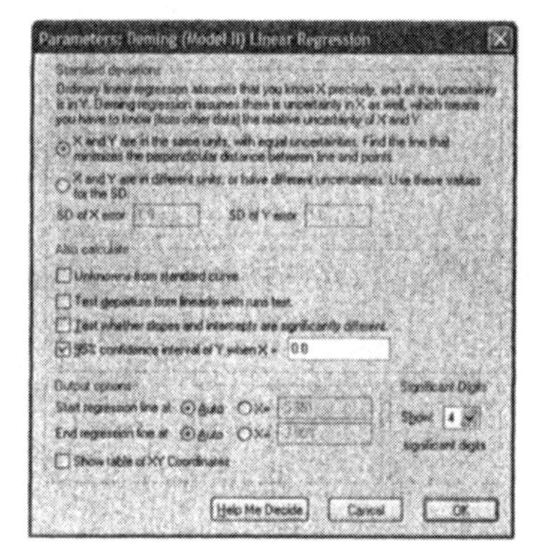

Deming regression can also be used when the X and Y are both subject to error, but the errors are not the same. In this case, you must enter the error of each variable, expressed as a standard deviation.

How do you know what values to enter? To assess the uncertainty (error) of a method, collect duplicate measurements from a number of samples using that method. Calculate the standard deviation of the error using the equation below, where each d_i is the difference between two measurements of the same sample (or subject), and N is the number of measurements you made (N equals twice the number of samples, since each sample is measured twice).

$$SD_{error} = \sqrt{\frac{\sum d_i^2}{N}}$$

Repeat this for each method or variable (X and Y), enter the two SD_{error} values into the Deming regression analysis dialog, and Prism will fit the line for you. If the X variable has a much smaller SD than the Y value, the results will be almost identical to standard linear regression.

If you try to compare Prism's results with those of another program or book, you may encounter the variable λ (lamda), which quantifies the inequality between X and Y errors.

$$\lambda = \left(\frac{SD_{X\ error}}{SD_{Y\ error}} \right)^2$$

Prism requires you to enter individual SD values, but uses these values only to calculate λ, which is then used in the Deming regression calculations. If you know λ, but not the individual SD values, enter the square root of λ as the SD of the X values, and enter 1.0 as the SD of the Y error. The calculations will be correct, since Prism uses those two values only to compute λ.

Inverse linear regression with Prism

Inverse linear regression is used rarely. It is appropriate when you know the variable plotted on the Y axis with great precision, and all the error is in the variable plotted on the X axis. You might want to plot your data this way if the independent variable is depth, so it logically belongs on the vertical (Y) axis. In this case, you want to minimize the sum-of-squares of the horizontal distances of the points from the line.

Prism does not have a command or option for inverse linear regression. But you can tweak the Deming regression analysis to perform inverse regression. To do this, choose Deming regression and enter a tiny value for the SD of the Y variable (say 0.0001) and a huge value for the SD of the X variable (say 10000). Since the ratio is so huge, Prism assumes that virtually all the error is in X and very little is in Y, so the calculations are the same as inverse linear regression.

52. Reading unknowns from standard curves

Introduction to standard curves

Standard curves are used to determine the concentration of substances. First you perform an assay with various known concentrations of a substance you are trying to measure. The response might be optical density, luminescence, fluorescence, radioactivity, or something else. Graph these data to make a standard curve – concentration on the X axis, and assay measurement on the Y axis.

Also perform the same assay with your unknown samples. You want to know the concentration of the substance in each of these unknown samples.

To analyze the data, fit a line or curve through the standards. For each unknown, read across the graph from the spot on the Y axis that corresponds to the assay measurement of the unknown until you intersect the standard curve. Read down the graph until you intersect the X axis. The concentration of substance in the unknown sample is the value on the X axis.

In the example below, the unknown sample had 1208 counts per minute, so the concentration of the hormone is 0.236 micromolar.

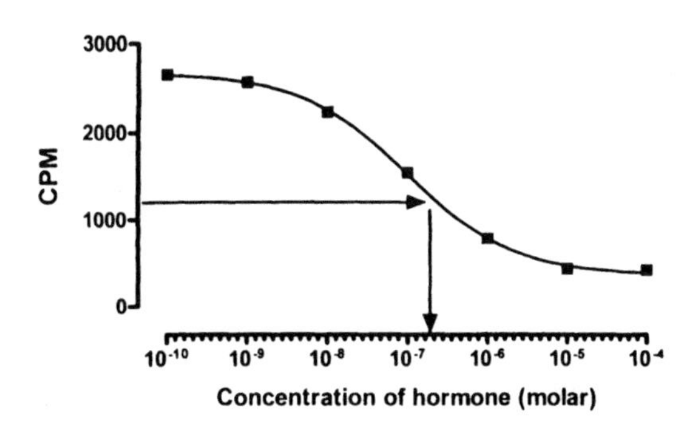

Prism makes it very easy to fit your standard curve and to read (interpolate) the concentration of unknown samples.

How to fit standard curves

Before you can read unknown values, you first must fit a line or curve through your standard points. Prism lets you fit a standard curve with one of these methods:

Creating a standard curve with linear regression

Standard curves are often nearly linear, at least within a certain range of concentrations. If you restrict your standard curve values to a linear region, you can analyze the curve with linear regression. This will be a useful analysis, even if the overall standard curve is not quite straight, so long as you choose a reasonable range. The standard curve should start a little below your lowest unknown value and extend to a little beyond your highest unknown value. There is no benefit to continuing the standard curve far above or below the range of your unknowns.

Creating a standard curve with cubic spline (or Lowess)

The easiest way to fit a curve is to create a cubic spline or Lowess curve. They are easier than nonlinear regression, because you don't have to choose an equation. Spline and Lowess curves tend to wiggle too much, so are not often used as standard curves. See chapter 54.

Creating a standard curve with polynomial regression

Polynomial regression is a convenient method to create a smooth curve. With Prism, you perform polynomial regression by choosing a polynomial equation from the nonlinear regression dialog. Try a second-, third-, or fourth-order polynomial equation. Higher order polynomial equations generate standard curves with more inflection points.

Creating a standard curve with nonlinear regression

Nonlinear regression is often used to fit standard curves generated by radioimmunoassay (RIA) or similar assays (ELISA). These assays are based on competitive binding. The compound you are assaying competes with a labeled compound for binding to an enzyme or antibody. Therefore the standard curve is described by equations for competitive binding. Try the one-site competitive binding curve. If that doesn't fit your data well, try the sigmoid dose-response curve with variable slope. When fitting sigmoid curves, enter the X values as the logarithms of concentrations, not concentrations.

Ordinarily, the choice of an equation is very important when using nonlinear regression. If the equation does not describe a model that makes scientific sense, the results of nonlinear regression won't make sense either. With standard curve calculations, the choice of an equation is less important because you are not interested in the best-fit values of the variables in the equation. All you have to do is assess visually that the curve nicely fits the standard points.

Determining unknown concentrations from standard curves

To read values off the standard curve:

1. Enter the unknown values on the same table as your standard curve. Just below the standard curve values, enter your unknowns as Y values without corresponding X values. This example shows X and Y values for five standards and Y values for four unknowns.

	X	Y
1	0.0	0.00
2	2.0	0.12
3	4.0	0.21
4	6.0	0.29
5	8.0	0.57
6		0.14
7		0.48
8		0.09
9		0.36

2. Click on the *Graphs* yellow tab (or use the Navigator to go to the graph) and look at the graph of your standard curve.

3. Click the *Analyze* button and choose how to fit the standard curve. Choose either linear or nonlinear regression, or create a LOWESS, spline, or point-to-point curve.

4. On the *Parameters* dialog, check the option for *Unknowns from standard curve.*

5. Go to the *Results* section to see the tabular results. Choose the subpage with interpolated values using either the explorer or the toolbar.

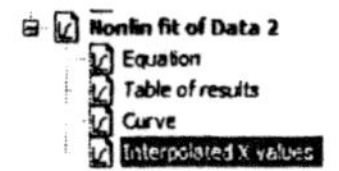

6. Look at the table of X and Y values. The Y column contains values you entered, and the X column shows the calculated concentrations in the same units that you used for the X axis.

7. If necessary, transform the results to antilogs. Click the *Analyze* button. Choose to analyze the data you are looking at. From the *Analyze Data* dialog, choose *Transforms* from the *Data manipulation* menu. Finally, in *Parameters: Transforms*, choose to transform X using *X=10^X.*

Standard curves with replicate unknown values

Prism's nonlinear regression analysis can interpolate from a standard curve, even if you have replicate unknown values.

Enter the data with all the replicates as shown below. The top part of the table is the standard curve. Below that are the unknown values. The standards and the unknowns do not need to have the same number of replicate determinations.

	X Values	A	
	X Title	Data Set-A	
	X	Y1	Y2
1	-9.0	1597	1531
2	-8.0	1453	1471
3	-7.0	1314	1245
4	-6.0	751	771
5	-5.0	336	306
6	-4.0	328	212
7	-3.0	207	307
8		1123	1085
9		1345	1298
10		1456	1421
11		987	

When you fit the standard curve with nonlinear regression, select the option to calculate unknowns from a standard curve.

The linear regression only can interpolate the mean of replicates unknown. If you want to interpolate individual values, use the nonlinear regression analysis and choose a linear (polynomial, first-order) model.

The interpolated values are shown on two output pages. The subpage labeled "Interpolated X mean values" shows the average of the replicate unknown Y values you entered in the Y column, with the corresponding interpolated X values in the X column.

Each value in "Interpolated X replicates" is a concentration corresponding to one of the replicate values you entered, and is expressed in the same units as the X axis of your standard curve. Because Prism cannot deal with replicate X values, Prism places these unknown X values in a Y column on the results table. Think of them as X values on your standard curve. But think of them as Y values when you want to do further analyses (such as a transform).

To calculate the mean and SD (or SEM) of the replicate values, press Analyze and choose row statistics.

Potential problems with standard curves

Reading unknown values from a curve or line is not always completely straightforward.

Problems when interpolating within the range of your standards

If you calculate X from Y, beware of a possible ambiguity. It is possible that two or more points on the curve have identical Y values but different X values. In this situation, Prism will report the lowest of the X values within the range of the standard curve and will not warn you that other answers exist.

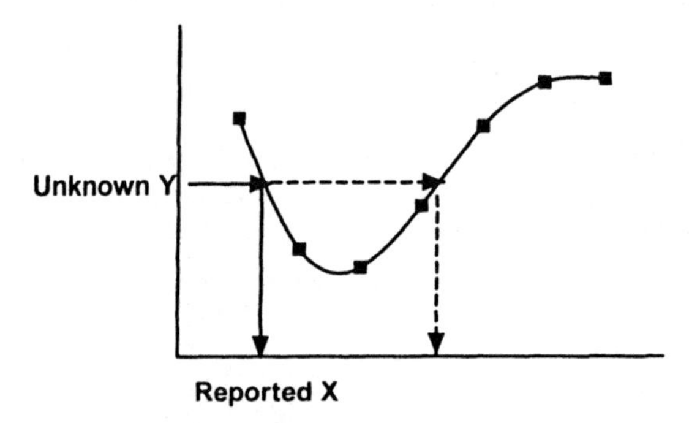

Prism defines a curve as a large number of points. To find unknown values, Prism linearly interpolates between the two points on either side of the unknown value. If you define the curve with more line segments, the interpolation will be more accurate. To increase the number of line segments, go to the *Output* tab on the nonlinear regression parameters dialog.

Problems extrapolating nonlinear regression beyond your standards

Prism can only read unknowns off standard curves generated by nonlinear regression within the range of the standard curve. If you enter an unknown value that is larger than the highest standard or smaller than the lowest standard, Prism will not try to determine the concentration. You can extend the curve in both directions, to include lower and higher X values, by settings on the Range tab in the Nonlinear regression parameters dialog.

Beware of extrapolation beyond the range of the data. Be especially wary of polynomial equations which will sometimes veer far from your data when you go outside the range of your data. The graph below shows the same data shown on page 339 fit to a fourth order polynomial equation (dashed curve). The curve fits the data well, but the path it takes outside the range of the data is not what you'd expect. Extrapolating beyond the range of the standards (the data points) would not yield useful values.

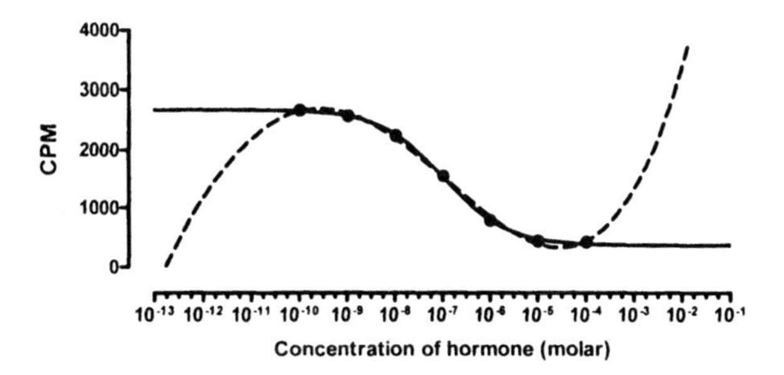

Problems extrapolating linear regression beyond your standards

With linear regression (in contrast to nonlinear regression), Prism will extrapolate the line to determine the unknown values, no matter how far they are from the knowns.

Mark Twain pointed out the folly of extrapolating far beyond your data:

> "In the space of one hundred and seventy six years the Lower Mississippi has shortened itself two hundred and forty-two miles. That is an average of a trifle over a mile and a third per year. Therefore, any calm person, who is not blind or idiotic, can see that in the Old Oölitic Silurian Period, just a million years ago next November, the Lower Mississippi was upwards of one million three hundred thousand miles long, and stuck out over the Gulf of Mexico like a fishing-pole. And by the same token any person can see that seven hundred and forty-two years from now the Lower Mississippi will be only a mile and three-quarters long, and Cairo [Illinois] and New Orleans will have joined their streets together and be plodding comfortably along under a single mayor and a mutual board of aldermen. There is something fascinating about science. One gets such wholesale returns of conjecture out of such a trifling investment of fact. "

53. Graphing a family of theoretical curves

Creating a family of theoretical curves

You'll often find it useful to graph a theoretical curve, or a family of theoretical curves, as a way to understand the properties of the model. This is sometimes called "plotting a function". This is different from fitting a curve with nonlinear regression. With nonlinear regression, you choose the equation and Prism finds the values of the parameters that make that equation fit your data the best. When plotting theoretical curves, you choose both the equation and the values of the parameters, and Prism graphs the function. You don't need to enter any data, and there is no curve fitting involved.

This chapter explains a special analysis that Prism offers to graph a family of theoretical curves. This is different from checking an option in the nonlinear regression dialog to plot the curve defined by the initial values. See page 305.

To simulate a family of curves, start from any data table or graph and follow these steps.

1. Click *Analyze*, select built-in analysis, and then select *Create a family of theoretical curves* from the list entitled "Simulate and generate".

2. The curve will be defined by a number of line segments. Choose how many segments you want. The default value of 150 will suffice for most purposes.

3. Choose a starting and ending value for X.

4. Select an equation.

5. Choose how many curves you wish to generate. All will be generated from the same equation, over the same range of X values. You can enter different parameters for each.

6. Enter the parameters for the first data set.

7. If you are creating more than one curve, go to the next one using the drop one (choose Curve B or Curve C ...). Click *Copy Previous Parameters* to copy the parameters from the previous curve, then edit as needed.

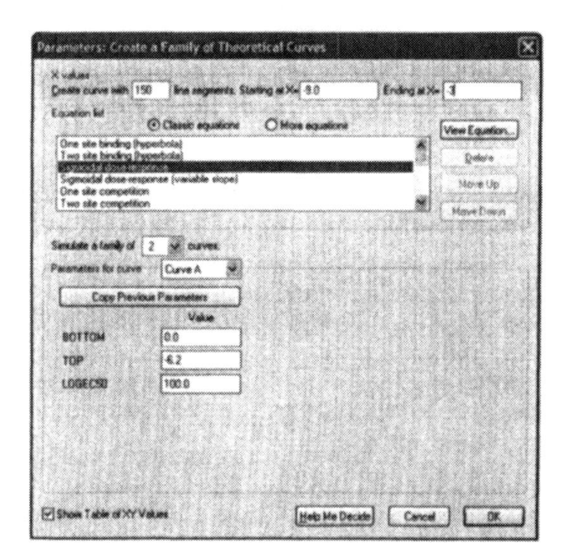

Note: Prism has a separate analysis to simulate data sets with random scatter. After choosing this "analysis", you'll need to define the arithmetic or geometrical series used to generate X, the equation used to generate Y, and the distribution used to add random scatter. See Chapter 10 in the Prism 4 User's Guide (pdf file free at www.graphpad.com) for details.

54. Fitting curves without regression

Introducing spline and lowess

Prism provides two approaches for fitting a curve without selecting a model. A ***cubic spline curve*** goes through every data point, bending and twisting as needed. A ***Lowess curve*** follows the trend of the data. LOWESS curves can be helpful when the data progresses monotonically, but are less helpful when there are peaks or valleys. Prism lets you choose between fine, medium and course LOWESS curves. The fine curve reveals the fine structure of the data, but tends to wiggle a lot. The coarse curve shows only the general trend, but obscures the detail.

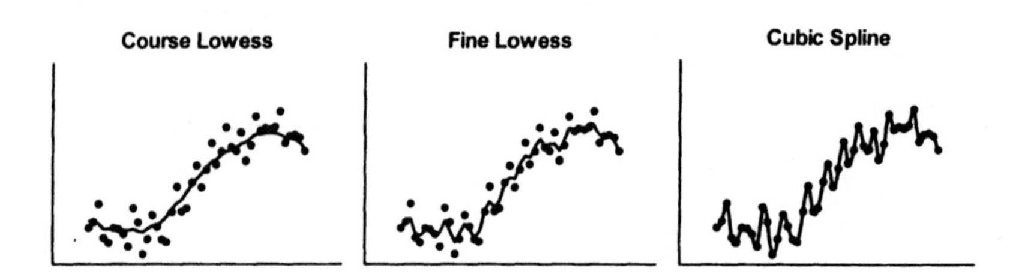

Spline and lowess with Prism

To create a Lowess or spline curve with Prism, click the *Analyze* button and choose *Fit spline/Lowess* from the list of curves and regressions to open the parameters dialog.

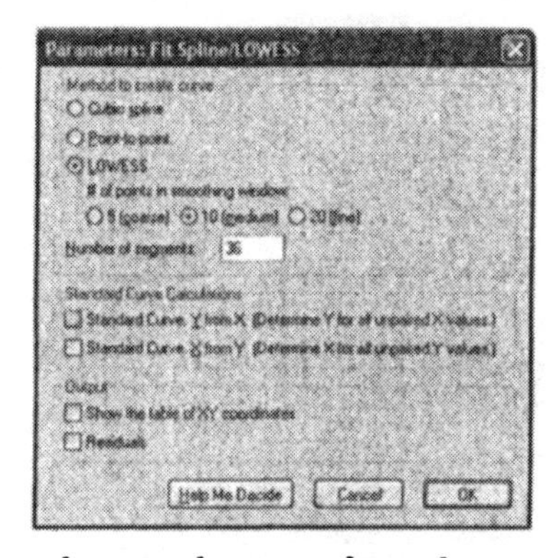

A cubic spline curve is a smooth curve that goes through every point. If your data are scattered, the spline will wiggle a lot to hit every point.

Select a Lowess curve only if you have well over twenty data points. Prism generates Lowess curves using an algorithm adapted from *Graphical Methods for Data Analysis*, John Chambers et al. (Wadsworth and Brooks, 1983). The smoothing window has 5, 10, or 20 points depending on whether you choose coarse, medium or fine smoothing.

Prism can also create a point-to-point "curve" -- a series of line segments connecting all your data. Don't create a point-to-point curve just so you can connect points with a line on the graph. You can do that by checking an option on the Format Symbols & Lines dialog from the Graphs section of your project. Only select the point-to-point analysis if you want to use the point-to-point line as a standard curve or to calculate area under the curve

Prism generates the spline, Lowess, or point-to-point curve as a series of line segments. Enter the number of segments you want, and check the option box if you want to view the table with the XY coordinates of each point.

Consider alternatives since spline curves wiggle so much, and Lowess curves can be jagged. To get a smoother curve, consider using nonlinear regression and pick a model empirically. You don't have to pick an equation that corresponds to a sensible model, and you don't have to interpret the best-fit values. Instead, you can use nonlinear regression simply as a way to create a smooth curve, rather than as a method to analyze data. Polynomial equations are often used for this purpose.

Annotated bibliography

DM Bates and DG Watts, *Nonlinear Regression Analysis and Its Applications*, Wiley, 1988.

A comprehensive book on nonlinear regression written by statisticians. Most biologists will find this book hard to follow.

PR Bevington, DK Robinson, *Data Reduction and Error Analysis for the Physical Sciences*, second edition, McGraw-Hill, 1992.

Despite the title, it is quite useful to biologists as well as physical scientists. It is a very practical book, with excellent practical and mathematical discussion of linear and nonlinear regression.

L Brand and ML Johnson, *Numerical Computer Methods*, Methods in Enzymology volume 210, Academic Press, 1992.

Thirty chapter by different authors on various aspects of numerical data analysis. Most deal with nonlinear regression.

KP Burnham and DR Anderson, *Model Selection and Multimodel Inference -- A Practical Information-theoretic Approach*, second edition, Springer, 2002.

A terrific resource to learn about model comparisons. Discusses the information theory (AIC) approach in detail, and barely mentions the extra sum-of-squares method. Very well written, and very persuasive about the advantages of the information theory approach.

A Christopoulos (editor). *Biomedical Applications of Computer Modeling*, CRC Press, 2001.

Detailed explanations of various models of signal transduction in G-protein linked receptors.

NR Draper and H Smith, *Applied Regression Analysis*, Wiley-Interscience, 1998.

Comprehensive resource about linear and multiple regression. Also has one chapter about nonlinear regression.

J. Gabrielsson, D Weiner, *Pharmacokinetic* and *Pharmacodynamic Data Analysis: Concepts and Applications*, 3rd edition, Garnder's UK, 2002.

As its title suggests, this book focuses on pharmacokinetics, and compartmental models. But the first section of the book is an excellent discussion of models and curve fitting, and will be useful to anyone who wants to master curve fitting, even if they have no interest in pharmacokinetics.

SA Glantz and BK Slinker, *Primer of Applied Regression and Analysis of Variance*, second edition, McGraw-Hill, 2001.

This is a terrific book about multiple regression and analysis of variance. The word "primer" in the title is not quite apt, as the explanations are fairly deep. The authors are biologists, and their examples discuss practical problems in data analysis. Chapter 11 explains nonlinear regression from a biologist's perspective, but includes a description of the mathematical basis of nonlinear regression.

H Gutfreund, *Kinetics for the Life Sciences. Receptors, transmitters and catalysts.* Cambridge University Press, 1995.

Enzyme and receptor kinetics, from first principles to fancy complexities.

R Hilborn and M Mangel, *The Ecological Detective: Confronting Models with Data,* Princeton University Press, 1997.

As the title suggests, this book was written for ecologists and population biologists. But it can be appreciated by any scientist. The authors approach all of statistics from the point-of-view of comparing models, and include clear explanations of maximum likelihood, probability and Bayesian thinking.

ML Johnson and L Brand, *Numerical Computer Methods (Part B),* Methods in Enzymology volume 240, Academic Press, 1994.

Thirty chapters by different authors on various aspects of numerical data analysis. Most deal with nonlinear regression.

LE Limbird, *Cell Surface Receptors: A Short Course On Theory And Methods,* Martinus Nijhoff Publishers, second edition, 1996.

Comprehensive book on receptors, including experimental design.

WH Press, et al, *Numerical Recipes in C,* second edition, Cambridge Press, 1992.

This book explains all sorts of algorithms for analyzing data. Chapter 15 explains nonlinear regression in plain language, but with mathematical rigor. If you want to write your own nonlinear regression program, this is the first place to look. It also is a great resource for understanding the mathematical basis of nonlinear regression by reading clear explanations that don't rely on heavy-duty math. The entire contents of this book are available on line at www.nr.com.

JW Wells, Analysis and interpretation of binding at equilibrium. Chapter 11 of EC Hulme editor, *Receptor-Ligand Interactions. A Practical Approach.* Oxford University Press, 1992.

This lengthy chapter gives detailed math for describing binding at equilibrium, including multiple sites (perhaps with cooperativity) and multiple ligands. It also has a great mathematical description of nonlinear regression.

Index

CPSIA information can be obtained at www.ICGtesting.com
Printed in the USA
BVOW051103160812

297966BV00004B/2/P

9 780195 171808